SEMICONDUCTOR OPTICAL MODULATORS

SEMICONDUCTOR OPTICAL MODULATORS

by

Koichi Wakita
NTT Opto-Electronics Laboratories

KLUWER ACADEMIC PUBLISHERS
Boston / Dordrecht / London

Distributors for North America:
Kluwer Academic Publishers
101 Philip Drive
Assinippi Park
Norwell, Massachusetts 02061 USA

Distributors for all other countries:
Kluwer Academic Publishers Group
Distribution Centre
Post Office Box 322
3300 AH Dordrecht, THE NETHERLANDS

Library of Congress Cataloging-in-Publication Data

A C.I.P. Catalogue record for this book is available
from the Library of Congress.

Printed on acid-free paper.

Printed in the United States of America

To my family

CONTENTS

Preface

The introduction of GaAs/AlGaAs double heterostructure lasers has opened the door to a new age in the application of compound semiconductor materials to microwave and optical technologies. A variety and combination of semiconductor materials have been investigated and applied to present commercial uses with these devices operating at wide frequencies and wavelengths.

Semiconductor modulators are typical examples of this technical evolutions and hsve been developed for commercial use. Although these have a long history to date, we are not aware of any book that details this evolution.Consequently, we have written a book to provide a comprehensive account of semiconductor modulators with emphasis on historical details and experimantal reports. The objective is to provide an up-to-date understanding of semiconductor modulators. Particular attention has been paid to multiple quantum well (MQW) modulators operating at long wavelengths, taking into account the low losses and dispersion in silica fibers occuring at around 1.3 and 1.55 mm.

At the present time, MQW structures have been investigated but these have not been sufficiently developed to provide characteristic features which would be instructive enough for readers. One problem is the almost daily publication of papers on semiconductor modulators. Not only do these papers provide additional data, but they often modify the interpretations of particular concepts. Almost all chapters refer to the large number of published papers that can be consulted for future study.

This book is organized as follows. The first four chapters introduce the basic convepts and provide mathematical derivations useful in understandingthe operation of semiconductor modulators. Chapters 5 and 6 describe various device structures and characteristics. Various modulators arecompared in Chapter 7. Chapter 8 considers monolithic integration with laser diodes and a surface normal switch is discussed in Chapter 9. The evaluation of modulator characteristics, such as waveguide loss, polarization sensitivity, and chirping are discussed in Chapter 10. Chapter 11 describes epitaxial growth and device fabrication techniques. The final Chapter (Ch.12) discusses new applications.

Acknowledgements

I wish to thank numerous colleagues both at NTT Opto-electronics Laboratories, and elsewhere, who greatly aided me in the organization and prepararation of this book. In particular, J. E. Zucker of Lucent Technologies provided me with data on optical switching and their monolithic integration. In Chapter 2, I used the effective index method developed by Yuzo Yoshikuni in the initial stage of wave propagation derivation and my discussions with Kenji Kawano were helpful in establishing the limit of applicability for this theory and the effectiveness of the beam propagation method. Naoto Yoshimoto, Thomas L. Koch of Lucent Technologies, and R. Zengerle of University Kaiserslautern provided me with data and graphs of mode-spot size converters. Prof. R. V. Ramaswamy of Univ. of Florida, S. Y. Wang of Hewlett Packard Lab., Profs. C. S. Tsai of Univ. of California at Irvine and T. Suhara of Osaka Univ. permitted me to use Figs. 3.3, 3.6, 3.14 and 3.15, respectively. My discussions with Takayuki Yamanaka were usuful in the description of strong and weak quantum confinement (Ch.4). Prof. D. A. B. Miller of Stanford Univ. permitted me to use Figs. 4.8, 4.9, and 4.12. The work of many people in the various groups at NTT was usuful: Yuzo Yoshikuni for direct modulation, Satoshi Oku for laser diode switching, and Koji Nonaka for wavelength conversion(Ch.5). J. F. Vincant of Alcatel Alshom, M. N. Khan of Lucent Technology, Hiroaki Inoue of Hitachi Lab., Prof. Hiroyuki Sakaki of Univ. of Tokyo, N. Agrawal of Heinrich-Herz-Institute permitted me to use Figs. 5.12, 5.13, 5.14, 5.15, and 5.18, respectively. In Chapter 6, I used many results obtained by NTT's optical switching group (Hiroaki Takeuchi, Kenji Kawano and Toshio Ito) as well as Profs. Kunio Tada of Univ. of Tokyo(Fig. 6.3), Hitoshi Kawaguchi of Yamagata Univ.(Fig. 6.4), Keiro Komatsu of NEC Opto-electronics Research Labs(Fig. 6.7). My discussions with Shunji Nojima were fruitful on optimizing the initial stage of the MQW structure and Takayuki Yamanaka calculated thick quantum well effectiveness(Ch.7). M. Erman of Alcatel Alshom and Tomoki Saito of NEC Opto-electronics Research Labs. permitted me to use Figs. 7.5 and 7.9. My discussions with Yuichi Kawamura (Osaka Prefecture University) on the initial monolithic integration of DFB lasers and modulators were also usuful(Ch.8). Masayuki Yamaguchi of NEC Corp., Hiroaki Inoue of Hitachi Central Research Lab., Masatoshi Suzuki of KDD Labs., Haruhisa Soda of Fujitsu Ltd. permitted me to use Figs. 8.2, 8.3, 8.4, and 8.5, respectively. Takashi Kurokawa and Chikara Amano provided much data including EARS(Ch.9). Profs. L. A. Coldren of Univ. California, Santa Barbara, D. A. B. Miller of Stanford Univ.,Kenichi Kasahara of NEC Opto-electronics Research Labs. and P. H. Shen of U. .S. Army Research Lab. permitted me to use Figs. 9.2, 9.4, 9.5, 9.6, 9.13, and 9.14, respectively. The discussions with Kazuo Kasaya aided my presentation of propagation loss and the discussions with Osamu Mitomi were helpful in describing the nonlinear effect on chirping(Ch.10). Prof. H. Melchior of Swiss Federal Institute of Technology., and T. H. Wood of Lucent Technologies permitted me to use Figs. 10.12, and 10.21, respectively. The discussions with Yoshio Noguchi and Susumu Kondo aided my description of crystal growth, Masashi Nakao helped me clarify the evaluation of the epitaxial layer, and Yasuhiro Suzuki provided me the disordering data(Ch.11). Isamu Kotaka contributed to the device processing and Kaoru Yoshino was very helpful in describing the module(Ch.11). Kenji Sato provided me with data on optical short pulse generation, modulation and multiplexing(Ch.12), and Tatsumi Ido of Hitachi Central Research Lab. permitted me to use Figs. 12.15 and 12.16.

I would like to thank Yoshihisa Sakai, Kiyoyuki Yokoyama and Masashi Nakao for their suggetions and comments on using word processor. Last,but not least, I would like to thank J. Yoshida, H. Tsuchiya, M. Naganuma and Y. Itaya for their encouragement throughout this work.

Chapter l

Introduction

l.1 Historical Perspectives

l.1.1 The First Ten Years

The electro-optic effect is one of the most useful physical phenomena for external modulators, and coefficients of this effect had been measured for various materials before the invention of the laser. However, the most considerable interest in the electro-optic effect has been taken specifically because the effect can be used for the modulation and polarization of laser light. Since the first demonstration of laser oscillation (with a ruby) was reported in 1960, many reports on such oscillation using gases, liquids, solids, glass and semiconductors have followed. This experimental success has opened the door to a new age in science and technology research.

In the first ten years since 1960, laser oscillation wavelengths ranged from the ultraviolet to the far-infrared, i .e. at least 3 orders of magnitude in frequency range. Modulator development proceeded in 3 phases: the first ten years, the last six years and the mid-range coresponding to Section 1.1.1, 1.1.2 and 1.1.3. At first, external modulators were necessary to produce any laser light, and many ferro-electrics and anti-ferroelectrics such as KDP and ADP, and perovskites such as $BaTi0_3$ and $LiNb0_3$ were used. With a few exceptions such as ZnO, ZnS, and GaAs, semiconductor materials were not used. All of these materials have large indices of refraction near the band edge: from 2 for ZnO to 3.5 for GaAs. As a result, the important product $n^3\gamma$ may be relatively large even though the electro-optic coefficient γ is not in itself large. The maximum range of transparency is fixed on the short wavelength end by the band edge and on the long wavelength end by Reststrahlen absorption. The transparency is also limited, however, by impurities and dislocations. In the long wavelength region free carriers, too, cause appreciable absorption.

Over three decades ago, modulators were developed using both lithium niobate ($LiNbO_3$ or LN) and GaAs [1-5]. However, due to the poor GaAs material quality resulting in excessively high insertion losses, the material of choice was LN. With numerous refinements, such as single mode waveguide creation by Ti-doped diffusion and traveling-wave electrode configuration, LN modulators and switch technologies have progressed to produce directional couples and Mach-Zehnder interferometers [6-11] with bandwidths in excess of 40 GHz[12] and 70 GHz [13], respectively. Figure 1.1 shows the progress of 3 dB bandwidth LN-modulator development.

Meanwhile, the Franz-Keldysh effect had been reported, describing the absorption-coefficient change near the bandgap of the semiconductors due to application electric fields. This electro-absorption effect can be used to construct an intensity modulator. The applied electric field shifts the band-edge, increasing the absorption of light. This phenomenon was reported in 1958 [14,15]. In the first stage, the change for bulk semiconductor materials was not so large that high voltage was necessary

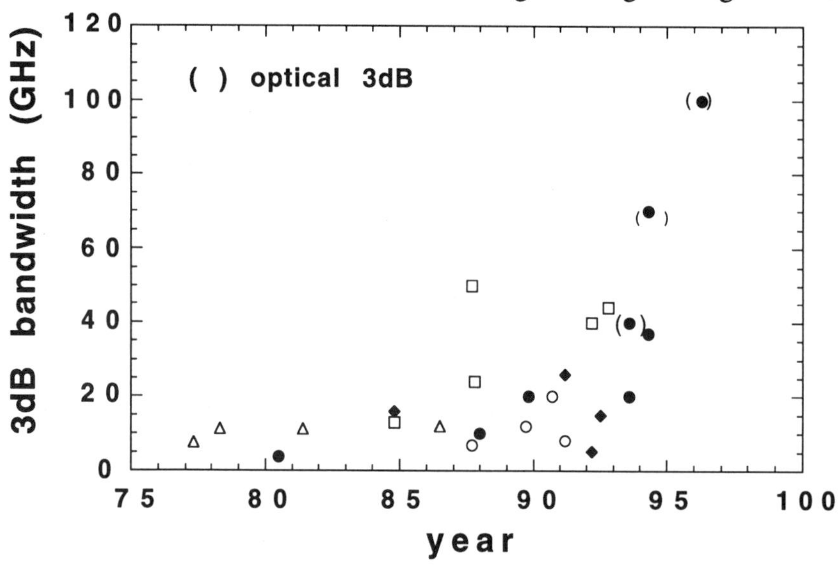

Fig. 1.1. Development of 3-dB bandwidth of $LiNbO_3$ modulators.

to obtain large extinction ratios.

The development of double heterostructure lasers [16,17] has changed this situation: the principal materials for semiconductor lasers were the direct energy-gap compounds from groups III through V of the periodic table. In general, increasing the atomic weight of a group III or group V element decreases the energy gap E_g, increases the refractive index n, and /with the exception of GaAs/AlGaAs and GaP/AlP, increases the lattice parameter a_0. Heterostructure lasers are layered semiconductor structures in which the lattice parameter a_0 is usually held constant from layer to layer, while E_g and n vary. The solid solutions provide continuously variable Eg and n for various compositions at a constant a_0 and permit additional possible heterostructure semiconductors. The waveguide theory and developments in crystal-growth techniques enable us to design and control various optical devices with external semiconductor modulators.

GaAs/GaAlAs double heterostructures were fabricated with band-edges close to the sources of the light to be modulated. Their changes in absorption were two orders of magnitude for a small (less than 10 V) applied voltage. For short-wavelength region, these modulators had relatively efficient characteristics [18]. Moreover, an electro-optic directional-coupler switch, first demonstrated with LN, was tried using GaAs and GaAs/AlGaAs and efficient electro-optical switching was demonstrated [19, 20] .

The direct modulation of laser diodes has been investigated by many groups. Such modulation has many advantages: modulation is very easy, being obtained by changing the injection current, and is free from the insertion loss associated with the use of external modulators. Such systems, being free from optical coupling are simple and compact in configuration, and this results in stability. Therefore, with the introduction of optical fiber transmission systems into commercial products, which began in the mid-1970s, direct modulation came into use, and external modulators were never used again.

High bit-rate and long haul optical fiber transmission systems have been developed and the bit rate has improved with the direct modulation speeds of semiconductor lasers. As technology evolved from low bit-rate short-haul systems to 1.6 Gbit/s systems, light sources, too, have changed from multi-mode, short-wavelength GaAs/AlGaAs laser diodes to single-mode, long-wavelength (1.3 μm and 1.55 μm) InGaAsP/InP lasers. Figure 1.2 shows the development of the capacity of optical-fiber transmission systems and optical components over time. In the mean time, the laser

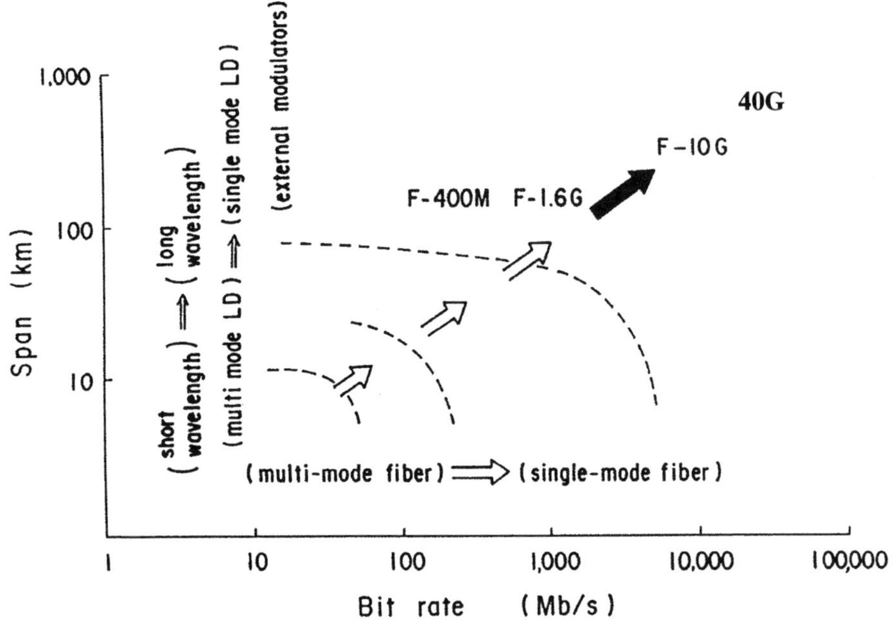

Fig. 1.2 Development of optical fiber transmission systems and optical semiconductor devices.

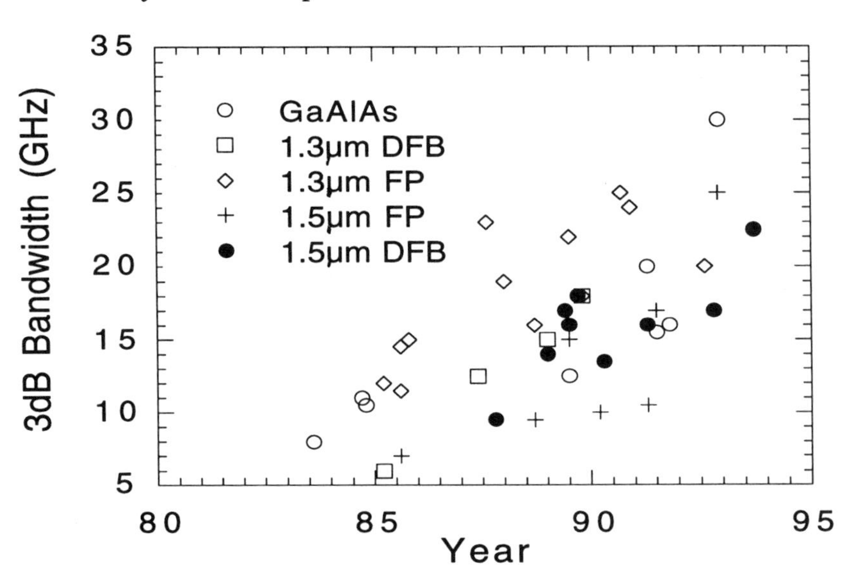

Fig. 1.3. Development of 3-dB bandwidth of semiconductor laser diodes.

diodes have progressed from Fabry-Perot types to distributed feedback (DFB) or distributed Bragg reflection (DBR) types and the dynamic single mode oscillation has been achieved [21,22].

Figure 1.3 shows the bandwidths of 3-dB laser diodes by year. Many efforts have been made to improve this bandwidth. Figure 1.4 shows the evolution of high-speed photodetector development, for reference. Figure 1.3 indicates a plateau in 3 dB bandwidth development for laser diodes, due to the nonlinear gain of such diodes. Countermeasures are now being tested but have not proven successful yet.

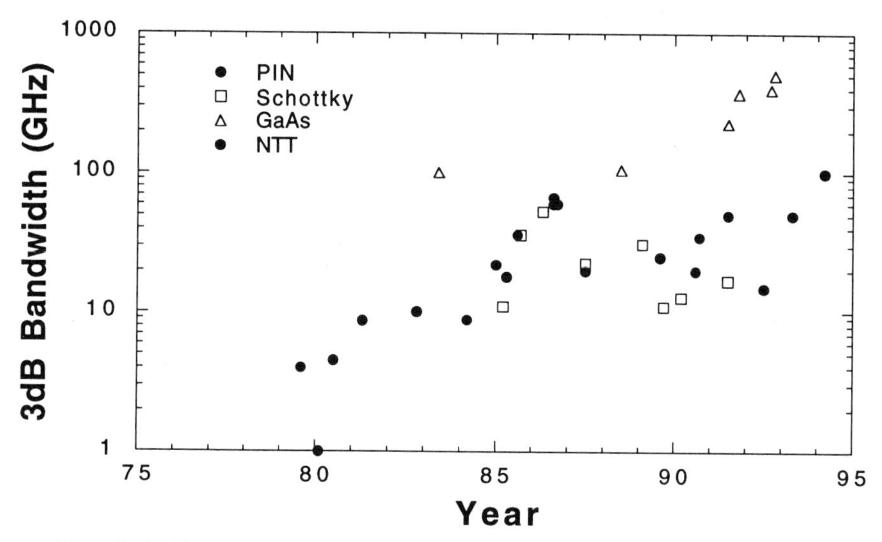

Fig. 1.4. Development of 3-dB bandwidth of photodetector.

1.1.2 The Last Six Years

Ultra-high speed optical components are necessary to exploit the wide bandwidths available when using light as a carrier for telecommunication, information processing and instrumentation applications. The frequency-chirping associated with the high-speed direct modulation of laser diodes above a few Gbit/s has become a clear problem and results in serious problems in high-speed long-distance optical-fiber transmission systems. Other uses of lasers, such as interconnecting and signal processing systems, in addition to the mainstay optical fiber transmission systems have been investigated in connection with a newly completed network running throughout Japan, from Asahikawa (in Hokkaido) to Okinawa. Moreover, the development of low propagation-loss optical waveguides has reached loss levels of 0.1 dB/cm [23].

At the same time, Er-doped optical fiber amplifiers have been developed, enabling us to overcome previous optical fiber transmission-length limits which were due to the large insertion-loss inherent to external modulators [24]. Advances in semiconductor crytal-growth techniques as well as in the quantum size effect theory, have made it possible to fabricate thin semisonductor layers with atomically smooth interface and superlattice structures. Prototypes of applications have been demonstrated for many devices. Since the first proposal of the supperlattices [25], a great deal of research and development into optical devices using the structures has been conducted with the goal of improving device characteristics. In particular, AT&T's groups have contributed to the development of the physics of the multiple quantum well (MQW) structure and preliminary applications thereof [26] .

These techniques have opened up new technical frontiers for external modulators with high-bit rates and low-voltage operation. At present, driving voltages as low as 2 volts with better than 20 dB extinction ratios and 20-Gbit/s modulation rates have been achieved in many laboratories [27, 28]. Moreover, optical-fiber transmission experiments have been reported using semiconductor modulators over 100 km of the 20 Gbit/s modulation speeds [29, 30], and over 400 km at 2.5 Gbit/s [31]. Figure 1.5 shows the evolution of the high-speed modulator and its integrated light source development.

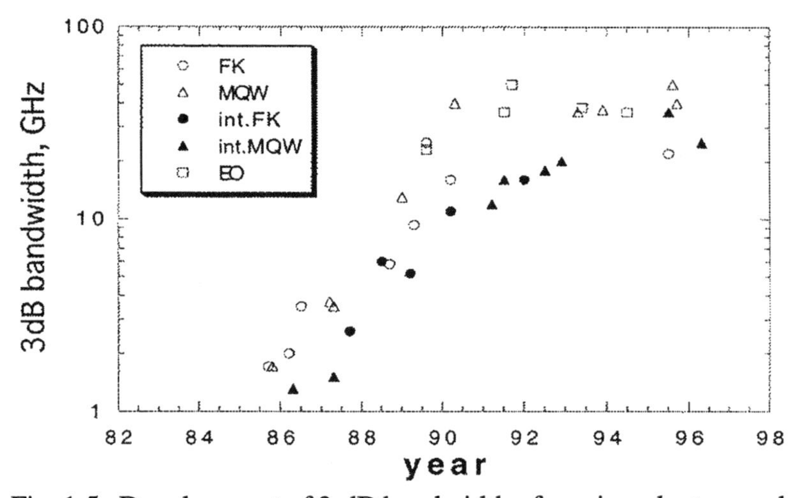

Fig. 1.5. Development of 3-dB bandwidth of semiconductor modulator. FK:Franz-Keldysh, EO:Electrooptic, int.: integrated modulator.

Monolithic integration of semiconductor modulators with laser diodes, optical amplifiers, and detectors has recently been developed. Semiconductor modulators have become indispensable for optical transmission systems as well as optical signal processing and instrumentation applications.

1.1.3 Mid-Range and Future Prospects

As mentioned in the previous section, adolescent-stage progress in semiconductor modulators and switches was not very remarkable because they were still in the shadow of laser diode development. However, crystal-growth and device-processing techniques as well as device design and evaluation methods used for laser diodes served to prepare for the subsequent innovation of semiconductor modulators and switches. In particular, the development of low propagation-loss optical waveguides was imperative for the success of optics integrated on III-V semiconductor materials. Uniformity of thickness and composition of epitaxial layers and low defect density was achieved through the use of metal-organic-vapor-phase epitaxial techniques (MOVPE). The propagation loss was reduced to 0.2 dB/cm [32] for single-hetero-junction ridge waveguides and below 0.7 dB/cm for dielectrically-loaded double-heterostructure optical guides [33]. The details will be discussed in Chapter 10.

Background impurity is one of the most important factors for reverse-biased devices, so high-purity epitaxial growth was investigated for use in photodetectors such as p-i-n photodiodes and avalanche photodiodes. This technique produces high quality modulators operating in the reverse-bias condition.

Dry etching techniques are also important for integrated optics. For small-area devices wet chemical etching was used, but such techniques are simply unable to achieve the kind of anisotropic removal of material necessary to attain target dimensions. In integrated devices, in particular, it is often necessary to epitaxially regrow traces on the etched surface and ohmic contacts are fabricated after etching. These processes are adversely affected by residue remaining after exposure to reactive discharge, and by near-surface plasma-induced damage. At first, GaAs/AlGaAs systems were investigated and remarkable results were obtained. Contrariwise, for InGaAsP/InP and InGaAs/InAlAs systems used in the long wavelength region were reported to give rise to rough surface morphologies. This problem arises because these materials consist of a low evaporating-pressure material (InP) and chemically active material (Al). There have been several demonstrations of the smooth, yet residue-free etching of InP and related compounds in CH_4/H_2 and C_2H_6/H_2 Plasma [34]. This is discussed in Chapter 11.

The design of integrated devices requires a detailed understanding of optical coupling mechanisms between waveguides and optical devices. Modal analysis and beam propagation methods have been developed and these provide a useful basis for simulating waveguide/optical device coupling [35, 36].

With this last section, we will discuss future prospects. There are three main trends: high speed, low power requirements and large-scale integration. As for the speed, $LiNbO_3$ modulators are in the forefront and techniques for their use have matured. For example, traveling wave structure is one of the next targets. They should help eliminate the device capacitance that limits speed. Trials have been run for bulk modulators, but not using MQW structures. Optical switches, which operate free from capacitance, driven by short optical pulses are another target.

For higher efficiency, modification of MQW structures, such as parabolic and two or three step structures, as well as quantum wire or quantum dot stuctures, is the next target for MQW structures. These have already been proposed but not demonstrated yet.

The large scale integration is another upcoming target. It is necessary to build up the fabrication techniques and crystal growth. Simultaneous selective area growth, which has been reported recently by NEC [37] and the Hitachi [38] group in Japan, and more recently by Alcatel [39], will be useful for semiconductor photonic integrated circuits. This issue is described in Chapter 11.

As for other applications or trends, the electroabsorption modulator will be used in a short optical pulse generator with a high repetition rate by using its nonlinear response to applying voltage. The repetition frequency can be easily changed and high duty ratio of pulse width to repetition period will be useful for optical soliton source. At the same time, monolithic integration with other functional devices encoding generated optical pulses will be the next target. This issue is discussed in the last chapter.

modulators	modulation	mechanism	structure
intensity	absorption(gain)	Franz-Keldysh	s-WG (a)
phase	index	QCSE	M-Z (b)
		Wannier Stark	DC (c)
		EO	IS (d)
		exciton bleaching	
		plasma	

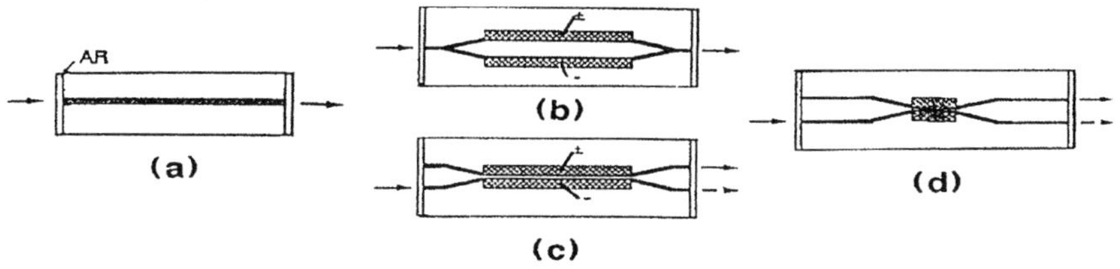

Fig. 1.6. Modulator structures and their operation principles. QCSE and EO stand for quantum-confined-Stark effect and electrooptic effect. s-WG, single waveguide; M-Z, Mach-Zehnder interferomeric modulator; absorption and DC, directional coupler; IS, intersectional type.

1.2 Optical Modulator Operation Principle

In this section, we introduce the operation principle of optical modulators for a guide. For easy understanding, we limit our discussion to a simple straight-waveguide structure. There are various operation principles and modulator structures as shown in Fig. 1.6. The modulation makes use of the change in absorption or refractive index, which is based on Pokels effect, Franz-Keldysh effect, quantum confined Stark effect (QCSE), exciton-bleaching or tunable carrier density effect, and plasma or injected carrier effect.

At first, we focus on a waveguided intensity modulator based on the electroabsorption effect and design an intensity modulator concretely. As for the waveguide structures, there are many types reported: single waveguide, Mach-Zehnder, directional coupler and intersectional, as shown in Fig. 1.6. There are many combinations. Our discussion will focus on single waveguided planar structures. Other waveguided structures will be described in detail in other sections.

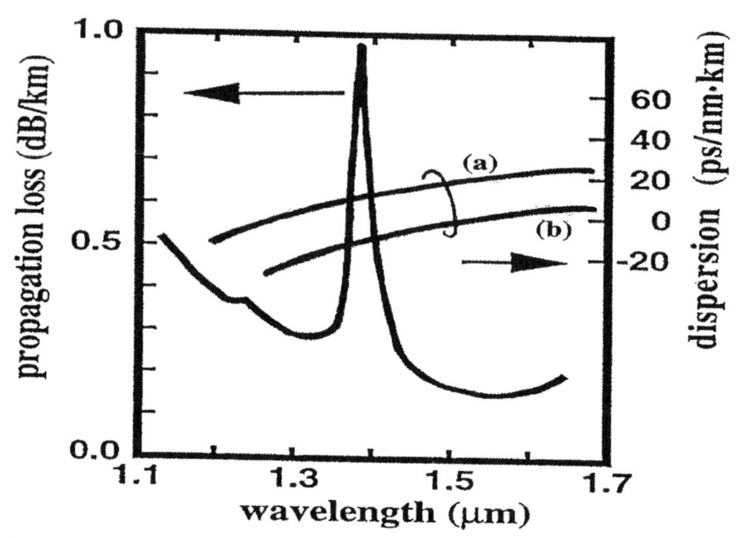

Fig. 1.7. Propagation loss and dispersion of single-mode fiber.

1.2.1 Design Procedures

At the first stage, we have to know the absolute value of the absorption coefficient and refractive index and their change with applied voltage or carrier injection. However, at the initial stage, we do not know any parameters and have to use some reported or supposed values. Usually, mixed crystals such as ternary or quaternary are often used for semiconductor waveguides and we can obtain the necessary parameters based on the linear or quadratic approximation of binary materials by extrapolating the fractional content values.

First the operating wavelength is to be determined. As shown in Fig. 1.7, propagation loss and dispersion of silica-based single-mode fiber depend on the wavelength. We have two kinds of optical fibers whose dispersion-minimum wavelength is 1.3 μm or 1.55 μm. The latter is called dispersion-shifted fiber, whereas the former is called normal fiber.

Generally the operating wavelength and the materials are not independent. Fig. 1.8 shows this situation. For 0.8 μm, we use GaAs/AlGaAs and for 1.3 or 1.55 μm, we use the InGaAsP/InP or InGaAs/InAlAs material systems. Then we determine the waveguide thickness and its fractional content of mixture. Waveguide characteristics are strongly dependent on the waveguide core and cladding materials, because the optical confinement depends on the refractive index profile that will

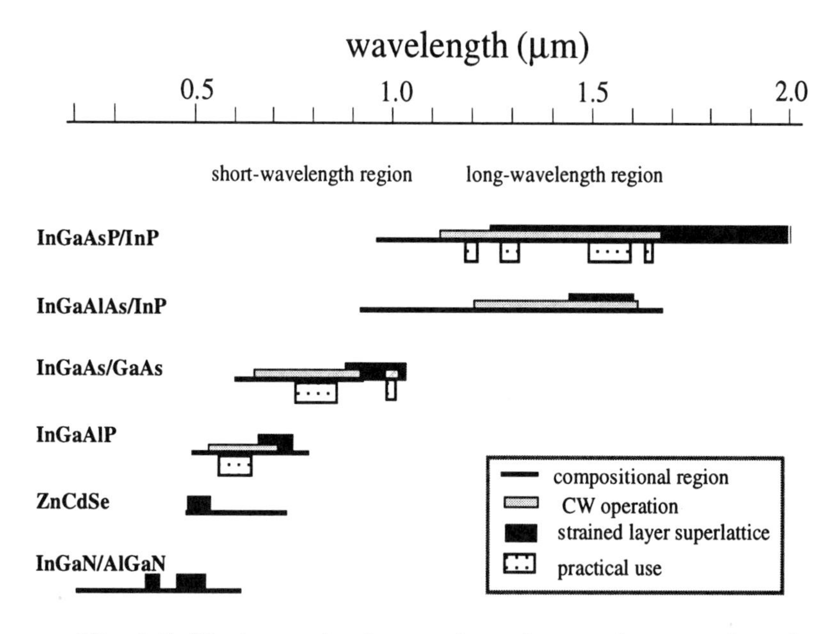

Fig. 1.8. Various mixed crystals and operating wavelength

be discussed in Chapter 2. Based on the extrapolation, we determine the absorption energy. As discussed in Chapter 7 in detail, the suitable detuning energy between the modulated light wavelength and the absorption band edge of electroabsorption modulators is about 40-50 meV. Therefore, the photoluminescence wavelength of InGaAsP is determined under the condition of lattice-matching with an InP substrate.

For MQW structures the thickness of an InGaAsP quantum well is about 7.5 nm when the well and barrier are lattice-matched to an InP substrate. If we use an InP barrier, the InGaAs well thickness is 5.0 nm when this modulator operates at 1.55 μm. Next we determine the total thickness of the waveguide core or the MQW layer. The barrier thickness must be thin to ensure large optical confinement and small driving voltage as far as the electron confinement is achieved. The thickness of the waveguide layer (summation of well and barrier thicknesses multiplied by their number) is determined from a practical point of view. That is, it depends on the operating voltage, 3-dB bandwidth, insertion loss, and so on. For usual devices the waveguide thickness is between 0.1 and 0.2 μm, so that the operating voltage is a few volts and the 3-dB bandwidth is around 10 GHz.

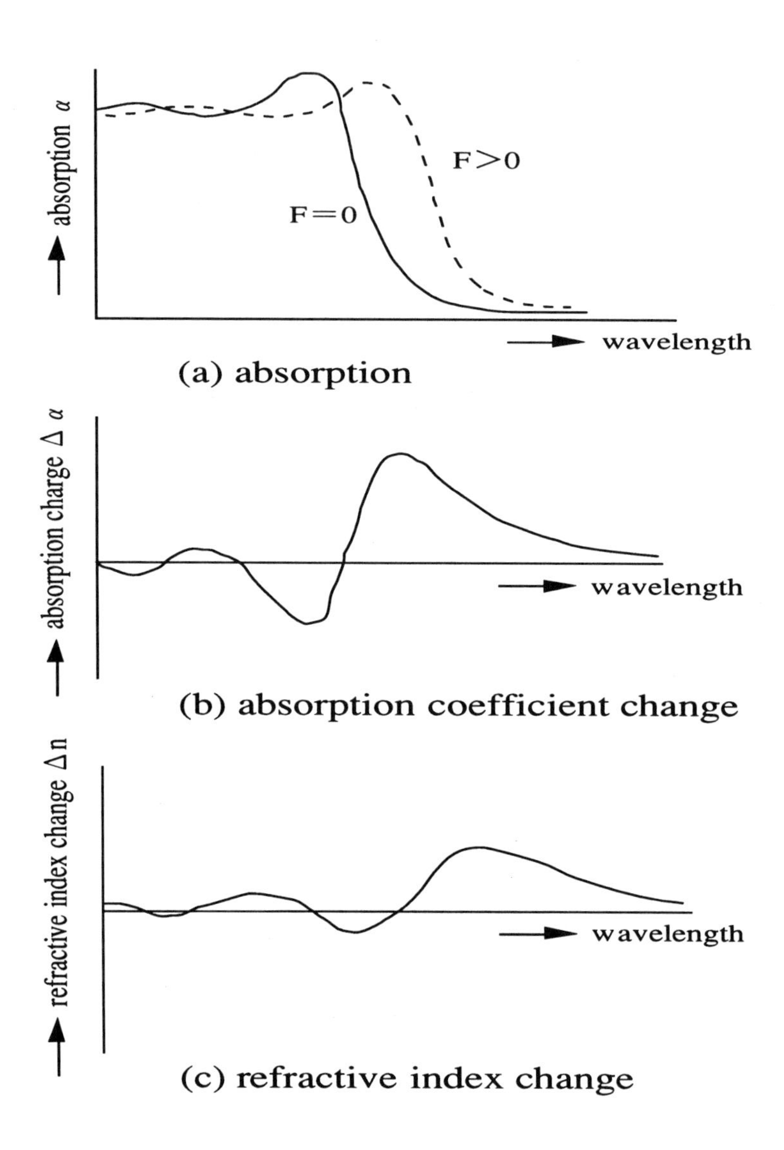

Fig. 1.9. Relationship between wavelength and the absorption coefficient (a), its change (b), and refractive index change(c) with applied bias.

1.2.2 Modulator Characteristics

There are five important factors in intensity modulators:

(1) on/off ratio,
(2) required voltage for required on/off ratio,
(3) 3-dB bandwidth,
(4) insertion loss, and
(5) chirping.

We discuss these factors as follows.

The on/off ratio is given by the product of absorption coefficient change and sample thickness or length for surface normal configuration or waveguide mode, respectively. The absorption coefficient and its change are to be measured before you design modulators. We have two candidates of modulator structures; surface normal and waveguide type. For surface normal modulators, the maximum extinction ratio is usually at most several dB, because of their short optical interaction length. As for waveguide structures, the optical confinement factor is to be taken into account. The maximum extinction ratio decreases because of the incomplete confinement of the optical field. The method to obtain the optical confinement factor is described in Chapter 2.

A. On/off Ratio. The most important parameter is the on/off ratio for the intensity modulator. Usually, 15 or 20 dB is necessary from the system experiment. The on/off ratio is defined as the ratio of incident light intensity P_{in} to transmitted light intensity P_{out} and is given as follows.

$$[on/off] = -10\log 10(P_{out}/P_{in}) \quad [dB].$$

In electroabsorption materials, the transmitted light intensity is given by using the absorption coefficient with the incident light intensity as follows.

$$P_{out}/P_{in} = \exp(-\Gamma\, \Delta\alpha L),$$

where Γ, $\Delta\alpha$ and L are the optical confinement factor, absorption coefficient change, and the sample length, respectively. Therefore, on/off ratio is given as

$$[on/off]=0.434\, \Gamma\Delta\alpha L \qquad [dB].$$

B. Required Voltage. As shown in Fig. 1.9, the absorption coefficient change $\Delta\alpha$ strongly depends on the applied voltage and the used wavelength. Based on the QCSE, the absorption coefficient peak shifts in proportion to the square of the applied electric field intensity and approximately to the fourth power of the well thickness. On the contrary, the well thickness increase results in the decrease in oscillator strength of the exciton absorption. Therefore, there is an optimum condition for the well thickness. The detailed discussion on the optimum condition will be discussed in Chapter 7.

From practical use, the smaller the applied voltage, the better for driving electronics circuits. Up to now, 2-V peak-to-peak operation is necessary for high-speed (more than 10 Gb/s) operation. The applied voltage is determined by the total thickness of the i-layer and the barrier thickness is also important.

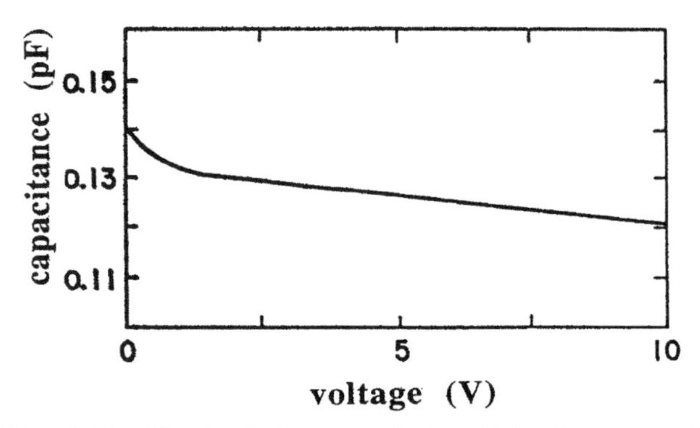

Fig. 1.10. Typical characteristic of device capacitance.

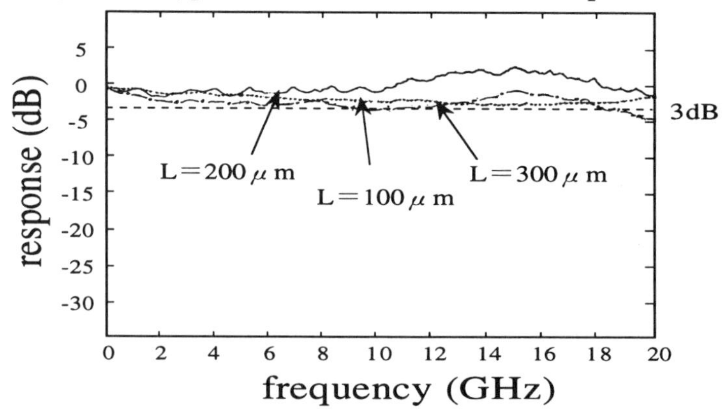

Fig. 1.11. Frequency response and its capacitance dependence.

C. 3-dB Bandwidth. The bandwidth is usually determined by the device capacitance when the device is operated in the reverse bias condition, except when it is operated by carrier injection, which is operated with very slow speed such as a few ns. When the speed is limited by the device capacitance, the 3-dB bandwidth is given as follows.

$$[f_{3dB}]_1 = 1/\pi RC,$$

where R and C are the load resistance and capacitance, respectively. The device capacitance is proportional to sample length L and width W, and it is inversely proportional to i-region thickness d when the stray capacitance is neglected. Figures 1.10 and 1.11 show device capacitance with L of 100 μm and frequency response of electroabsorption modulators with the same W and d and different L. As the capacitance decreases, the frequency response increases. Usually, stray capacitance induced from the bonding pad is not neglected and some countermeasures, such as polyimide with small dielectric constant, are used to reduce the stray capacitance.

The above discussion is based on the assumption that the sample is short enough and the transit time of light through the sample is very short. When we use long samples, the bandwidth is limited by the transit time τ of the light through the sample and the bandwidth is given as follows.

$$[f_{3dB}]_2 = 1.39/\pi\tau = 1.39c_0/\pi n_g L,$$

where n_g is the refractive index of the waveguide and c_0 is the speed of light in a vaccum. As an example, we calculate this value for InGaAsP with a refractive index of 3.24 as

$$[f_{3dB}]_2\, L = 4.13 \text{ GHz cm}.$$

That is, the 3-dB bandwidth for the lumped InGaAsP device is limited to above $[f_{3dB}]_2$ even if $[f_{3dB}]_1$ is increased.

In order to improve this limitation, a traveling-wave-type structure is proposed and demonstrated. In this structure, a traveling line is arranged along the optical waveguide to match the transmission velocity of microwave and optical waves. In this case, the bandwidth is [40]

$$[f_{3dB}] = 1.39c_0/\pi[n_g-n_m]L,$$

where n_m is the refractive index of the transmission line for the modulated microwave. Based on this equation, [f_{3dB}] seems to increase monotonously if we increase L when the velocity matching condition is fulfilled. Practically, the losses for optical and microwave transmission are not neglected, however, the long waveguide and the maximum value in the reported data is 350 GHz, which was obtained by using superconducting materials for the ohmic contact [41].

D. Insertion Loss. Insertion loss consists of transmission loss, reflection loss, and coupling loss. The transmission loss consists of residual absorption loss of the material, free carrier loss, and scattering loss. The coupling loss is due to the mode-spot size mismatch between the incident light and the guided light. The reflection loss is at most 3 dB and is eliminated by anti-reflection coating.

The transmission loss consists of the inherent absorption α_g due to the energy difference between the incident light and the absorption

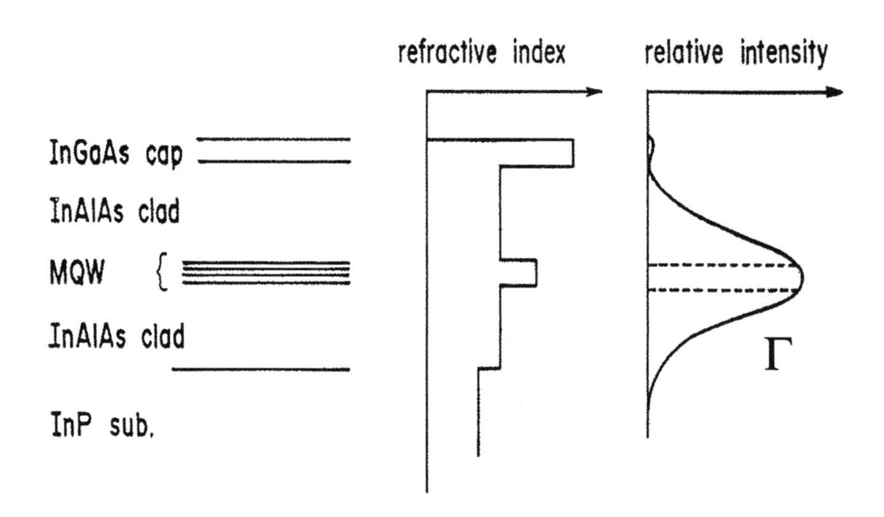

Fig. 1.12. Electric field distribution and optical confinement factor Γ for an InGaAs/InAlAs MQW modulatorthree-layer waveguide.

edge of the waveguide material, free carrier absorption α_{fc} due to the free carrier in the waveguide and in the cladding layer, and light scattering α_s as shown in Fig. 1.12 and given by

$$\alpha_{tran} = \Gamma\alpha_g + (1-\Gamma)\alpha_{fc} + \alpha_s,$$

where Γ is the optical confinement factor. This value depends on the refractive index profile. The free carrier absorption in the p-doped cladding layer is much larger than that in the n-doped cladding layer by as much a 1 order of magnitude for InP. The doping level is required to be small in order to provide the low transmission loss, while high doping is required so that the applied voltage is biased only to the i-region resulting in small bias voltage. Therefore, the p-doping is chosen to be 5×10^{17} to 1×10^{18} cm^{-3}.

In a semiconductor waveguide, the mode spot size is generally very small, especially in the y-direction (perpendicular to the layer) due to the large refractive index difference between those of the guide layer and cladding layer. This strong optical confinement produces small driving voltage, on the contrary, causing effective overlap between the optical field and the electric field which results in small spot-size. Therefore, the large insertion loss in semiconductor waveguides is mainly due to the coupling loss and some counterplans, such as monolithic integration of laser diodes or mode-spot size tansformers are tried.

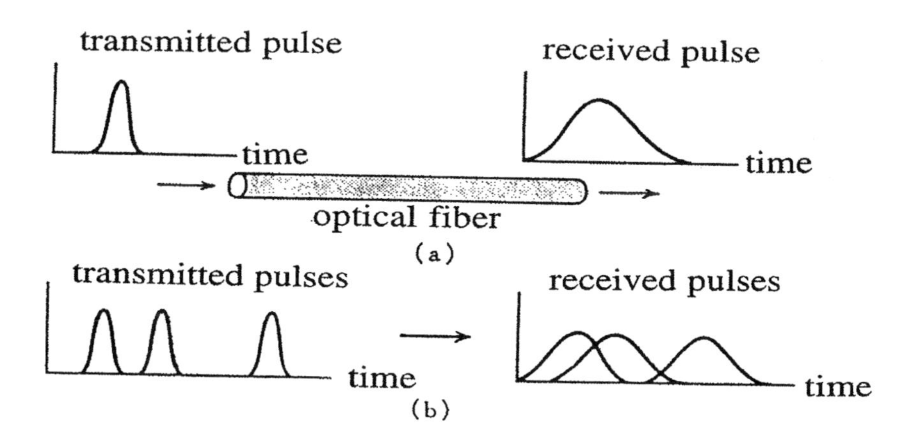

Fig. 1. 13. Schematic of optical pulse transmission through a dispersive fiber.

E. Chirping. Any change of absorption coefficient in a material structure will be accompanied by a phase shift of light, since the real and imaginary parts of dielectric constant constitute a Kramers-Kronig transform's pair. This causes frequency chirping in intensity modulation and intensity fluctuation in phase modulation. When the chirping is induced, the transmitted optical pulses are broadened through the fiber due to the dispersion. Figure 1.13 schematically shows pulse broadening and its overlapping through the optical fiber. Though the chirping for external modulators is thought to be much smaller than that for direct-modulated laser diodes, it is a limiting factor for the capacity of long-haul high-bit-rate optical-fiber communication systems due to the fiber dispersion as shown in Fig. 1.14.

The magnitude of chirping is defined as the ratio of the change of refractive index n to the change of the extinction coefficient k. It

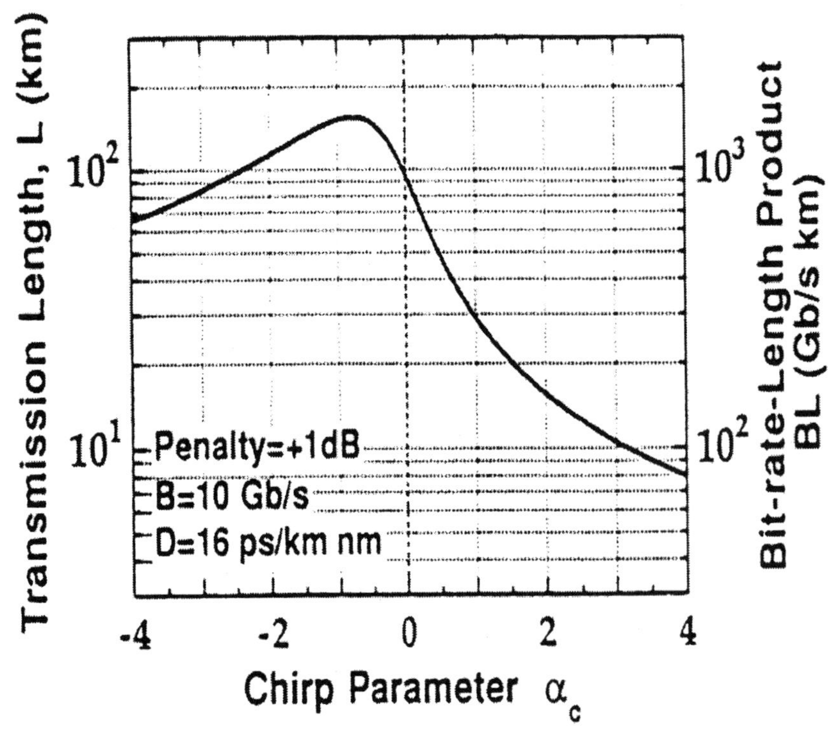

Fig. 1.14. Calculated transmission length versus chirping parameter. Modulation speed, power penalty, and fiber dispersion are assumed to be 10 Gbit/sec, 1 dB, and 16 ps/km nm, respectively.

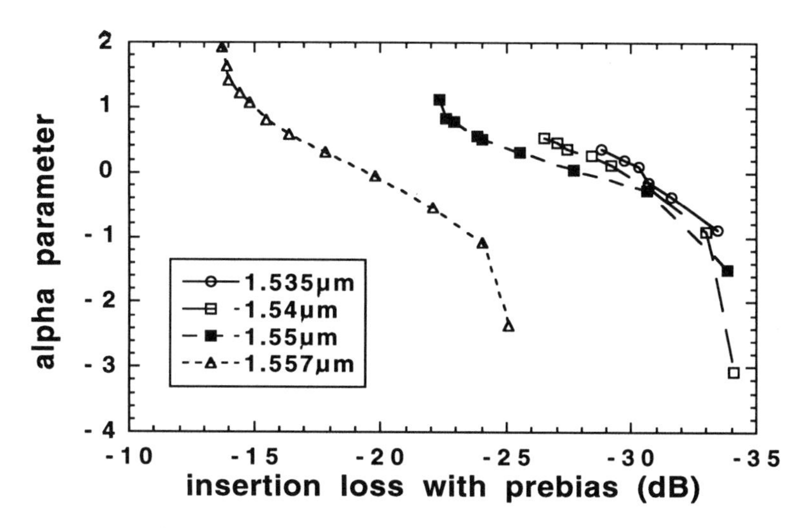

Fig. 1.15. Chirp parameter versus insertion loss.

depends on the device structure and operating condition and modulation mechanism. It has been proposed that waveguide Mach-Zehnder-amplitude modulators may be operated in a perfectly chirpless mode and devices designed to operate chirp-free have been reported, whereas the chirping exists more or less for electroabsorption modulators but its magnitude is at most 1.5 in previously reported data.

However, the lowest fiber-dispersion penalty is generally not obtained for a frequency chirp parameter identically equal to zero. Choosing a non-zero value for the chirp parameter, depending on the fiber dispersion coefficient and distance, so as to provide some amount of pulse compression can be advantageous.

The chirp parameter for the electroabsorption modulator is shown in Fig. 1.15 where the transverse axis indicates the transmission loss with pre-bias. It is noticeable that the chirp parameter decreases as the operating wavelength approaches the absorption edge energy, while the absorption at zero bias (transmission loss) increases. That is, the chirp parameter and the transmission loss contradict each other. Therefore, some compromise is necessary. In general, the transmission loss is too large to be used under the condition of the small chirp parameter. Recent advanced technologies produce highly efficient optical amplifiers and monolithically integrated light sources with electroabsorption modulators and DFB lasers, which enable us to use electroabsorption modulators with large propagation loss.

The details of the chirping in the external modulators will be discussed in Chapter 10.

F. Figure of Merit. As described in the previous sections, there are trade-offs among the above five parameters. For example, on/off ratio is proportional to the product of absorption coefficient change and sample length L, whereas 3-dB bandwidth f_{3dB} is inversely proportional to L and proportional to i-region thickness d, and the operating voltage is proportional to d. Therefore, the figure of merit is defined as the following and comparison of modulator characteristics can be discussed. The details will be given in Chapter 7 .

$$[\text{figure of merit}] = 2R\, f_{3dB}\, \lambda/(50+R)\, V\pi.$$

1.3 References

[1] A. Ashkin and M. Gerhenzon, J.Appl. Phys., 34, pp. 2116-2119, 1963.
[2] D. F. Nelson and F. K. Reinhart, Appl. Phys. Lett., 5,pp. 148-150, 1964.
[3] W. L. Walters, J. Appl. Lett., 37, pp. 916-918, 1966.
[4] I. P. Kaminow and E. H. Turner, Proc. IEEE, 54, pp. 1374-1390, 1966.
[5] F. S. Chen, Proc. IEEE, 58, pp. 1440-1457, 1970.
[6] K. Kubota, J. Noda, and O. Mikami, IEEE J. Quantum Electron., QE-19, pp. 754-760,1980.
[7] M. Izutsu and H. Haga, and T.Sueta, Trans. IECE Japan, E63, pp.817-818, 1980.
[8]M. Izutsu and T. Sueta, IEEE J. Quantum Electron., QE-19, pp. 668-674, 1983.
[9]P. S. Cross, R. Baumgartner, and B. H. Kplner, Appl. Phys. Lett., 44, pp. 486-488,1984.
[10] D.W.Dolfi, Appl. Opt., 25, pp.2479-2480,1986.
[11] M. Nazarathy, D. W. Dolfi, and R. L. Jungerman, J. Opt. Soc. Amer., 4, pp. 1071-1079, 1987.
[12] S. K. Korotky, G. Eisenstein, R. S. Tucker, J. J. Veselka, and G. Raybon, Appl. Phys. Lett., 50, pp. 1631-1633, 1987.

[13] K. Noguchi, H. Miyazawa, and O. Mitomi, Electron. Lett., 30, pp.949-950, 1994.
[14] W. Franz, Z. Naturforsch., 13, 484, 1958.
[15] L. V. Keldysh, Zh. Eksp. Teor. Fiz., 34, 1134, 1958 [Sov. phys.-JETP 7, 788, 1958].
[16] Zh. I. Alferov, V. M. Andrev, D. Z. Garbuzov, Yu. V. Zhilyaev, E. O. Morozov, E. L. Portnoi, and V. G. Trofim, Sov. Phys. Semicond., 4, 1573, 1971. (Translated from Fiz. Tekh. Poluprovodu., 4, 1826, 1970).
[17] I. Hayashi, M. B. Panish, P. W. Foy, and S. Sumski, Appl. Phys. Lett., 17, pp. 109-111,1970.
[18] F. K. Reinhart, Appl. Phys. Lett., 22, 372, 1973.
[19] K. Tada and K. Hirose, Appl. Phys. Lett., 25, 561, 1974.
[20] J. C. Campbell, F. A. Blum, D. W. Show, and K. L. Lawley, Appl. Phys. Lett., 25, pp. 202-204, 1976.
[21] T. Matsuoka, H. Nagai, Y. Itaya, Y. Noguchi, Y. Suzuki, and T. Ikegami, Electron. Lett., 18, pp. 27-28, 1982.
[22] Y. Abe, K. Kishino, and Y. Suematsu, Electron. Lett., pp. 410-411, 1982.
[23] R. J. Deri and E. Kapon, IEEE J. Quantum Electron., 27, pp. 626-640, 1991.
[24] K. Hagimoto, K. Iwatsuki, A. Tanaka, M. Nakazawa, M. Saruwayari, K. Aida, K. Nakagawa, and M. Horiguchi, OFC'89, Post dead paper, PD15, 1989.
[25] L. Esaki and R. Tsu, IBM J. Res. Devlop., 14, 61,1970.
[26] S. Schmitt-Rink, D. S. Chemla, and D. A. B. Miller, Advances Physics, 38, pp. 89-188, 1989. References are therein.
[27] I. Kotaka, K. Wakita, K. Kawano, M. Asai, M. Naganuma, Electron. Lett., 27, pp. 2162-2163.
[28] F. Devaux, F. Dorgeuille, A. Ougazzaden, F. Huet, M. Carre, A. Carenco, M. Henry, Y. Sorel, J. -F. Kerdiles, and E. Jeanney, Photn. Technol. Lett., pp. 1288-1290, 1993.
[29] T. Kataoka, Y. Miyamoto, K. Hagimoto, K. Wakita, and I. Kotaka, Electron. Lett., 28, pp. 897-898, 1992.
[30] T. Kataoka, Y. Miyamoto, K. Hagimoto, K. Sato, I. Kotaka, and K. Wakita, Electron. Lett., 30, pp. 872-873, 1994.
[31] O. Gautheron, G. Grandpierre, P. M. Gabla, J. -P. Blondel, E. Brandon, P. Bousselet, P. Garabedian, and V. Harvard, ECOC'94, pp. 15-18, 1994.

[32] H. Inoue, K. Hiruma, K. Ishida, and H. Matsumura, IEEE Trans. Electron. Devices, ED-32, pp. 2662-2668, 1985.
[33] S. H. Lin, S. Y. Wang, S. A. Newton, and Y. M. Young, Electron. Lett., 21, pp. 597-598, 1985.
[34] S. J. Pearton, W. S. Hobson, F. A. Baiocchi, and A. B. Emerson, J. Vac. Sci. Technol., B8, pp. 57-67, 1990.
[35] A. W. Snyder and J. D. Love, Optical Waveguide Theory, London, UK: Chapman and Hall, 1983.
[36] M. D. Feri and J. A. Fleck, Jr., Appl. Opt., 19, pp. 1154-1160, 1980.
[37] T. Kato, T. Sasaki, N. Kida, K. Komatsu, and I. Mito, ECOC'91, WeB7-1.
[38] M. Aoki, M. Takahashi, M. Suzuki, H. Sano, K. Uomi, T. Kawano, and T. Takai, Photon. Technol. Lett., 4, pp. 580-582, 1992.
[39] H. Haisch, W. Baumert, C. Hache, E. Kuhn, M. Klenk, K. Satzke, M. Schilling, J. Weber, and E. Zielinski, ECOC'94, pp. 801-804, 1994.
[40] A. Yariv, *Quantum Electronics*, 2nd Ed., pp.347-351, John Wiley & Sons, 1975.
[41] J. Nees, S. Williamson, and G. Mourou, Appl. Phys. Lett.,54, pp. 196-1964, 1989.

Chapter 2

Analysis of Slab Waveguide

2.1 Introduction

Optical waveguides are the structures that are used to confine and guide the light in the guided-wave devices and circuits of integrated optics. This chapter is devoted to the theory of these waveguides. The guides to integrated optics are usually planar structures such as planar films or strips.

The simplest waveguide is the planar slab guide shown in Fig. 2.1, where a planar guide of refractive index n_1 is sandwiched between a substrate and a cover material with lower refractive indeces n_0 and n_s. Often the cover material is air, in which case $n_0=1$. As an illustration, we have listed in Table 2.1 the refractive indeces of some semiconductor waveguide materials used in integrated optics. Typical differences among the indeces of the guide were formed by three-step epitaxial growth, impurity diffusion and strip loaded.

Dielectric waveguides have already been the subject of textsbooks [1-3] and we can refer the reader to these for a history on the subject as well as for a more complete list of references. In this chapter we hope

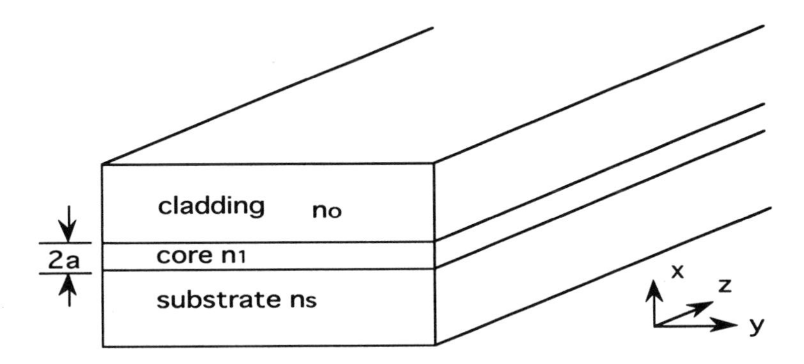

Fig. 2.1. Schematic of a planar slab waveguide.

Table 2.1. Indices of semiconductors

materials	refractive index	wavelength (μm)
GaAs	3.7	0.85
AlAs	2.95	0.85
$Al_{0.2}Ga_{0.8}As$	3.4	0.85
InP	3.17	0.85
$In_{0.53}Ga_{0.47}As$	3.59	1.55
InGaAsP	3.17 - 3.45	1.55
$In_{0.52}Al_{0.48}As$	3.22	1.55

to give both an introduction to the subject as well as a collection of important results sufficiently detailed to be of use to the experimenter.

Figure 2.2 shows the main optical waveguide structures. Figure 2.1 is planar type or two-dimensional waveguide and 2.2 is channel (stripe) or three-dimensional waveguide. The planar guides provide no confinement of the light within the film plane, i.e., the y-z plane, and confinement takes in the x-dimension only. Strip waveguides can provide this additional confinement, which we assume to be in the y-dimension. In several active integrated optics devices such as lasers or modulators, this additional confinement can help to bring about desirable savings in drive voltage and drive power for example.

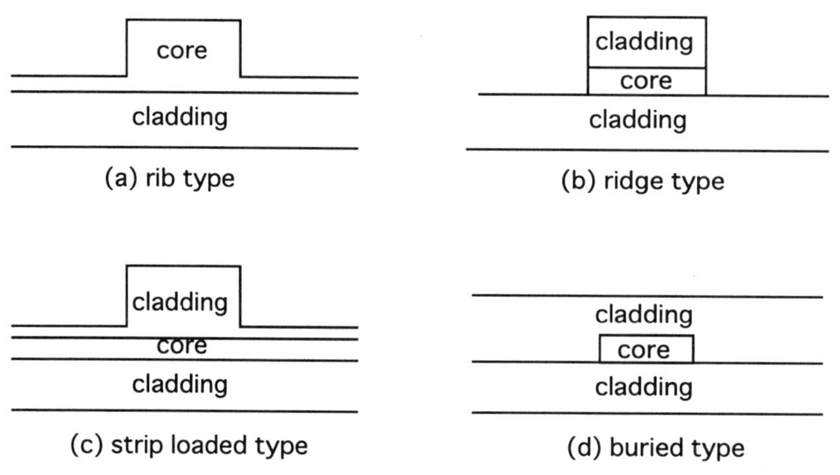

Fig. 2.2. Various waveguide structures.

2.2 Slab Waveguide

For the most fundamental three-layer-slab waveguide structures (Fig. 2.2(a)), we can analyze Maxwell's equations and the wave equations as the following, assuming a lossless medium with a scalar dielectric constant e(w) and a scalar magnetic permeability m.

$$\mathbf{E} = \mathbf{E}(x,y)\exp\{j(\omega t-\beta z)\}$$

$$\mathbf{H} = \mathbf{H}(x,y)\exp\{j(\omega t-\beta z)\},$$

where **E**, **H** are the vectors of the electric and magnetic field, respectively, and ω and β are the angular frequency, and the propagation constant of the mode. In planar guides, the light is confined in one dimension only, which we choose to be in the x-direction. The refractive index n(x) of a planar guide and the corresponding modal fields are functions of only this coordinate. One can then simplify the modal differential equations by setting $\partial E/\partial y=0$ and $\partial H/\partial y=0$. A planar guide supports transverse electric (TE) modes with zero longitudinal electric field ($E_z=0$) and transverse magnetic modes (TM) with zero longitudinal magnetic field ($H_z=0$).

For TE-modes we set $H_y=0$ and get $E_z=0$ and $E_x=0$.

$$E_y = A\cos(\kappa a+ \phi)\exp\{-\sigma(x-a)\} \qquad (x>a)$$

$$= A\cos(\kappa x+\phi) \qquad (-a<x<a)$$

$$= A\cos(\kappa a-\phi)\exp\{\xi(x+a)\} \qquad (x<-a)$$

$$\kappa = (k^2 n_1{}^2-\beta^2)^{1/2}$$

$$\sigma = (\beta^2-k^2 n_0{}^2)^{1/2}$$

$$\xi = (\beta^2-k^2 n_s{}^2)^{1/2}$$

Moreover, the application of the boundary conditions yields the formula

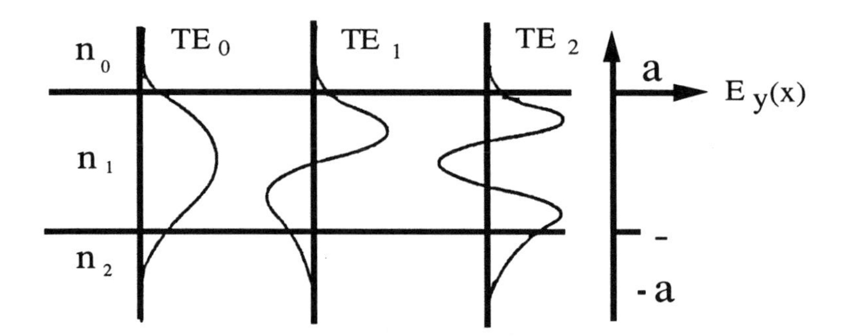

Fig. 2.3. Electric field distribution for TE_0, TE_1, and TE_2 mode.

for the phase shifts and the dispersion relation

$$2\kappa a=(m+1)\pi-\tan^{-1}(\kappa/\sigma)-\tan^{-1}(\kappa/\xi) \quad (m=0,1,2,-----),$$

where the mode label m is an integer. The equivalent refractive index N has discrete values in the range of $n_s< N <n_1$ ($n_0<n_s$ is assumed). Figure 2.3 shows the electric field Ey distribution in the x-direction for TE_0, TE_1, and TE_2 mode. The number of the maximum or minimum points are (m+1) and the larger m, the larger the guided light spreads into the substrate and cover. As for the TM mode, the following relation is yielded.

$$2\kappa a=(m+1)\pi - \tan^{-1}(n_0^2\kappa/n_1^2\sigma) - \tan^{-1}(n_s^2\kappa/n_1^2\xi) \quad (m=1,2,----).$$

In addition to the guided modes, radiation modes also exist and these are continuous and the latter N for the radiation modes is also continuous.

In order to investigate the guided modes more in detail, it is usual to use the normalized relation from the above relation.

$$V(1-b_E)^{1/2}=(m+1)\pi - \tan^{-1}\{(1-b_E)/b_E\}^{1/2}$$

$$-\tan^{-1}\{(1-b_E)/(b_E+\gamma_E\}^{1/2}.$$

where $V=ka(n_1^2-n_s^2)^{1/2}$:normalized frequency, $b_E=(N^2-n_s^2)/(n_1^2-n_s^2)$:normalized equivalent refractive index, $\gamma_E=(n_s^2-n_0^2)/(n_1^2-n_s^2)$:asymmetry measure.

Figure 2.4 shows the dispersion relation for the normalized ω–β-diagram, which is taken from Kogelnik and Ramaswamy[4] and where the guide index b_E is plotted as a function of the normalized frequency

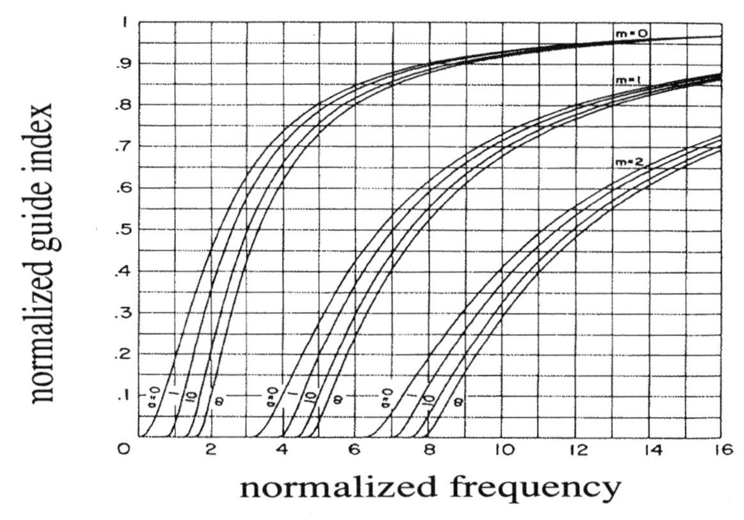

Fig. 2.4. Normalized ω–β-diagram of a planar slab waveguide showing the guide index b as a function of the normalized thickness V for various degrees of asymmetry.

V for four different values of the asymmetry measure and for the mode orders m=0, 1, and 2. From these relations, it is clear that more mode number can be guided as the thickness of the guide or the normalized frequency is increased and the equivalent refractive index for each mode results in an increase. The cutoff is defined as the condition where a certain mode m is not guided due to the small V. The normalized cutoff frequency is

$$V_m = m\pi + \tan^{-1}(\gamma_E)^{1/2}.$$

For the symmetry waveguide, i.e., $n_0=n_s$ ($\gamma_E=0$) and $V_m=m\pi$. The fundamental mode (m=0) is not cutoff.

For the TM mode, we get cutoff conditions of the same form as for TE-mode and ω–β diagrams that are very similar. The normalized dispersion relation is

$$V(q_s)^{1/2} n_1/n_s(1-b_M)^{1/2}$$

$$= (m+1)\pi - \tan^{-1}\{(1-b_M)/b_M\}^{1/2}$$

$$- \tan^{-1}\{(1-b_M)/b_M + \gamma_M(1-b_M d))\}^{1/2}$$

where

$$b_M = (n_1/n_s q_s)^2 (N^2 - n_s^2)/(n_1^2 - n_s^2),$$

$$q_s = (N/n_1)^2 + (N/n_s)^2 - 1,$$

$$\gamma_M = (n_1/n_0)^4 (n_s^2 - n_0^2)/(n_1^2 - n_s^2),$$

$$d = (1 - n_s^2/n_1^2)(1 - n_0^2/n_1^2).$$

The relations are more complicated compared with those for TE-modes. But the results shown in Fig. 2.4 can be used by changing the parameter γ_E into γ_M as $q_s=1$, $d=0$, $b_M=b_E$ when $n_1-n_s << 1$.

In the real planar waveguide structures, over three layers are often used and the equations are very complicated. It is simpler to solve Maxwell's equation numerically under the given boundary conditions when the planar waveguides consist of multi-layers.

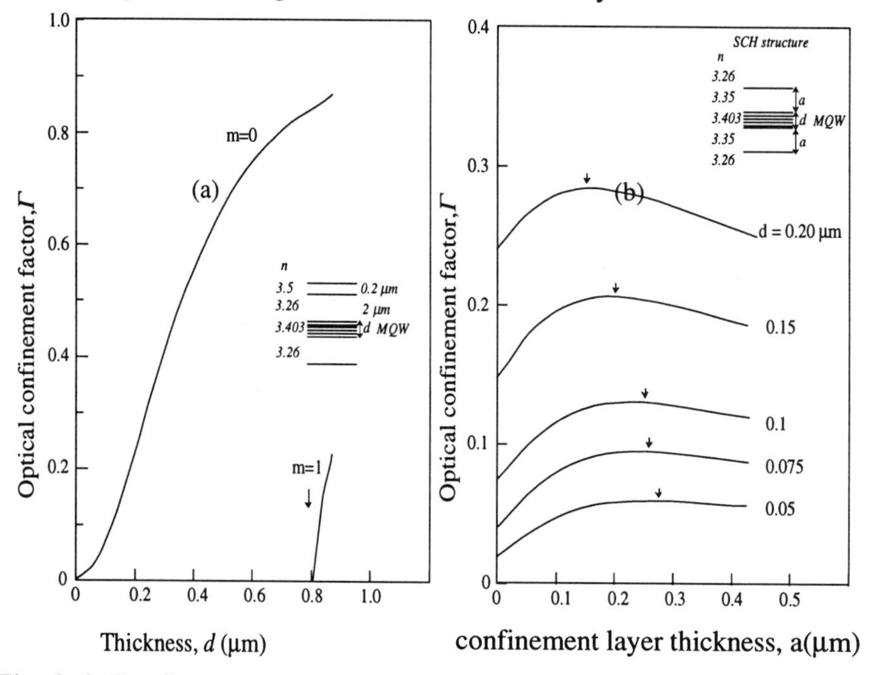

Fig. 2.5. Confinement factor vs. guide and confinement layer thickness for an electroabsorption MQW modulator operating at 1.55μm wavelengths.

2.3 Optical Confinement Factor

Thin waveguide structures are usually used for several active integrated optics such as lasers or modulators; in this situation the confinement factor, which is defined as the ratio of the guided power in the guide to the whole power in a certain mode, is very important. For example, the confinement factor for TE_0 mode of a symmetric waveguide is given as the following.

$$\Gamma = \{V+2(b_E)^{1/2}\}/\{V+2(1/b_E)^{1/2}\}.$$

As shown in Fig. 2.4, b_E decreases and Γ results to go to zero as the thickness of the guide, i. e., V decreases. Figures 2.5(a) and (b) show the relationship between the optical confinement factor and guide layer thickness, and confinement layer thickness with guide layer thickness. The arrow in the Fig. 2.5 indicates the condition of a single mode operation.

2.4 Channel Waveguide

The analysis of dielectric strip guides is much more complex than that of planar waveguides, and no exact analytic solutions for the modes of strip guides are available. Numerical calculations have been made for guides with a core of rectangular cross-section embedded in a uniform surround of lower index. Approximate calculations are usually employed for practical applications. The following two methods are popular.

2.4.1 Marcatili Method

This method is applicable to a large class of strip guides whose cross-sections are indicated in Fig. 2.2. When the optical field is confined and transmitted enough, the electromagnetic field abruptly decreases as it goes away from the boundary. Then the optical energy intensity in the part of the oblique line is negligible and we can solve the electromagnetic field component by separating the variables. When the refractive index changes to a step-index, E_x mode for H_x=0 and E_y mode for H_y=0 are obtained, respectively. For example, for each region,

$$H_y = A\cos(k_x x - \phi)\cos(k_y y - \psi) \qquad \text{region I}$$

$$A\cos(k_x a - \phi)\exp(-\gamma_x(x-a))\cos(k_y y - \psi) \qquad \text{region II}$$

$$A\cos(k_x x - \phi)\exp(-\gamma_y(y-b))\cos(k_y b - \psi) \qquad \text{region III,}$$

where $\beta^2 = -k_x^2 - k_y^2 + k^2 n_1^2$ region I

$$\beta^2 = \gamma_x^2 - k_y^2 + k^2 n_0^2 \qquad \text{region II}$$

$$\beta^2 = -k_x^2 + \gamma_y^2 + k^2 n_0^2 \qquad \text{region III}$$

and

$$\phi = (p-1)\pi/2 \qquad (p = 1,2,3,------)$$

$$\psi = (q-1)\pi/2 \qquad (q = 1,2,3.------).$$

The condition that E_z $\partial Hy/\partial x$ at x=a and H_z $\partial Hy/\partial y$ at y=b are continuous result in the following dispersion equations for $E_{pq}{}^x$ mode.

$$k_x a = p\pi/2 - \tan^{-1}(n_0^2 k_x / n_1^2 \gamma_x)$$

$$k_y b = q\pi/2 - \tan^{-1}(k_y/\gamma_y),$$

where $\gamma_x^2 = k^2(n_1^2 - n_0^2) - k_x^2$ and $\gamma_y^2 = k^2(n_1^2 - n_0^2) - k_y^2$ are used. Using above equations, we can obtain k_x and k_y, and

$$\beta^2 = k^2 n_1^2 - (k_x^2 + k_y^2).$$

The dispersion equation for $E_{pq}{}^y$ is also obtained according to the above procedure.

2.4.2 Effective Index Method

As an example we take the E_{pq}^{x} mode for this method as shown in Fig. 2.6. First of all we divide the whole into three sections and each section is assumed to be a planar waveguide with the infinite lateral width. Then we obtain the effective refractive indices N_1, N_2, N_3 for TE mode. Then we assume three layers stacking in the x-direction and extending in the y-direction. The effective index of the TM mode for such a planar waveguide is approximately that of the initial channel waveguide. This approximation offers good physical meaning and is a help to understand the rib waveguide.

The wave equation for E_{pq}^{x} is given as the following.

$$\partial^2 H_y/\partial x^2 + \partial^2 H_y/\partial y^2 + [k^2 n^2(x,y) - \beta^2]H_y = 0$$

The fundamental assumption of the effective index method is to separate the variables in the field distribution. That is

$$H_y(x,y)=X(x)Y(y).$$

Inserting the above equation, we obtain

$$(1/X)d^2X/dx^2 + (1/Y)d^2Y/dy^2 + [k^2n^2(x,y) - \beta^2] = 0$$

This equation is divided into the following two equations.

$$(1/Y)d^2Y/dy^2 + [k^2n^2(x,y) - k^2n_{eff}^2(x)] = 0$$

$$(1/X)d^2X/dx^2 + [k^2n_{eff}^2(x) - \beta^2] = 0,$$

where n_{eff} is called as effective index. Based on the continuous condition of H_z at the boundary, dY/dy is also continuous. The eigen value equation for the four layers slab waveguide as shown in Fig. 2.6 is given as the following.

$$\sin(\kappa d-2\phi) = \exp(-2(\sigma S+y))\sin(\kappa d),$$

where

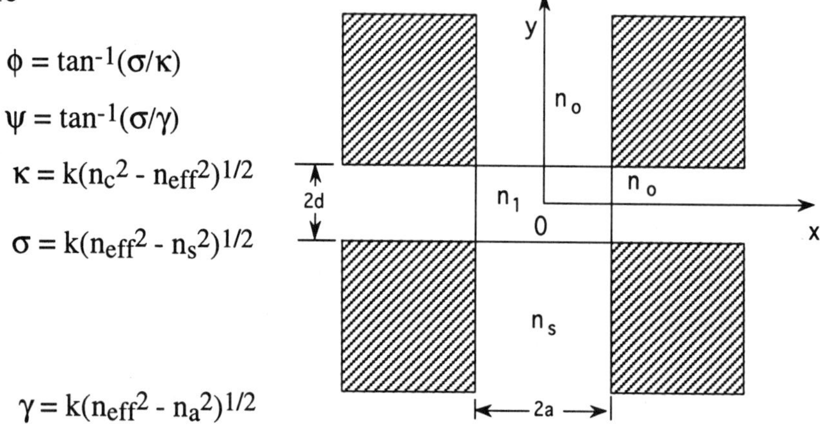

$$\phi = \tan^{-1}(\sigma/\kappa)$$

$$\psi = \tan^{-1}(\sigma/\gamma)$$

$$\kappa = k(n_c^2 - n_{eff}^2)^{1/2}$$

$$\sigma = k(n_{eff}^2 - n_s^2)^{1/2}$$

$$\gamma = k(n_{eff}^2 - n_a^2)^{1/2}$$

Fig. 2.6. Two dimensional buried rectangular waveguide for Marcatili method.

2.5 Three Dimensional Channel Waveguides

In the two dimensional waveguides discussed in the previous section, only one dimension is confined. In the practical application, it is necessary to develop the optical waveguide into two dimensions. Various analytical methods have been developed. Variational analysis techniques [5] are very attractive since they can treat structures with high index differences along a nonplanar contour, as is the case for instance with rib waveguides [6]. Another technique, beam propagation method (BPM)[7-9], is a perturbation method and therefore only structures with small index differences can be analyzed in principle. The advantage of the BPM is that the propagation of any beam in the waveguide structure can be computed. In this section, we only introduce reported structures and resultant waveguide characteristics for optical modulators and we will not discuss the details of analytical methods at any length here, concentrating instead on the experimental data. Figure 2.2 shows four types of typical waveguide channels, ribed and ridged (lying on top of the substrate), imbedded in the substrate, by strip-loaded planar guides. Various modified structures based on these structures have been reported but the fundamental way of thinking is the same.

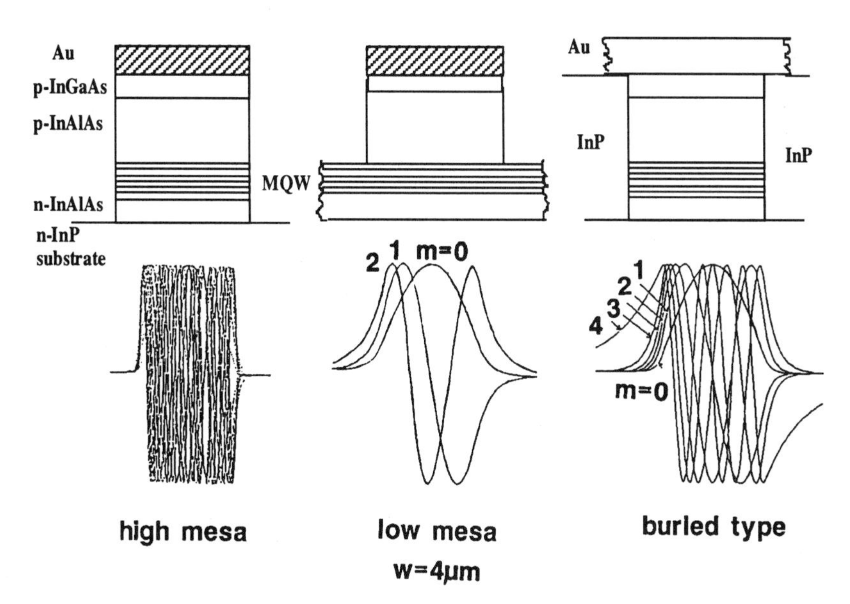

Fig. 2.7. Calculated field distributions for MQW cores associated with high mesa, low mesa and buried type waveguide.

Fig. 2.2 (a) and (b) are called rib type and ridged type, respectively, and have air on either side of the waveguide channel, which results in highly multimode guided light propagation because of the large dielectric discontinuity in the transverse direction. The guided light profile typically fills the channel and no field penetrates into the air. Therefore, the rib sidewall roughness severely affects the propagation and the loss is relatively large. However, the overlap field between the optical and the electric is effective and a single mode is launched for a relatively short waveguide when a single mode is incident as will be described later on.

By contrast, the strip-loaded guide (Fig. 2.2(c)) typically supports only one mode and the tail of the light distribution outside the guide can be large or small, depending on the amount of the loading. The imbedded guide (Fig. 2.2(d)) has properties between those of the ridged and strip-loaded guides and the number of propagating modes depends on the dielectric discontinuity between the channel guide and the imbedding medium.

Figure 2.7 shows the sectional views of InGaAs/InAlAs MQW optical modulators and their field distributions. Based on the usual waveguide theory, this guide structure may be multimode. Figure 2.7

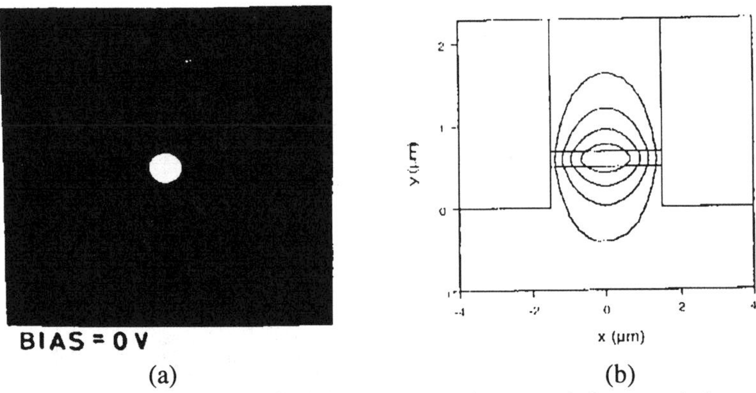

Fig. 2.8. Near field pattern observed for a slab waveguide MQW modulator with no applied bias(a), and its calculation by finite-element method (b).

shows the field distribution for MQW cores, which is calculated by the slab waveguide theory. Certainly, multimode is launched even for a narrow strip width. However, in the experiment, we can observe quasi-single mode when we lauch the waveguide by a single mode fiber. Figure 2.8 shows the near field pattern for this waveguide with a strip width of 3 μm and the length of 100 μm.

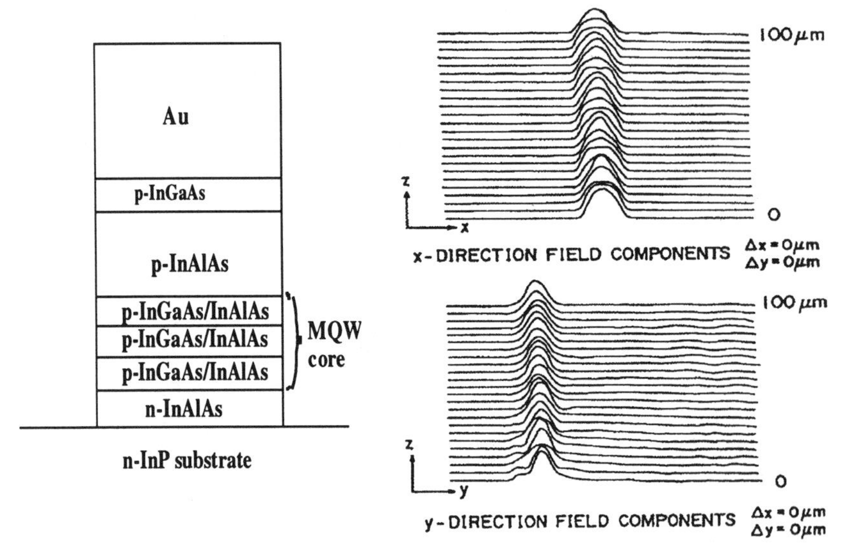

Fig. 2.9. Calculated field distribution for an MQW core with the cross-section shown in the left using three dimensional BPM [10].

Figure 2.9 shows the x-and y-directed propagating field distributions calculated using the three demensional BPM for the MQW waveguide where the core thickness and waveguide length are 1.0 and 100 μm, respectively[10]. The input and output of the single-mode fibers are assumed to be Gaussian beams of 1.5 μm spotsize. The calculated total insertion losses are 3.2 and 4.6 dB for thin and thick core structures, respectively. These values do not include the Fresnel reflections. Note that quasi-single mode propagation can be observed even for the thick core structure.

The strip-loaded waveguide has been also investigated in order to reduce propagation loss [11], especially on the rib sidewall scattering, rib etch depth, and etch damages.

The imbedded waveguide structure has been done in the buried heterostructure using liquid phase epitaxy (LPE) at first and recenly metal organic vapor phase epitaxy (MOVPE). At present, semi-insulating InP is used for the buring layer in order to improve overlap field between the optical and the electric, and to reduce device capacitance for high speed operation. This structure also enables to operate stable single-mode and reduce the coupling loss between the optical fiber.

2.6 Mode Spot Size

Highly efficient and reliable fiber modules for semiconducor devices are necessary to construct optical communications systems and networks. Spot-size matching of optical fibers and semiconductor devices is key technology. When the optical modes of typical semiconductor waveguide devices such as modulators and switches are directly coupled to flat-end single mode fibers, the losses easily reach as much as 10 dB per facet. Even if we use a micro-optical setup with specially prepared fiber ends, the coupling losses are relatively large in spite of the critical alignment tolerances. This is due to the large mode mismatch between single-mode fibers and semiconductor devices. In particular, the large insertion loss inherent in semiconductor modulators and switches, which is their only drawback compared with $LiNbO_3$ modulators, is due to this coupling loss mainly. Besides, micro-optical setups are not suitable for mass production and long-term stability.

Various types of integrated optical modeshape adapters that are directly integrated on semiconductor chips have been proposed since the first report by Koch [12]. Vertical-taper types[12,13,15,18], two layer lateral-taper types [14], and the selectively grown vertical and lateral types [16,17] have been demonstrated in which the spot size of the semiconductor devices is converted to the spot size of the optical fiber. They are designed for low-loss optical coupling to single-mode fibers while maintaining relatively large alignment tolerances.

Recenly a semiconductor phase modulator [19] and 2x2 directional coupler waveguide switch modules [20] integrated with spot size converters have been reported.

2.7 Concept of Integrated Modeshape Adaptors

Figure 2.10 shows how ultra-thin etch-stop fabrication techniques were first applied to adiabatically expand the vertical optical mode size. The principle of this expansion of mode size is the stepped reduction in thickness of a waveguide core to obtain a narrow divergence output

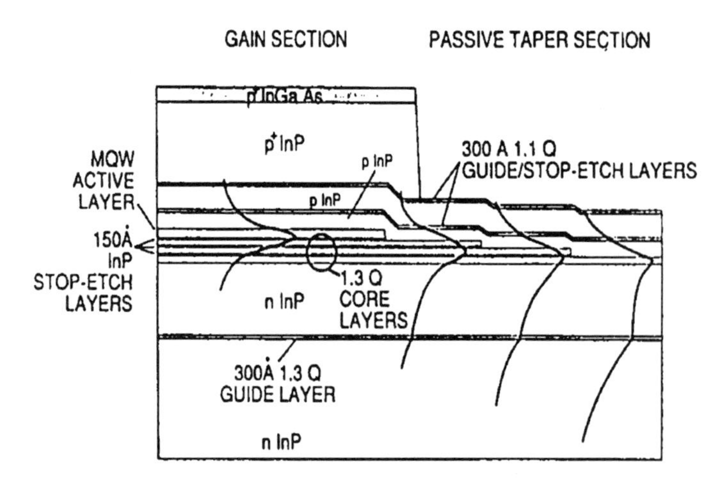

Fig. 2.10. Longitudinal structure through the buried heterostructure core for a simple two-step passive adiabatic mode expansion taper [12].

beam.

Fig. 2.11 shows the schematic view of a spot size converter with a lateral taper where no thickness variation of any of the waveguides forming the beamwidth transformer is needed [14]. The spot-size transformer consists of three sections; Section I contains a small-spot waveguide for optical interconnections on the chip itself. The spot size transformation is performed by a two-layer taper in section II, where the width of the upper waveguiding layer decreases linearly to a fine wedge-like end. The width of the lower waveguide increases to that of

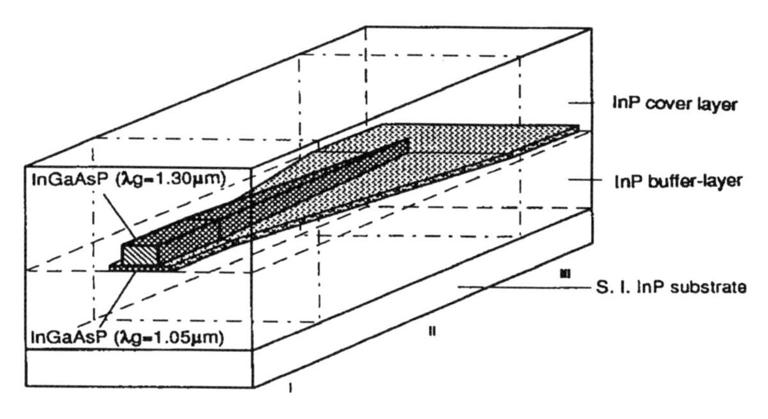

Fig. 2.11. Schematic drawing of the laterally tapered spot-size transformer [14].

the output waveguide in section III.

A phase modulator with spot converters has been reported [18]. This device is composed of a phase modulator section and a spot size convertion section as shown in Fig. 2.12. It has a primary core and a secondary core. The former consists of 28 pairs of InGaAlAs (9nm)/ InAlAs(5nm) MQW that are 0.4 μm thick. Its exciton absorption peak wavelength is 1.44 μm. The equivalent refractive index of this core is 3.45, which gives a small spot size. To gradually make the spot size as large as that of optical fibers, the core is laterally tapered in the spot-size -conversion section. The secondary core is located under the primary core and consists of a 3.1-μm-thick InP(240 nm)/InAlAs(20 nm) MQW. The refractive index difference between the secondary core and the InP clad layer was adjusted to 0.1% by controlling the ratio of InP and InAlAs layer thickness in the MQW.

The total insertion loss (measured from fiber to fiber) is 12.5 dB, and that of the phase modulator without the spot size conversion structure is 17.2 dB. Loss reduction definitely improves fiber coupling properties with spot size conversion. The propagation loss and fiber coupling loss per facet is 5.5 and 0.8 dB, respectively. Residual loss of 5.4 dB is considered to be due to mode conversion loss.

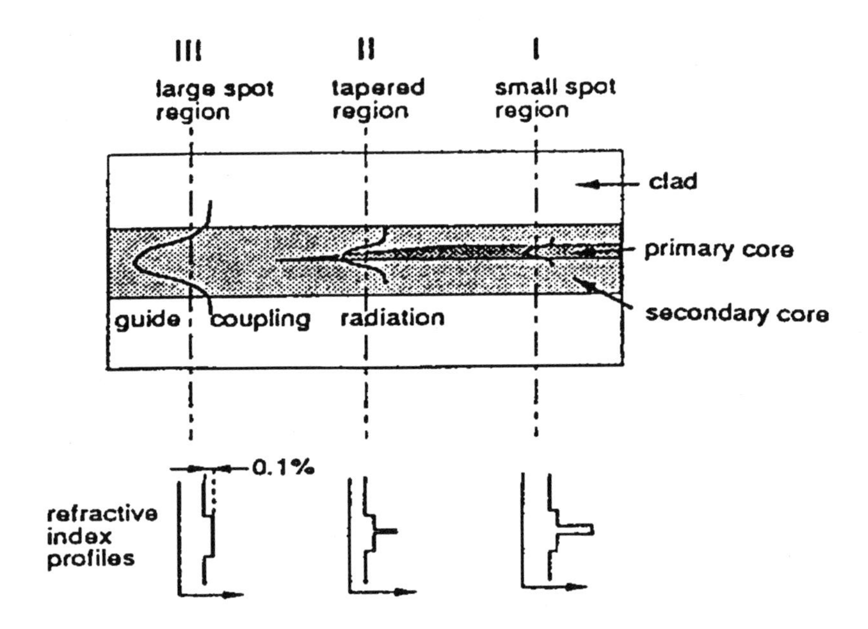

Fig. 2. 12. Cross sectional view of the spot-size converter for a phase modulator.

For 2x2 directional coupler waveguide switch modules integrated with spot size convertors, the connector-to-connector total insertion loss and crosstalk are, on average, 16 and -15 dB, respectively [20]. The spot size conversion loss, including coupling loss to arrayed polarization-maintaining (PANDA) single mode fibers is, on average, 2.3 dB per port.

2.8 References

[1] E. A. J. Marcatili, Bell Syst. Tech. J., 48, 2071, 1969.

[2] T. Tarmir (ed.), *"Integrated Optics*", Springer-Verlag, Chap. 2, 1975.

[3] K. Okamoto, *"Theory of Optical Waveguides"*, CORONA PUBLISHING CO. LTD (in Japanese), 1992.

[4] H. Kogelnik and V. Ramaswamy, Appl. Opt., 13, 1857, 1974.

[5] M. Matsuhara, J. Opt. Soc. Amer., 63, 1514, 1973.

[6] W. Austin, J. Lightwave Technol., LT-2, 688, 1984.

[7] M. D. Feit and J. A. Fleck, Jr., Appl. Opt., 17, 3990,1978.

[8] M. D. Feit and J. A. Fleck, Jr., Appl. Opt., 19, 2240,1980.

[9] M. Erman, N. Vodjdani, P. Jarry, D. Graziani and H. Pinhas, J. Lightwave Technol., LT-4, 1524, 1986.
[10] K. Kawano, K. Wakita, O. Mitomi, I. Kotaka, and M. Naganuma, IEEE J. Quantum Electron., 28, 224, 1990.
[11] R. Deri and E. Kapon, IEEE J. Quantum Electron., 27, 626, 1991 and detailed references are therein.
[12] T. L. Koch, U. Koren, G. Eisenstein, M. G. Young, M. Oron, C. R. Giles, and B. I. Miller, Photon. technol. Lett., 2, pp.88-90, 1990.
[13] G. Mueller, B. Stegmueller, H. Westermeier, and G. Wenger, Electron. Lett., 27, 1836, 1991.
[14] R. Zengerle, H. Bruckner, H. Olzhauzen, and A. Kohl, Electron. Lett., 28, 631, 1991.
[15] T. Brenner, W. Huziker, M. Smit, M. Bachmann, G. Guekos, and H. Melchior, Electron. Lett., 28, pp. 2040-2041, 1992.
[16] N. Yoshimoto, K. Kawano, H. Takeuchi, S. Kondo, and Y. No guchi, Electron. Lett., 28, pp. 1610-1611, 1992.
[17] R. J. Deri, C. Caneau, E. Colas, L. M. Schiavone, N. C. Andreadakis, and G. H. Song, Appl. Phys. Lett., 61, 952-954, 1992.
[18] T. Brenner and H. Melchior, Photon. Technol. Lett., 5, 1053-1056, 1993.
[19] N. Yoshimoto, K. Kawano, Y. Hasumi, H. Takeuchi, S. Kondo, and Y. Noguchi, Photon. Technol. Lett., 6, 208-210, 1994.
[20] K. Kawano, M. Kohtoku, N. Yoshimoto, S. Sekine, and Y. No guchi, Electron. Lett., 30, pp. 353-354, 1994.

Chapter 3

Electrooptic Modulation

3.1 Modulation Configuration

It is always necessary to use modulators in order to modulate the lightwave with signals including information. There are three main methods in modulation; direct, internal modulation, and external modulation. The direct modulation method is represented by semiconductor laser diodes, where injection current for the laser diode is changed by signals. This has been used for a long time for lightwave transmission systems. The second method is used for changing the Q in the cavity of solid state lasers by inserting the modulators. The lasing condition is very sensitive to losses in the cavity and only a low percentage modulation index results to sufficiently operate the lasers because a small change in the loss can stop even lasing. This method is highly efficient, but it is difficult to obtain stable lightwave signals due to the relaxation oscillation of lasers associated with modulation. However, this method is necessary to use a laser as actively mode-locking that genarates ultrashort and repetitive optical pulses fixed by the cavity length. The third method is very popular and various types are used when its large insertion loss is not a serious problem. The structure of internal modulators is also similar to that of external modulators. Therefore, in this chapter we focus mainly on the external modulators.

Light modulation is defined as the change of (1) the amplitude, (2) phase, (3) polarization, or (4) frequency of the incident light wave by changing the refractive index, absorption coefficient, or the direction of the transmitted light in the medium where a laser light is incident. This situation is shown in Fig. 3.1. In these categories, frequency modulation is very difficult to create due to the great energy needed and not used for normal use. Therefore, the above (1), (2), and (3) methods are used for modulation. However, frequency modulation is attractive for wavelength-multiplexing, and is relatively easy for diode lasers. Therefore some descriptions of the frequency modulation will be given in Chapter 5, where optical switching is discussed.

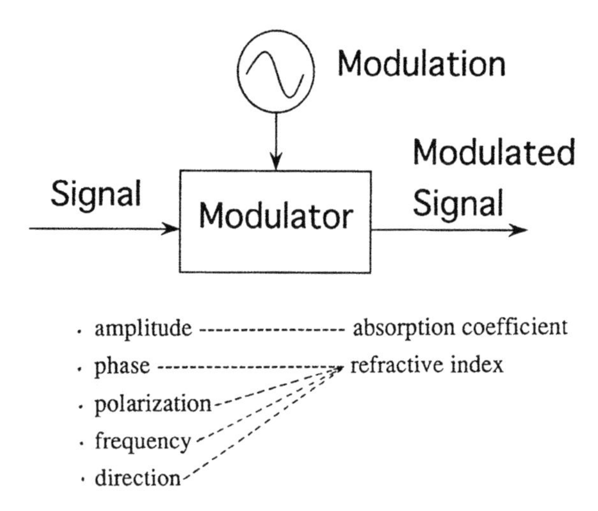

Fig. 3.1. Principle of modulation.

Figure 3.1 shows the various external forces and physical effects or phenomena interacting with the light. Modulation of the direction change of the light beam by a mechanical method is simple and easy, but its frequency is very low and is not used except in special uses. The absorption change is easily done for semiconductors, but usable wavelengths are limited. The method of refractive index change is efficient and there are many kinds of modulators using this method. Electro-optic, acousto-optic, and magneto-optic effects are methods using a refractive index change, which is used for practical modulators.

3.2 Electrooptic Effect

The optical properties of a crystal are frequently described in terms of the index ellipsoid (or indicatrix). The index of refraction of a crystal is described for the three arbitrary Cartesian coordinates (x_1, x_2, x_3) as the following ellipsoid surface:

$$x_i x_j / n_{ij}^2 = 1,$$

where i and j indicate the numbers 1, 2, and 3 and $n_{ij}=n_{ji}$. This ellipsoid is called an index ellipsoid or indicatrix. When we describe the principle axes by x, y, and z, the index ellipsoid is expressed as follows.

$$x_1^2/n_1^2 + x_2^2/n_2^2 + x_3^2/n_3^2 = 1,$$

where the coordinates x_i are parallel to the axes of the ellipsoid and n_i are the principal refractive indices. The properties of the indicatrix can be seen from a simple example. If a wavefront has its normal in the x_3 direction, then we consider the ellipse formed by the indicadrix and the x_3=0 plane. The wave has components polarized along x_1 and x_2 and indeces given by the semi-axes of the ellipse (n_1 and n_2). In a more general case, the direction of normal wave can be chosen arbitrarily and the two indices are obtained as the semi-axes of the elliptical section perpendicular to the arbitrary direction; That is, the concrete quantity of refractive index is not generally determined unless the incident light direction and polarization plane are designated for the principal axes of the index ellipsoid. However, their changes are described based on the changes of $(1/n_{ij}^2)$ under the external forces.

When the external forces of electric field intensity $E(E_1, E_2, E_3)$ and stress $T(T_{11}, T_{22}, T_{33}, T_{23}, T_{31}, T_{12})$ are applied to a medium, the refractive index changes can be written as

$$1/n_{ij}^2=(1/n_{ij}^2)_0 + \gamma_{ijk}E_k + 1/2\, g_{ijkl}E_kE_l + \pi_{ijkl}T_{kl},$$

where i=j indicates the change along the principal axes and i=j not along it. E_k is a vector, T_{kl} is the 2nd tensor, γ_{ijk} is the 3rd tensor, g_{ijkl}, and π_{ijkl} are the 4th tensor components, respectively. The indices i and j can be interchanged, as can k and l, so the usual contraction can be made: γ_{mk}—-$\gamma_{(ij)k}$ and g_{mn}—$g_{(ij)(kl)}$, where m and n run from 1 to 6 and m is related to (ij) and n to (kl) as follows: 1--->11, 2--->22, 3--->33, 4--->23, 5--->13, 6--->12; for example, γ_{13}---γ_{52}, g_{2322}--g_{42}. The number of matrix elements are E(3x1), T(6x1), γ(6x3), g(6x6), π(6x6), respectively. n_{ij0} indicates the refractive index at zero bias condition.

The γ_{mk} are linear electro-optic tensor components and called Pockels constants. If the material has a center of symmetry, reversing the sense of the applied field E does not change the physical situation and, in particular, $1/n^2$ will be independent of the sign of E. Terms of odd power in E in the above equation will change sign, however, so the coefficients of these terms must vanish in centrosymmetric materials. Only noncentro-symmetric crystals can produce a linear effect. There is, of course, no restriction on the terms of even powers of E; that is,

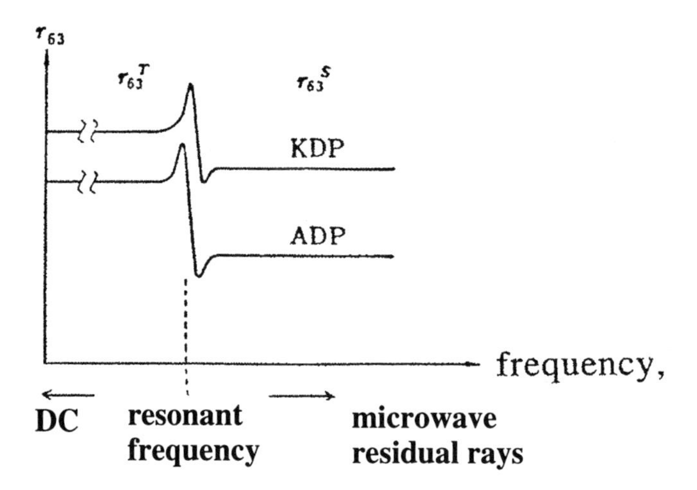

Fig. 3.2 The relation between Pockels constant $\gamma_{63}{}^{T}$ and γ^{S} versus electric field frequency.

Pockels effect exists in 20 crystal classes of 32 classes to which all crystals belong to and are classified into. These crystals also have piezoelectric effect. On the other hand, the crystals have centro-symmetry in some directions and the independent coefficients decrease in 18 numbers.

The g_{mn} are quadratic electro-optic tensor components and called Kerr constants. The Kerr constants are the 4th tensor and do not vanish under the symmetrical operation described above. Therefore, they exist in any symmetrical crystals such as isotropic materials as glass. The p_{mn} is a coefficient related to the photoelastic constant and exists even in an isotropic materials such as the g_{mn}. The Kerr effect is a second order effect and therefore its coefficients are generally very small. This results in a large modulating voltage when its effect is used for modulators. Consequently, this effect is not used except in special uses such as for materials that have no or a very small Pockels effect.

If the crystal develops macroscopic strain under the influence of the electric field, there will be a change in index through the elastooptic effect. All solids exhibit an elastooptic effect and all solids are strained by an electric field, either through the converse piezoelectric effect (linear effect) or through electrostriction effect (quadratic effect). When these effects are considered, the electrooptic effect γ and g are given as the following:

$$\gamma_{mn} = \gamma_{mn}{}^S + \pi_{mn} e_{kn}$$

$$g_{mn} = g_{mn}{}^S + \pi_{mn} Q_{kn}$$

$$m,n=1,----,6, \ k=1,2,3.$$

where e_{mn} and Q_{kn} are a piezoelectric constant and an electrostriction constant, respectively. The resultant secondary effect can be shown to depend on crystal symmetry in the same way as the direct effect. The above equations mean that if the driving field frequency is sufficiently higher than that corresponding to the acoustic resonance of the material, a change in the material cannot follow the frequency and the strain is zero (S=0). In this condition, γ^S is used. On the other hand, if the frequency is lower than that of the piezoelectric resonance frequency ($f_r=v_A/d$; v_A is the sound velocity, d is the contact distance), the crystal change can follow the modulation frequency and stress is zero (T=0). In this condition, γ^T is used. The optical modulation index changes for the modulation frequency as shown in Fig. 3. 2 and the suppression of piezoelectric resonance in the modulator is important. However, the crystals used in the practical condition are sufficiently fixed and the influence from the piezoelectric effect is negligible.

It may be useful to consider an example of the use of the indicatrix in describing the electrooptic effect.

3.2.1 T_d = 43m (ZnS) Type Crystals

The semiconductor materials commonly used in optoelectronics are GaAs and InP as substrate, and the alloys which can be lattice matched to these substrates, GaAlAs for GaAs and GaInAsP or InGaAlAs for InP. These types belong to the cubic and the indicatrix is spherical when the electric field is not applied($n_x=n_y=n_z=n_0$).

In T_d type crystals, the only nonzero electrooptic coefficients are the γ_{41} [1-4]:

$$\gamma_{ij} = \begin{pmatrix} 0 & 0 & 0 \\ 0 & 0 & 0 \\ 0 & 0 & 0 \\ \gamma_{41} & 0 & 0 \\ 0 & \gamma_{41} & 0 \\ 0 & 0 & \gamma_{41} \end{pmatrix}$$

For crystal in which only the linear effect is appreciable, if the applied field is (E_x,E_y,E_z), then the equation of the indicatrix is

$$1/n_0^2(x^2 + y^2 + z^2) + 2\gamma_{41}(E_x yz + E_y zx + E_z xy) = 1$$

A.1. When the electric field is applied along the {001} direction and light propagates along the {110} direction, the above equation is

$$1/n_0^2(x^2 + y^2 + z^2) + 2\gamma_{41}Exy = 1,$$

where the electric field E_x=E_y=0, E_z=E. By rotating the coodinates 45 deg. around the z-axis, the above equation can be diagonalized, and it becomes

$$(1/n_0^2 + \gamma_{41}E)X^2 + (1/n_0^2 - \gamma_{41}E)Y^2 + 1/n_0 Z^2 = 1,$$

where the principal refractive indices n_1 and n_2 are

$$n_1 = n_0 + (1/2)\gamma_{41}n_0^3E$$

$$n_2 = n_0 - (1/2)\gamma_{41}n_0^3E$$

$$n_3 = n_0$$

A.2. When the electric field is applied along the {110} direction and light propagates along the {110} direction, the indicatrix equation becomes

$$1/n_0^2(x^2 + y^2 + z^2) + \sqrt{\ } 2\gamma_{41}E(xy + zx) = 1,$$

where the electric field E_y=E_z=E/$\sqrt{2}$, E_x=0. In this case the principal refractive index is

$$n_1 = n_0 + (1/2)\gamma_{41}n_0^3E$$

$$n_2 = n_0 - (1/2)\gamma_{41}n_0^3E$$

$$n_3 = n_0$$

A.3. When the electric field is along the {111} direction, and $E_x=E_y=E_z=E/\sqrt{3}$, the indicatrix equation becomes

$$1/n_0^2(x^2 + y^2 + z^2) + 2/\sqrt{3}\gamma_{41}E(yz + zx + xy) = 1$$

The principal refractive index becomes

$$n_1 = n_0 + (1/2\sqrt{3})\gamma_{41}n_0^3E$$

$$n_2 = n_0 + (1/2\sqrt{3})\gamma_{41}n_0^3E$$

$$n_3 = n_0 - (1/\sqrt{3})\gamma_{41}n_0^3E$$

3.2.2 C_{6m}=6mm Type Crystals

Popular semiconductors such as ZnO, ZnS and CdS belong to this type and the only nontrivial electrooptic coefficients are

$$\gamma_{ij} = \begin{pmatrix} 0 & 0 & \gamma_{13} \\ 0 & 0 & \gamma_{13} \\ 0 & 0 & \gamma_{33} \\ 0 & \gamma_{51} & 0 \\ \gamma_{51} & 0 & 0 \\ 0 & 0 & 0 \end{pmatrix}$$

The indicatrix becomes

$$(1/n_0^2 + \gamma_{13}E_z)(x^2 + y^2) + (1/n_e^2 + \gamma_{33}E_z)z^2$$
$$+2\gamma_{51}E_y yz + 2\gamma_{51}E_x xz = 1,$$

where n_e is extraordinary refractive index.

B.1 $E_x=E_y=0, E_z\neq0$. In this case, the principle axis does not change but the above equation is changed to the following:

$$(x^2 + y^2)/(n_0^2(1 - 1/2n_0^2\gamma_{13}E_z)^2) + z^2/(n_e^2(1 - 1/2n_e^2\gamma_{33}E_z)^2) = 1.$$

That is,

$$n_x = n_y = n_0 - 1/2n_o^3\gamma_{13}E_z$$

$$n_z = n_e - 1/2n_e^3\gamma_{33}E_z.$$

When the incident light is parallel to the x-axis, the phase difference is given as the following:

$$\Gamma_x=(2\pi L/\lambda)\{(n_o - n_e) + 1/2n_e^3[\gamma_{33} - (n_o/n_e)^3\gamma_{13}]V_z/d\},$$

where d is the thickness of the crystal along the z-axis, L is the length of the sample along the x-axis direction and V_z is the applied voltage.

B.2 $E_x=E_z=0, E_y\neq0$. The above indicatrix is given as the following:

$$(x^2 + y^2)/n_o^2 + z^2/n_e^2 + 2\gamma_{51}E_y yx = 1$$

The priciple axis changes 45 degrees along the z-axis; x' and y'. Then,

$$x'^2/n_o^2(1 + 1/2n_o^2\gamma_{51}E_y)^2 + y'^2/n_o^2(1 - 1/2n_o^2\gamma_{51}E_y)^2 + z^2/n_e^2 = 1$$

That is,

$$n_{x'} = n_o + 1/2n_o^3\gamma_{51}E_y$$

$$n_{y'} = n_o - 1/2n_o^3\gamma_{51}E_y$$

$$n_{z'} = n_z = n_e$$

When the thickness of the y-direction and of the z-direction are d and L, respectively, the phase difference along the z-axis is given as the

Table 3.1. Electrootic coefficient for several III-V semiconductors.

$T_d = \bar{4}3m$ Pockels constant

Material	Symmetry	γ_{nl}	λ_1 (μm)	n	λ_2 (μm)	ε
ZnS	$\bar{4}3m$	(T) $\gamma_{41}=1.2$	0.40	$n_0=2.471$	0.45	(T) 16
		2.0	0.546	2.364	0.60	(S) 12.5
		2.1	0.65	2.315	0.8	8.3
ZnSe	$\bar{4}3m$	(T) $\gamma_{41}=2.0$	0.546	$n_0=2.66$	0.546	9.1
						8.1
ZnTe	$\bar{4}3m$	(T) $\gamma_{41}=4.55$	0.59	$n_0=3.1$	0.57	10.1
		3.95	0.69	2.91	0.70	
		(S) $\gamma_{41}=4.3$	0.63			
CuCl	$\bar{4}3m$	(T) $\gamma_{41}=6.1$		$n_0=1.996$	0.535	(T) 10
		(T) $\gamma_{41}=1.6$		1.933	0.671	(S) 8.3
						(S) 7.7
CuBr	$\bar{4}3m$	(T) $\gamma_{41}=0.85$		$n_0=2.16$	0.535	
				2.09	0.656	
GaP	$\bar{4}3m$	(S) $\gamma_{41}=0.5$		$n_0=3.4595$	0.54	10
		(S) $\gamma_{41}=1.06$	0.63	3.315	0.60	12
GaAs	$\bar{4}3m$	(T) $\gamma_{41}=0.27\sim1.2$	$1\sim1.8$	$n_0=3.60$	0.90	(T) 12.5
				3.50	1.02	(S) 10.9
		(S−T) $\gamma_{41}=1.3\sim1.5$	$1\sim1.8$	3.42	1.25	(S) 11.7
		(S) $\gamma_{41}=1.2$	$0.9\sim1.08$	3.30	5.0	
		(T) $\gamma_{41}=1.6$	3.39, 10.6			

γ_{nl} : Pockels constant (10^{-12}m/V) λ_1 : γ_{nl} measurement wavelength (T) : stress constant (S) : strain constant n : refractive index
ε : dielectric constant λ_2 : n measurement wavelength

Table 3.2. Electrootic coefficient for several II-VI semiconductors.

$C_{6\nu}$ = 6mm Pockels constant

Material	Symmetry	γ_{nl}	λ_1 (μm)	n	λ_2 (μm)	ε
ZnO	6mm	(S) $\gamma_{33}=2.6$	0.63	$n_z=2.123$	0.45	$\varepsilon \approx 8.15$
		(S) $\gamma_{13}=1.4$	0.63	$n_y=n_x=2.106$	0.45	
		$\gamma_{33}/\gamma_{13}<0$		$n_z=2.015$	0.60	
				$n_x=n_y=1.999$	0.60	
ZnS	6mm	(S) $\gamma_{33}=1.85$	0.63	$n_z=2.709$	0.36	
		(S) $\gamma_{13}=0.92$	0.63	$n_y=n_x=2.705$	0.36	
		$\gamma_{33}/\gamma_{13}<0$		$n_z=2.368$	0.60	
				$n_y=n_x=2.363$	0.60	
CdS	6mm	(T) $\gamma_{51}=3.7$	0.589	$n_z=2.726$	0.515	(T) $\varepsilon_1=10.6$
		(T) $\gamma_c=4$	0.589	$n_y=n_x=2.743$	0.515	(T) $\varepsilon_3=7.8$
		(S) $\gamma_{33}=2.4$	0.63	$n_y=n_x=2.493$	0.60	(S) $\varepsilon_1=8.0$
		(S) $\gamma_{13}=1.1$	0.63			(S) $\varepsilon_3=7.7$
		$\gamma_{33}/\gamma_{13}<0$				

γ_{nl} : Pockels constant(10^{-12}m/V) λ_1 : γ_{ne} measurement wavelength n : fractive index ε : dielectric constant (T) : stress constant
(S) : starin constant λ_2 : n measurement wavelength

following:

$$\Gamma_z = (2\pi L/\lambda) n_o^3 \gamma_{51} L/dV_y$$

Tables 3.1 and 3.2 shows the electrooptic coefficients for various semiconductors as well as typical dielectric materials used for modulators. Usually the EO modulators are used under the condition that the phase change is maximum at the given voltage, where the incident light polarization affects on the characteritics. In some conditions, this polarization dependence is solved by using (111) substrate [5,6] or the waveguide at 45 degrees to the {011} cleavage plane on (100) substrate [7], though some more voltages are to be added and polarization insensitivity has been achieved. The detailed discussions for the polarization insensitivity will be carried out in Chapter 10.

3. 3 Phase Modulators

3.3.1 Introduction

Refractive index changes in semiconductors play an important role in optical devices such as couplers, switches, modulators, and lasers. In this section, we introduce and discuss the contributions to refractive index changes from various sources, e.g., the linear electrooptic (LEO), quadratic electrooptic (QEO), band-gap shift (BS), and free carrier plasma (PL) effects. As for the linear electrooptic effect, the details were described in the previous section and we compare this effect with other effects quantitatively in this section. At first we review the above effects for bulk materials. Next we introduce the effects for MQW structures.

Semiconductor waveguide phase modulators are better suited than the absorption modulators since they use only a refractive index change, without an absorption change. The combination of the phase modulation element with an optical interferometer or directional coupler geometry can provide amplitude modulation or optical switching. Absorption modulators, although somewhat simple in concept and low operating voltages, are not quite versatile, and potential problems with heating, wavelength sensitivity, and disposal of the generated carrier exist. Absorption modulators will be discussed in other sections.

By utilizing the various refractive index changes mentioned above, semiconductor phase modulators with efficient phase shifts and low drive voltages have been reported [8,9]. The phase modulators usually employ a p-n junction double heterostructure (DH) in order to inject or deplete free carriers from the junction and to guide light efficiently. They are often operated in a reverse biasd condition because of their beneficial high-speed characteristics with small device capacitance. On the other hand, forward-biased modulators provide much higher efficiencies, though their speed characteristics are limited by their carrier lifetime, which is usually a few ns. Under reverse bias, depletion of the free carriers away from the junction causes an electric field to develop across the depletion region, resulting in a local refractive index change and the phase of the propagating wave.

In the first study of modulators based on GaP p-n junction waveguides, Reinhart, Nelson and McKenna [10] showed that the dominant contribution to phase modulation is due to the LEO effect with possible free carrier and QEO effect contribution. The link between QEO effect and electroabsorption was demonstrated on DH waveguides [11].

The high efficiency of doped modulators was not recognized until recently [8,9,12], because usually birefringence measurements, rather than absolute phase measurements, are performed. The isotropic carrier contributions obviously disappear in birefringence measurements [13]. Both the electrooptic and free carrier effects are involved in the refractive index changes [13,14]. The electrooptic effect includes the LEO and QEO effects and the free carrier effect includes the PL and bandgap shift (BS) effects; the BS effect is the result of both the band-filling (BF) and bandgap shrinkage (BGS) effects. These effects have been analyzed in detail in refs 12-14. Though the reports are only for bulk or DH, and the best data are in the order of 60-100 degree/V mm, being lower than those of MQW or BRAQWETS which were recently developed. We give an overview of the various effects in this section.

3.3.2 Free Carrier Effects

Carrier and electric field effects are treated separately because the electric field vanishes in the regions occupied by free carriers. The refractive index change due to free carriers is associated with variations of two kinds of absorption phenomena. One is the change of free carrier absorption in the conduction band or valence band which leads to the PL effect. The other is the change of fundamental absorption edge which leads to the BS effect. These absorption changes are related to the refractve index change through the Kramers-Kronig dispersion relation.

A. Plasma effect. The plasma effect is due to scattering of light by free carriers (electrons or holes). The refractive index variation induced at frequency ω is given by

$$\Delta n = -(n/2)\,(\omega_p/\omega)^2,$$

where n is the refractive index of the material and

$$\omega_p^2 = e^2 N_{e,h} / \varepsilon m_{e,h}^*$$

is its plasma frequency. $N_{e,h}$ is the carrier concentration (electrons or holes), $m_{e,h}^*$ their corresponding effective mass, ε the dielectric constant of the material, and e the electric charge.

B. Bandgap shift. Since the band-filling and bandgap shrinkage effects occur simultaneously by the presence and removal of free carriers, the BS effect indicates both these effects. The refractive index change Δn_{BS} due to the BS effect has been studied extensively for doped GaAs [14,15] and the carrier injection in semiconductor diode lasers [16-18]. The semi-empirical expressions of Δn_{BS}, including the band-filling and bandgap shrinkage effects, have been reported [13]. A linear dependence of Δn_{BS} on carrier concentration is reported by Henry [18]. The empirical equation for the coefficient in $Al_xGa_{1-x}As$ was duduced by Lee et al.[12] as:

$$\Delta n_{BS} = B(\lambda)N$$

where

$$B(\lambda) = S_b / \{2n[E_b(x)^2-E^2]\}.$$

$B(\lambda)$ is the BS coefficient. $E_b(x)$ and S_b are the position and strength of the single-oscillator for the absorption change near the bandgap by free carriers. $E_b(x) = 1.45 + 1.247x$, where x is the AlAs mole fraction. $S_b = 1.37 \times 10^{-20}$ eV^2 cm^3 for n-type and 3.91×10^{-21}eV^2 cm^3 for p-type. This equation shows an excellent agreement with Henry's data [9] up to E = 1.4 eV. $B(\lambda)$ is a strong function of wavelength and increases drastically as the wavelength approaches the bandgap value due to the absorption change near the bandgap.

3.3.3 Comparison with Various Refractive Index Changes

Various mechanisms as discussed in the previous section contribute to changing the refractive index. For relatively longer wavelengths, the LEO effect is the major contribution to the phase change, followed by QEO, BS and PL in order of decreasing importance. However, for wavelengths very close to the absorption edge, the BS and QEO effects will dominate. For an electric field along the {100} direction and TE modes propagating along the $\{1\bar{1}0\}$ and {110} directions, the the total phase shift is obtained as a sum of the individual effects. The PL, BS, and QEO contributions have the same sign regardless of the propagation direction and polarization, whereas the LEO contribution is positive for propagating along the {1¯10} direction adding to the others, and is negative for propagating along the {110} direction. For TM modes, the total phase shift is a sum of the PL, BS, and QEO contributions without LEO contribution.

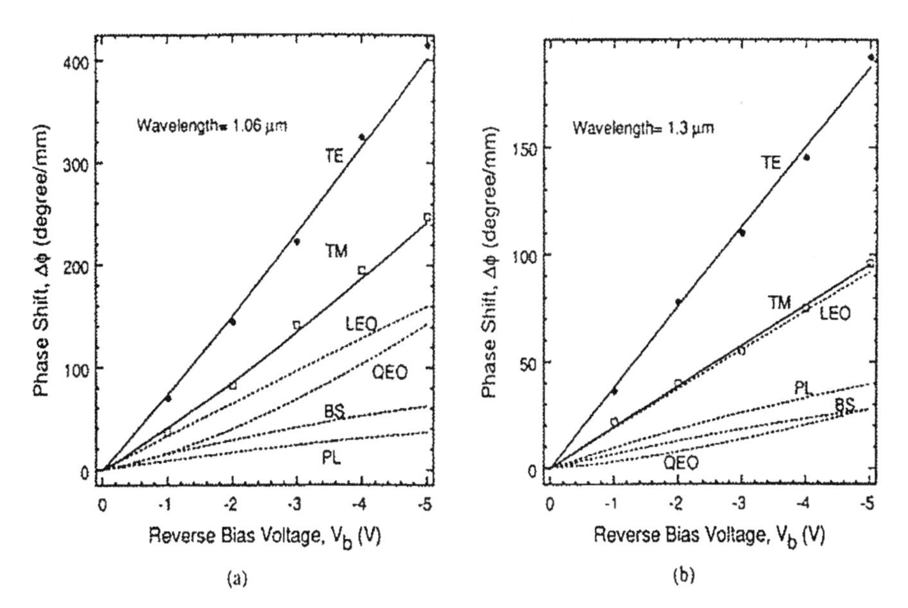

Fig. 3.3 Experimental and theoretical phase shifts as a function of reverse bias voltage at (a) λ=1.06 μm, and (b) λ=1.3 μm. The total phase shift for the TE mode is obtained by summing up individual contributions from the linear electroaptic (LEO), quadratic electrooptic (QEO), bandgap shift (BS), and free carrier plasma (PL) effects. For the TM mode, LEO contribution is not included [12].

Figure 3.3 shows the experimental and theoretical phase shifts as a function of reverse bias voltage at the wavelength of λ = 1.06 μm and λ = 1.3 μm [12]. The total phase shift for the TE mode appears to increase almost linearly with reverse bias voltage. This figure also shows individual contribution from the PL, BS, LEO, and QEO effects. The phase efficiency, defined as the phase shift per unit voltage per unit length, is estimated to be 82, and 37.5 °/V·mm for the TE mode at λ=1.06 μm and 1.3 μm, respectively.

3.3.4 EO of MQW Structures

The optical properties [20,21] of semiconductor quantum wells have received considerable attention over the past decade. A significant feature of quantum confinement is the change from a parabolic density of states for electrons and holes in bulk material to a stair-step density of states for quantum confined structures. Based on this quantum confinement, large absorption coefficient changes have been observed as described in the section 4.2. Associated with the large changes in the absorption coefficient are changes in the real part of the refractive index. From the measured electroabsorption spectrum, it is possible to calculate the electrorefraction spectrum through the Kramers-Kronig dispersion relation [22, 23]. In this section, we discuss the refractive index changes in MQW structures.

A. GaAs/AlGaAs MQW structures. Large changes of refractive index near the band edge have been reported for GaAs/AlGaAs MQW structures by direct measurement using the electroreflection method [24], which has made surface normal configuration and lacks in accuracy. Zucker and co-workers [23] used a waveguide containing only two GaAs quantum wells and having a small optical confinement factor to measure changes in the absorption coefficient $\Delta\alpha$ and Δn across the exciton resonance. Although large changes in the refractive index were observed at wavelengths close to the exciton resonances, they were accompanied by substantial absorption and electroabsorption losses. Zucker and co-workers [23, 25] have also demonstrated that the ratio $\Delta n/\Delta\alpha$ increases with the detuning from the exciton resonance and that $\Delta n/\Delta\alpha$ depends on the magnitude of the applied field.

Glick et al. [26-28] have measured electroabsorption and electrorefraction in GaAs waveguides at detunings of 30, 50, and 400 meV. At very large detunings of several hundred meV, the quantum wells behave essentially like bulk GaAs. For detunings of 30 and 50 meV, the measured phase shift is strongly quadratic with applied field. Zucker and co-workers [25] made a detailed study of electroabsorption and electrorefraction in GaAs MQW waveguides at a detuning of 65 meV from the exciton resonance. Phase shift of 50 degree/V mm were demonstrated for a 1 V swing with a dc bias of -25 V. Optimized structures, number of wells, the operating bias, the device length, and the background doping have been investigated for maximum phase change relative to the contrast and the efficiency of 27 degree/V·mm at 6 V has been reported [29].

The electrooptic coefficients of GaAs MQW waveguides at energies of 25-110 meV below the exciton resonance have been measured [30] and the quadratic electrooptic coefficient was observed to decrease rapidly with detuning. A comprehensive set of measurements on electroabsorption and electrorefraction as a function of polarization and crystal propagation direction in a region where the propagation losses are small (55-110 meV) has been reported [31], where the quadratic electrooptic coefficient is large ($1\text{-}7\times10^{-15}$ cm^2/V^2) in a wavelength bandwidth of ~25 nm and the magnitude of electroabsorption is < 0.5 dB/mm for applied fields of 0.9×10^5 V/cm. For TE polarization, the index change in a quantum well due to the QCSE is up to 5x larger than the bulk at electric fields of 1×10^5 V/cm. For TM polarization, this factor is much larger since the linear electrooptic coefficient is zero for bulk zinc-blende crystals.

B. Linear and quadratic electrooptic effect. The index change is related to the linear and quadratic electrooptic coefficients by

$$\Delta n = -n_0^3(\Gamma_{LEO}\, rF + \Gamma_{QEO}\, sF^2)/2$$

where r and s are the linear and quadratic electrooptic coefficients, respectively, n_0 is the zero-field index, Γ_{LEO} and Γ_{QEO} are the confinement factors for the linear and quadratic electrooptic coefficients, and F is the electric field. The confinement factor is the ratio of the light intensity within the active electrooptic layers to the sum of the total light intensity, as discussed in Chapter 2,

$$\Gamma = \int |E_y(x)|^2 dx \,/\, \{ \int |E_y(x)|^2 dx \},$$

where $E_y(x)$ is the field profile of the guided wave. If the depletion width is constant and the dc electric field in the depleted layer is uniform, the confinement factors are constant and independent of the applied bias voltage. Strictly speaking, the Γ_{LEO} consists of GaAs wells as well as AlGaAs cladding layers and barriers. The latter effects are not large compared with those in GaAs wells. The above equation depends on the well and barrier thickness as well as the cladding layer composition.

The relation between the applied bias voltage and electric field must be known. To calculate the electric field profile, the intrinsic layer carrier concentration must be known. As the doping is decreased below 10^{15} cm^{-3}, there is very little change in device performance. This is due to the fact that the electric field is very nearly flat at these dopings. When the doping is increased above 10^{16} cm^{-3}, the approximations begin to break down, especially for linear electrooptic coefficient [31]. The high background doping leads to an electric field profile that is not flat. Then each well has a different Δn and $\Delta\alpha$ for a given bias voltage, and so the magnitude of the overall changes induced will vary with doping of the intrinsic region. Therefore, an optimum intrinsic layer thickness and an optimum number of wells exist which are associated with the background doping level [32].

Fig. 3.4 shows the relation between the value of the electric field at the center of the guiding layer versus the applied reverse-bias voltage for background carrier concentrations of $1\text{x}10^{15}$cm^{-3} and $1\text{x}10^{16}$cm^{-3}. At low bias voltage, the linear relationship between the applied voltage and the electric field degrades, and up to certain voltages, the electric field is zero to deplete the guiding layer.

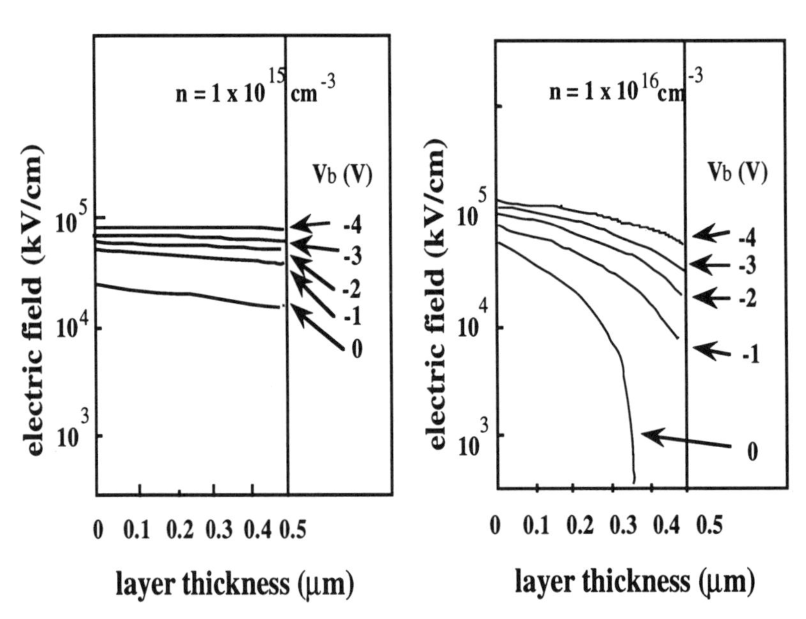

Fig. 3.4. Relation between the applied voltage and the electric field in the guide layer.

The quadratic electrooptic (QEO) coefficients in the MQW waveguide for both TE and TM polarization is on the order of 10^{-14} cm^2/V^2, which is dominant compared with the linear electrooptic (LEO) effect. This effect affects the modulation characteristics; the linear EO coefficient is enhanced by this term through the built-in potential [33]. The internal field in the depleted layer should be expressed by a summation of the F_B field from the built-in potential and the F_D field arising from the externally applied voltage. Hence, the induced refractive change is given by [33]

$$\Delta n = -n_0^3(rF + sF^2)\Gamma/2$$

$$= -n_0^3\{r(F_B + F_D) + s(F_B + F_D)^2\}\ \Gamma/2$$

$$= -n_0^3\{rF_B + sF_B^2 + (r + 2sF_B)F_D + sF_D^2\}\ \Gamma/2$$

Therefore, the induced refractive index change has an effective first order term on the applied field. The effective LEO ($r_{eff} = r+2sF_B$) is larger than the contribution from the second term, when the applied voltage is on the same order of the built-in potential. The amount of r_{eff} reaches excess on the order of 10^{-9} cm/V, which is more than one order of magnitude larger than that of the conventional Pockels effect. This means that even for TM polarization we can observe linear electrooptic effect. However, this effect is not larger than that for TE polarization.

Recently it has been reported that even for TM polarization a large electrooptic effect is obtained for thick InGaAlAs/InAlAs MQW structures with a large detuning energy condition [34]. This is considered to be due to valence band mixing where the light hole band interacts with the heavy hole band. Materials in the MQW structures, such as InGaAs, InAlAs and InGaAsP, operating at long wavelengths generally have small effective masses and their energy bands are more or less nonparabolic for energy-wavevector relations, resulting in the strong interaction. As a result, the refractive index change for TM polarization is enhanced.

Figure 3.5 shows the relationship between the refractive index change difference for TE and TM polarized light and the quantum well thickness for the same detuning energy [34]. In phase modulators the refractive

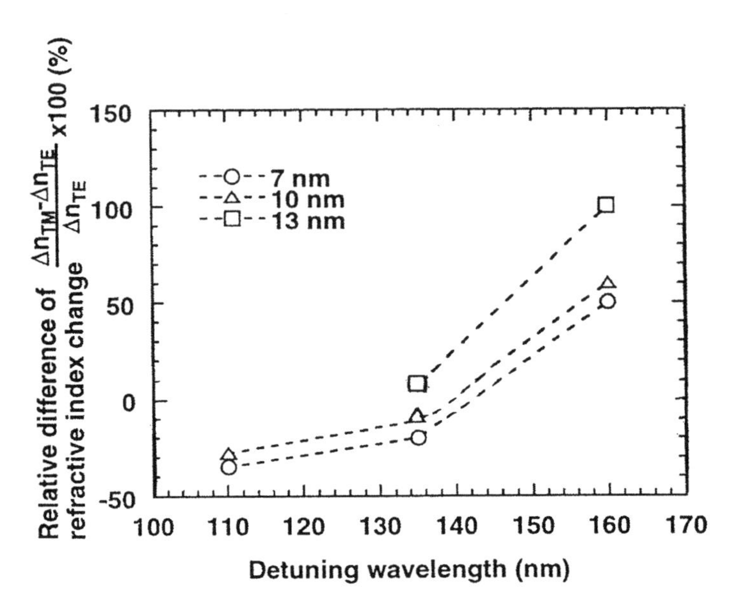

Fig. 3.5. Difference of refractive index change with applied voltage versus quantum well thickness.

index change induced by electric field intensity has a tendency to increase as the the detuning energy (energy difference between the operating wavelength and the absorption band edge) decreases, whereas the amplitude modulation increases. Usually the amplitude modulation is smaller for TM polarization than for TE polarization and, therefore, the use of TM polarization is better for phase modulators. In fact, amplitude modulation less than 0.5 dB at $V\pi$ (required voltage for π phase shift) has been achieved [34].

Based on this phenomenon and fabricating MQWs with a suitable well thickness, polarization insensitive optical switches have been fabricated [35] without any introductory tensile strain in the quantum wells. The introduction of tensile-strain in quantum wells to reduce polarization sensitivity is well-used as discussed in Chapter 10.

3.4 EO Modulators

As described in the previous section, phase modulators using bulk materials are inevitably long enough to obtain a required phase change because of their small electrooptic effects. Therefore, phase modulators

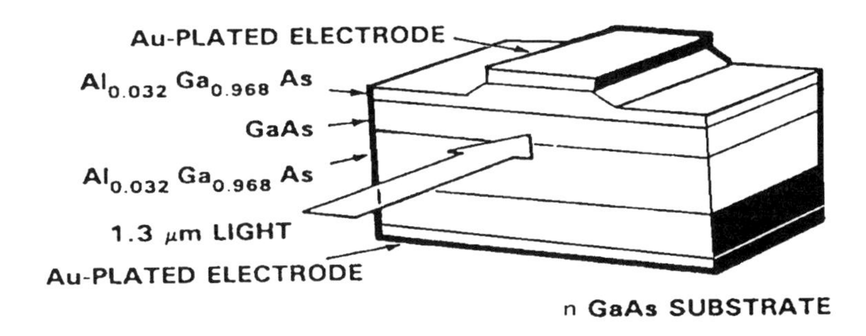

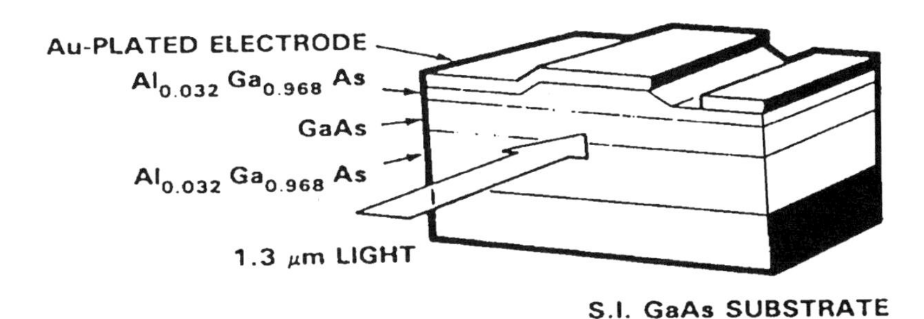

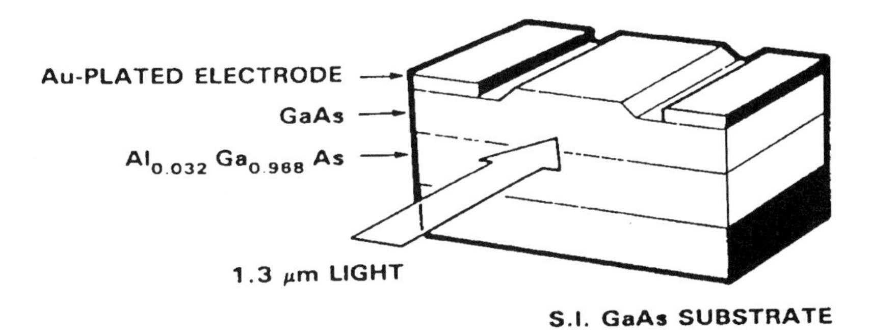

Fig. 3.6. Cross-section of three different configurations with applied electric field parallel and perpendicular to the substrate surface. (a) a p-i-n waveguide microstrip configuration, (b) a double heterostructure waveguide with coplanar strip electrodes, (c) a single heterostructure waveguide with electrodes on either side to provide electric field along {110} [36].

have large device capacitance as long as lump structures are used. As a result, though many waveguide phase modulators have been reported, almost all of the devices are preliminary and only dc operated, and lack in frequency response. Traveling wave structures, which are difficult to fabricate, are necessary. However, the traveling-wave approach to modulator design possibly represents the ultimate in modulator bandwidth engineering. Previously, traveling-wave modulators have exceeded 6 GHz·mm/V and >40 GHz bandwidths [36-39]. Up to now, only a few phase modulators have been remarkable from a practical viewpoint. In this section we briefly introduce these devices.

Figure 3.6 shows the typical structures for high speed phase modulators which have already been reported on [36]. Either a Mach-Zehnder or a directional coupler structure, or an external analyzer is necessary to complete the device of an amplitude modulator. However, as a vehicle to study fundamental characteristics, the simplest structure is a straight optical guide. Advanced interferometric-type modulators are discussed in other sections [37-39]. GaAs, or InP, is optically isotropic and belongs to the $4^{-}3m$ zincblende crystal symmetry group. As described in the previous section, two orientations are considered: 1) the applied electric field is applied along the {001} direction, and 2) the applied electric field is along the {110} direction. In both cases light propagates along the {11¯0} direction.

For the applied electric field along {001}, two electrode configurations are possible: one is the microstrip configuration using a p-i-n structure as shown in Fig. 3.6(a), and the other is a coplanar strip electrode configuration as shown in Fig. 3.6(b) and 3.6(c). In the Fig. 3.6(b) configuration, the optical guide is a double heterostructure, a dielectrically loaded structure where the metal electrode crosses over the guide providing field components along {001}. In Fig. 3.6(c) the electrodes are on either side of the guide and make a Schottky barrier contact with the semiconductor.

For the p-i-n microstrip modulator, $V\pi$ =8V and a 3-dB bandwidth of about 10 GHz have been achieved. For Schottky barrier symmetric coplanar strip modulators on semi-insulating GaAs substrates, bandwidths in excess of 20 GHz have been developed and a 37.5 GHz bandwidth has been predicted based on velocity mismatch and loss with $V\pi$= 38 V [36].

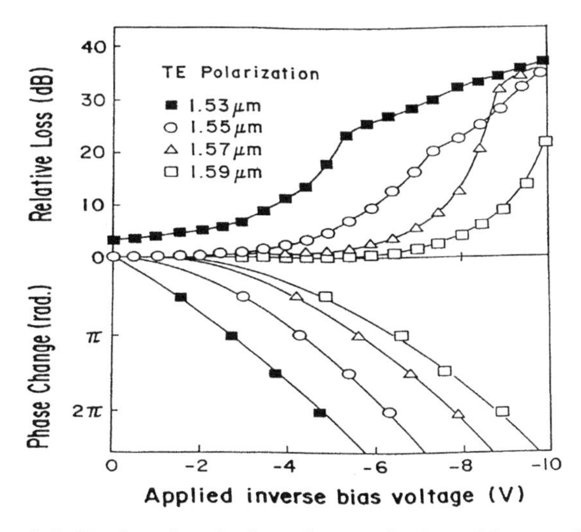

Fig. 3.8. Refractive index change induced by applied voltages and associated with the electroabsorption.

loss and is described in the section 10.1.6 to see how it might be reduced.

The frequency response for a phase modulator with length of 300 μm are shown in Fig. 3.9. The modulator was terminated with a parallel 50 Ω resistor. Polarization-maintaining optical (PANDA) fibers were used for the LD module's pigtail. Light from a pigtailed 1.554 μm LD is coupled to the MQW guides. The light output from the modulator was also coupled to the PANDA fiber, whose end face was formed hemispherically. The response was measured with a lightwave component

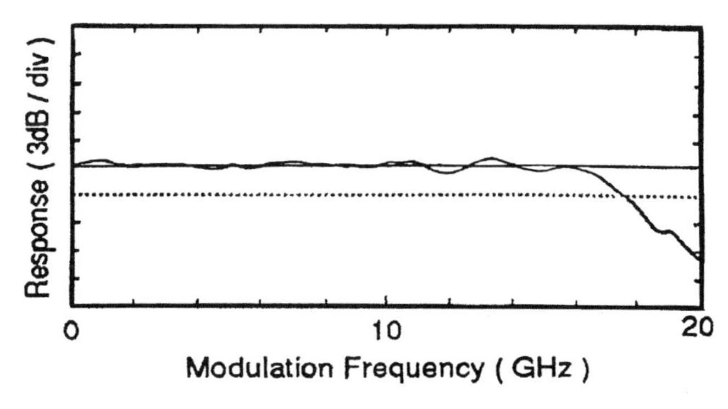

Fig. 3.9. Frequency response for 300 μm long phase modulators. The response is measured by intensity modulation using a Mach-Zehnder interferometer.

analyzer (HP 8703A) by using the intensity modulation signal that was obtained by Mach-Zehnder interferometer. The MQW modulator was driven by a zero dBm RF signal at a dc bias of -3 V. This voltage is nearly the same as $V\pi$ for a sample with a length of 300 μm. The measured electrical 3-dB bandwidth with 18 GHz and more than 20 GHz for a 300 and 200 μm-length modulator, respectively. This high-frequency response is limited by the RC time constant.

As a figure of merit for modulators, a device with a 20 GHz bandwidth and 3.8 V π-phase shift voltage results in a bandwidth to voltage ratio of 5.3 GHz/V. The low required power and high-frequency operation of this modulator will enable us to supply useful candidates as external modulators.

However, the driving voltage and insertion loss are relatively large. The driving voltage must be lower than 2 V because the electrical driver has very low output and needs a large bandwidth amplifier. As for the insertion loss, preceding modulators, such as LNs, operate about 5 dB from fiber to fiber. This is the pending problem for all semiconductor modulators. The reduction methods will be discussed in section 10.1.6 but here a phase modulator with spot size converters will be introduced.

3.5.3 Phase Modulator with a Spot-Size Converter

Figure 3.10 shows the structure of a spot-size converter with a MQW phase modulator [47], where spot-size converter consists of a core layer with 240 nm thick InP and 20 nm thick InAlAs. The taper length and

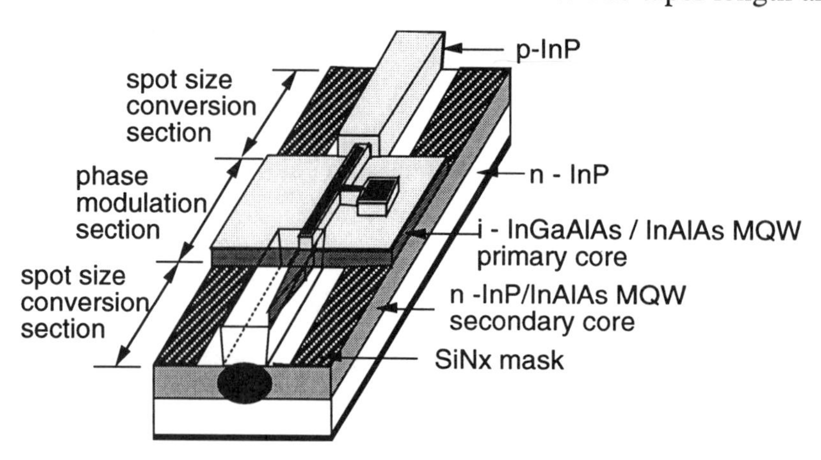

Fig. 3.10. An MQW phase modulator with a spot-size converter.

mesa stripe width are 900 μm and 14 μm, respectively. A small guided spot is enlarged through the spot-size converter as shown in Fig. 3.11. The insertion loss is reduced by 2/3 compared with that of straight waveguide and the off-axis tolerance for 1 dB loss increase is 2 mm. Mode-conversion using a spot-size converter has been discussed in the previous chapter.

Fig. 3.11. Observed near field pattern with and without a spot-size converter. (a) spot size converter section, (b) phase modulator section.

3.6 Acoustoelectric Effect

3.6.1 Introduction

The acoustooptic effect is based on the interaction of an optical wave with an acoustic wave through the photoelastic effect, which is defined as the refractive index change due to the strain. A traveling acoustic wave forms an optical phase grating through periodic refractive index changes associated with the acoustic strain field. This periodic fluctuation in density can be induced by a longitudinal acoustic wave, and a periodic distribution of sheer stress can be generated by a transverse acoustic wave.

A longitudinal acoustic wave consists of a uniformly spaced alternation of regions of higher and lower density. This periodic structure acts as a grating that moves with the velocity of sound. Light is scattered by this grating in the same way that X-rays are scattered by a lattice. Figure 3.12 shows the situation. The energy and momentum conservation law result in the following equations:

$$h\nu_s = h\nu_i - h\nu_A$$

$$\mathbf{k}_s = \mathbf{k}_i - \mathbf{k}_A,$$

where ν_s, ν_i, and ν_A are the frequencies of the scattered light, incident light wave, and sound wave, and $\mathbf{k}_s$, $\mathbf{k}_i$, and $\mathbf{k}_A$ are the wave vectors of the scattered light, incident light, and sound wave. Based on the above equations, the radiation reflects from the planar fronts of the acoustic wave at an angle θ such that

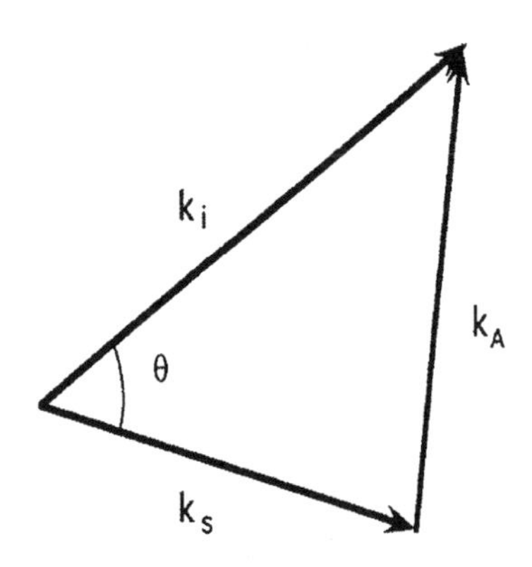

Fig. 3.12 The relation between wave vectors of incident and scattered light and the lattice vector of the ultrasound.

$$2\lambda_A \sin\theta = m\lambda_i,$$

where λ_A is the acoustic wavelength, m is an integer indicating the order of the scattering, and λ_i is the wavelength of the incident photon inside the semiconductor. This effect can also be thought of as light scattering by the acoustic phonon (Brilouin scattering).

Two regions of operation are possible as shown in Fig. 3.13 [48, 49]. When the angle θ = 0, the light of the diffraction order number corresponding to the sideband order number of phase modulation is diffracted in each direction. This diffraction is called Raman-Nath diffraction and is generated when the ultrasonic frequency is relatively low. When the incident light is parallel to the wavefront of the ultrasonic waves, the optical intensity corresponding to the sideband order is represented by a Bessel function and the diffraction intensity is given as

$$I_m = J_m^2(\xi),$$

where $\xi = (2\pi\Delta nL/\lambda)$, and J_m is the m-th-order Bessel function. This is approximate and the approximation holds when $L << \Lambda^2/2\pi\lambda$, where

L is the thickness of the periodic structure.

On the other hand, when θ is varied from 0, only the first-order mode is strengthened because of the phase being the same. Then the diffraction intensity is expressed by the approximation

$$I_1 = I_0 \sin^2(\xi/2).$$

This equation holds when $L >> \Lambda^2/2\pi\lambda$, and this diffraction is called Bragg diffraction. In the Bragg regime, one finds the highest modulation. The intensity of the first-order diffracted light is modulated by changing the intensity of the acoustic wave, and the diffractive angle of the first-order diffractive light is modulated by changing the frequency of the acoustic wave. The frequency of the incident light will differ from that of the first-order diffractive light by the frequency of the acoustic wave, so a frequency shifter will be obtained.

Bulk-type materials were initially used as the ultrasonic cell, but highly efficient surface acoustic waves (SAW) have been discovered recently. Up to the present, acoustooptic devices using $LiNbO_3$-based planar waveguides have been extensively studied [51].

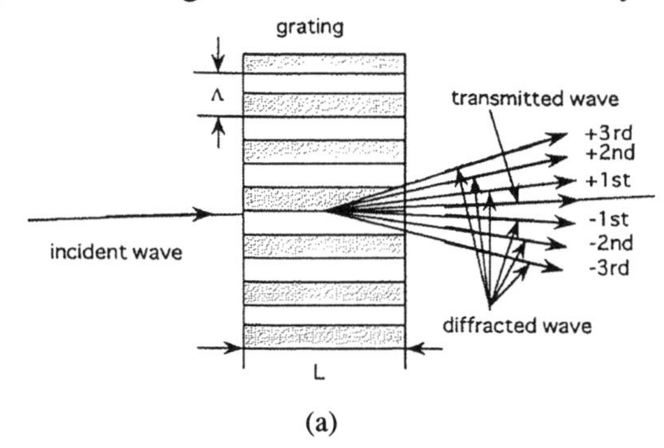

(a)

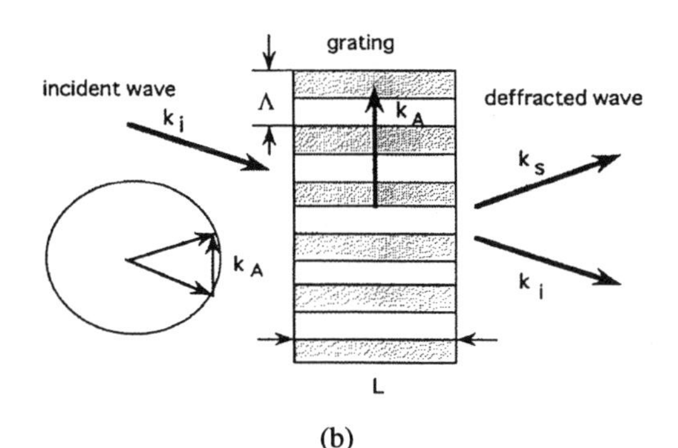

(b)

Fig. 3.13 Lightwave diffraction by an elastic wave:
(a) Raman-Nath diffraction
(b) Bragg reflection.

3.6.2 Figure of Merit for Acousooptical Devices

The usefulness of diffracting materials for application in ultrasonically driven light modulators and scanners is expressed by a figure of merit [48, 50]. There are at least three different criteria for judging a material's usefulness in optical information processing applications. When the intensity of the acoustic-beam is held constant, the following combinations of material parameters constitute figures of merit (M).

When both modulation bandwidth and diffracted light intensity are important, the figure of merit is given by

$$M_1 = (n^7 p^2) / (\rho v),$$

where n, p, and ρ are respectively refractive index, photoelastic constant, and density, and v is the sound velocity.

If only the scattered light intensity is important, the figure of merit is given by

$$M_2 = (n^6 p^2) / (\rho v^3).$$

Finally, in an acoustooptic device in which the intensity of the acoustic-beam is not otherwise constrained and may be made as small as the optical beam in the region of the light-sound interaction, a third generally useful figure of merit is given by

$$M_3 = (n^7 p^2) / (\rho v^2).$$

Figures of merit for several semiconductor materials are listed in Table 3.3 [50]. M_1, M_2, and M_3 are given for several typical acoustic- and optical-beam polarizations and propagation directions. It will be seen from the values listed in Table 3.3 that GaP and GaAs are exceptionally good modulating materials, and this is derived largely from their high refractive indices.

The high-frequency acoustic attenuations in GaP and GaAs are not extremely low, making these materials questionable for room-temperature applications at microwave frequencies, but light modulation near 500 MHz can be utilized without great difficulty. When very high acoustic frequencies must be used, low attenuation is another important factor and $LiNbO_3$ modulators appear most promising despite their relatively low figure of merit.

Table 3.3. Figures of merit for acoustooptic devices.

Material	λ (μ)	n	ρ (g/cm³)	Acous. wave polarization and direction	V (10^5cm/s)	Opt. wave polarization and direction	M_1 ($n^7p^2/\rho v$)	M_2 ($n^6p^2/\rho v^3$)	M_3 ($n^7p^2/\rho v^3$)	P / f (mW/MHz)
GaP	0.63	3.31	4.13	L[110]	6.32	∥	590	44.6	93.5	0.074
GaP	0.63			T[100]	4.13	∥ or ⊥ in[010]	137	24.1	33.1	0.21
GaAs	1.15	3.37	5.34	L[110]	5.15	∥	925	104	179	0.24
GaAs	1.15			T[100]	3.32	∥ or ⊥ in[010]	155	46.3	49.2	0.86
β−ZnS	0.63	2.35	4.10	L[110]	5.51	∥ in[001]	24.3	3.41	4.41	1.58
β−ZnS	0.63			T[110]	2.165	∥ or ⊥ in[001]	10.6	0.57	4.9	1.42
CdS	0.63	2.44	4.82	L[11-20]	4.17	∥	51.8	12.1	12.4	0.56
Te	10.6	4.8	6.24	L[11-20]	2.2	∥ in[0001]	10200	4400	4640	7.14
$LiNbO_3$	0.63	2.20	4.7	L[11-20]	6.57	(b)	66.5	6.99	10.1	0.69

3.6.3 Semiconductor AO Devices

Acoustooptic (AO) devices using semiconductors are little reported [52,53] because of their high losses and only applications for AO Bragg cells in a GaAs/GaAlAs waveguide have been reported [54]. Despite the various successes of the $LiNbO_3$-based AO Bragg devices, lasers and detectors have not been integrated. A semiconductor substrate, in

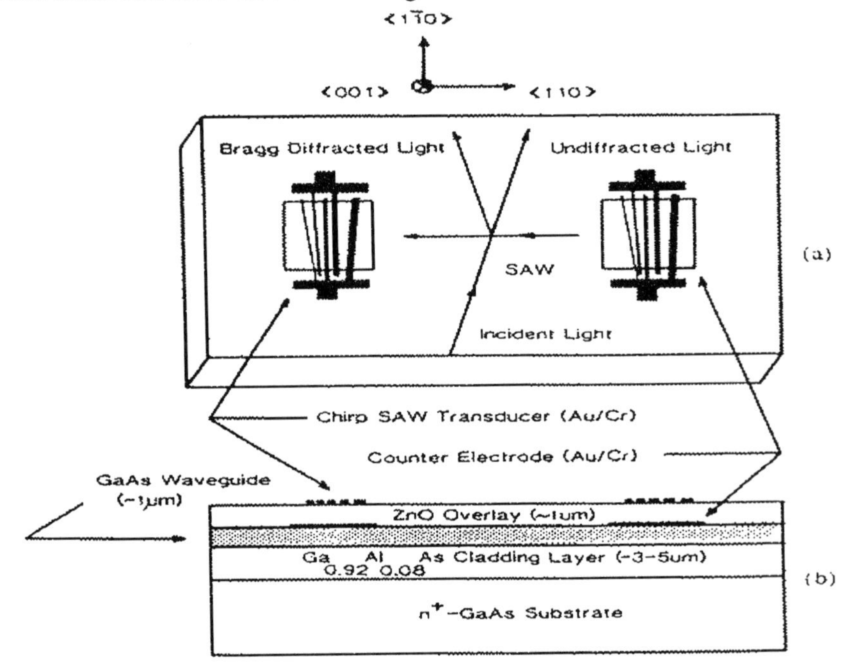

Fig. 3.14 Guided-wave acoustooptic Bragg cells: (a) top view, (b) side view [54].

contrast, can potentially provide the capability for total or monolithic integration because both the laser sources and photodetector arrays as well as electronic devices can be successfully integrated in the same substrate. Compact GaAs/GaAlAs planar waveguide AO Bragg cells that operates at a frequency between 190 and 625 MHz have been demonstrated [52].

Figure 3.14 shows a guided-wave AO Bragg diffractor in a Z-cut GaAs/GaAlAs planar waveguide with its cross-section and ZnO transducer geometry. Semiconductors generally have a small electromechanical transducing coefficient, and efficient zinc oxide (ZnO) transducers are used as an exciting source. The transducer counterelectrode, the ZnO overlay film, and the transducer electrodes were fabricated sequentially on the top of the optical waveguide for efficient excitation of the surface acoustic waves (SAW). To get a large bandwidth, a single tilted-finger chirp transducer [55] that coveres the frequency range from 250 to 500 MHz was designed and fabricated and the AO Bragg diffraction and acoustic propagation losses for four separate Bragg cells were measured.

As for Si-based devices, a waveguide was formed on a SiO_2 buffer

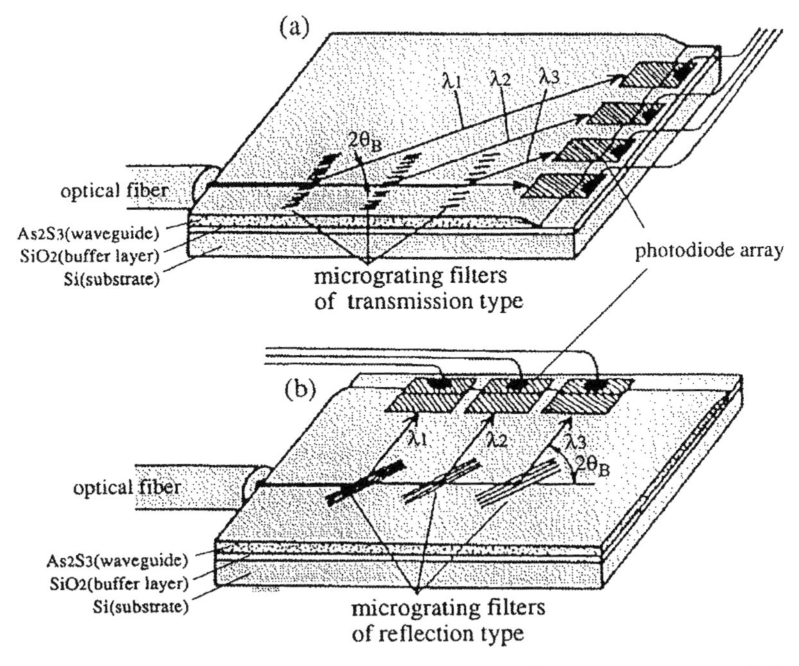

Fig. 3.15. Monolithic integrated wavelength-division-multiplexing receiver terminals: (a) transmission-grating type, (b) reflection-grating type [61].

layer that was a few micrometers thick and was grown by thermal oxidation. Such a layer has a smooth surface, and acts as a good buffer. Corning glass 7059 as well as Si_3N_4 and As_2S_3 are often used as waveguide materials. On the waveguide, an exciting material such as ZnO is formed. A ZnO film is fabricated by sputtering. The exciting characteristics of ZnO SAW have been investigated [56] and the AO properties in ZnO-on-glass waveguides have been reported [57]. Chirped

Bragg-grating lenses [58,59] and Fresnel waveguide lenses [60] made in $As_2/SiO_2/Si$ waveguides by EB (electron beam) direct writing have been reported. Figure 3.15 shows devices constructed by integrating a micrograting array demultiplexer fabricated by EB direct writing in an $As_2S_3/SiO_2/Si$ waveguide and a Schottoky-barrier photodiode array fabricated on a Si substrate [61]. An integrated optic Fourier processor using an AO Bragg cell and Fresnel lenses in an $As_2S_3/SiO_2/Si$ waveguide have been reported [62]. An AO Bragg cell using an amorphous As_2S_3 waveguide can be highly efficient and have low drive power since the AO figure of merit for As_2S_3 is very high. These devices were formed on semiconductor Si substrates, but the active layer is not semiconductor and is therefore outside the scope of this book. Semiconductor materials have not yet been generally used for acoustic devices.

As described above, periodic structures including static and dynamic gratings produced through the AO effect provide a variety of passive functions (grating deflectors, filters, lenses, couplers) and an effective means for guided-wave control (AO grating elements). Integrated optic devices (i.e., wavelength demultiplexers, RF spectrum analyzers, optical disk pickup, etc.) using these structures have been reported and discussed in detail [63, 64].

3.7 References

[1] S. Namba, J. Opt. Soc. Am., 51, 76, 1961.

[2] J. F. Nye, *Physical Properties of Crystals*, Oxford, England, Oxford Univ. Press, 1964.

[3] I. P. Kaminow, *An introduction to Electrooptic Devices*, New York, Academic Press, 1974. I. P. Kaminow and E. H. Turner, Appl. Opt., 5, 1612, 1965.

[4] A. Yariv and P. Yeh, *Optical Waves in Crystals*, New York, Wiley, 1984, Chap. 7 and 8.

[5] K. Tada and H. Noguchi, Trans. IEICE, E73, 88, 1990.

[6] K. Komatsu, K. Hamamoto, M. Sugimoto, Y. Kohga, and A. Suzuki, Top. Meet. on Photon. Switching, 1991, P.24.

[7] M. Bachmann, E. Gini, and H. Melchior, ECOC'92, TuB7.4, P.345, 1992.

[8] L. A. Coldren, J. G. Mendoza-Alvarez, and R. H. Yan, Appl. Phys.

Lett., 51, 792, 1987.
[9] J. Faist, F. K. Reinhart, and D. Martin, Electron. Lett., 23, 1391, 1987.
[10] F. K. Reinhart, D. F. Nelson, and J. Mckenna, Phys. Rev., 177, 1208, 1969.
[11] H. G. Bach, J. Krauser, H. P. Noting, R. A. Rogan, and F. K. Reinhart, Appl. Phys. Lett., 42, 692, 1983.
[12] S. S. Lee, R. V. Ramaswamy, and V. S. Sundaram, IEEE J. Quantum Electron., 27, 726, 1991.
[13] J. Faist and F.-K. Reinhart, J. Appl. Phys., 67, 6998, 1990 and ibid, 7006, 1990.
[14] G. Mendoza-Alvalez, L. A. Coldren, A. Alping, R. H. Yan, T. Hausken, K. Lee, and K. Pedrotti, J. Lightwave Technol., 6, 793, 1988.
[15] J. Zoroofch and J. K. Butler, J. Appl. Phys., 44, 3697, 1973.
[16] D. D. Sell, H. C. Casey, Jr., and K. W. Wecht, J. Appl. Phys., 45, 2650, 1974.
[17] J. G. Mendoza-Alvalez, F. D. Nunes, and N. B. Patel, J. Appl. Phys., 51, 4365, 1980.
[18] C. H. Henry, R. A. Logan, and K. A. Bertness, J. Appl. Phys., 52, 4457, 1981.
[19] J. Manning, R. Olshansky, and C. B. Su, IEEE J. Quantum Electron., QE-19, 1525, 1983.
[20] D. A. B. Miller, Opt. Photonics News 1, 7, 1990.
[21] H. Okamoto, Jpn. J. Appl. Phys., 26, 315, 1987.
[22] J. S. Weiner, D. A. B. Miller, and D. S. Chemla, Appl. Phys. Lett., 50, 842, 1987.
[23] J. E. Zucker, T. L. Hendrickson, and C. A. Burrus, Appl. Phys. Lett., 52, 945, 1988.
[24] H. Nagai,M. Yamanishi, Y. Kan, and I. Suemune, Electron. Lett., 22, 888, 1986.
[25] J. E. Zucker, T. L. Hendrickson, and C. A. Burrus, Electron. Lett., 24, 112, 1988.
[26] M. Glick, D. Pavuna, and F. K. Reinhart, Electron. Lett., 23, 1235, 1987.
[27] M. Glick, F. K. Reinhart, G. Weimann, and W. Schlapp, Appl. Phys. Lett., 48, 989, 1986.
[28] M. Glick, F. K. Reinhart, and D. Martin, J. Appl. Phys., 63, 5877, 1988.

[29] P. J. Bradley, G. Parry, J. S. Roberts, Electron. Lett., 25, 1349, 1989.
[30] A. Jenning, C. D. W. Wilkinson, and J. S. Roberts, Semicond. Sci. Technol., 7, 60, 1992.
[31] M. J. Bloemer and K. Myneni, J. Appl. Phys., 74, 4849, 1993.
[32] D. J. Newson and A. Kurobe, Electron. Lett.,23, 439, 1987.
[33] S. Nishimura, H. Inoue, H. Sano, and K. Ishida, IEEE Photon. Technol. Lett., 4, 1123, 1992.
[34] N. Yoshimoto, S. Kondo,Y. Noguchi, T. Yamanaka, and K. Wakita, App. Phys. Lett., 69, 1, 1996.
[35] N. Yoshimoto, Y. Shibata, S. Oku,S. Kondo,Y. Noguchi, K. Wakita, K. Kawano, and M. Naganuma, Intern. Top. Meeting on Photon. in Switching (PS'96), PWB3, Sendai, pp. 78-79, 1996.
[36] S. Y. Wang and S. H. Lin, J. Lightwave Technol., 6, pp. 758-771, 1988.
[37] R. G. Walker, IEEE J. Quantum Electron., 27, pp. 654-667, 1991.
[38] F. Kappe, G.G. Mekonnen, F. W. Reier, and D. Hoffman, Electron. Lett., 30, pp. 1048-1049, 1994.
[39] R. Spikermann, N. Dagli, and M. G. Peters, Electron. Lett., 31, pp. 915-916, 1995.
[40] U. Koren, T. L. Koch, H. Presting, and B. I. Miller, Appl. Phys. Lett., 50, 368, 1987.
[41] J. E. Zucker, I. Bar-Joseph, B. I. Miller, U. Koren, and D. S. Chemla, Appl. Phys. Lett., 54, 10, 1989.
[42] K. Wakita, O. Mitomi, I. Kotaka, S. Nojima, and Y. Kawamura, Photon. Technol. Lett., 1, 441, 1989.
[43] S. Nojima and K. Wakita, Appl. Phys. Lett., 53, 1958, 1988.
[44] S. Nojima, Appl. Phys. Lett., 55, 1868, 1989.
[45] K. Wakita, I. Kotaka, O. Mitomi, H. Asai, Y. Kawamura, and M. Naganuma, J. Lightwave Technol., LT-8, 1027, 1990.
[46] T. H. Wood, Z. Pastalan, C. A. Burrus, B. C. Johnson, B. I. Miller, J. L. deMiguel, U. Koren, and M. G. Young, Appl. Phys. Lett., 57, 1081, 1990.
[47] N. Yoshimoto, K. Kawano, Y. Hasumi, H. Takeuchi, S. Kondo, and Y. Noguchi, Photon. Technol. Lett., 6, 208, 1994.
[48] E. I. Gordon, Proc. IEEE, 54, 1391, 1966.
[49] A. Alphonese, RCA Rev., 33, 543, 1972.
[50] R. W. Dixon, J. Appl. Phys., 38, 5149, 1967.
[51] C. S. Tai, *Guided-Wave Acousto-Optics: Interactions, Devices and Applications*, Springer Series in Electronics and Photonics,

Vol. 23, Springer-Verlag (1990).
[52] K. W. Loh, W. S. C. Chang, W. R. Smith, and T. Grudkowski, Appl. Opt., 15, 156, 1976.
[53] T. W. Grudkowski, G. K. Montress, M. Gilden, and J. F. Black, in
1981 Ultrason. Symp. Proc., 88, 1981.
[54] C. J. Lii, C. S. Tsai, and C. C. Lee, IEEE J. Quantum Electron., QE-22, 868, 1986.
[55] C. J. Lii, C. S. Tsai, and C. C. Lee, IEEE J. Quantum Electron., QE-15, 1166, 1979.
[56] C. J. Lii, C. C. Lee, O. Yamazaki, L. S. Yap, K. Wasa, J. Merz, and C. C. Lee, and C. S. Tsai, 4th IOOC783, Tokyo, Japan, p. 252, 1983.
[57] F. S. Hickernell, R. L. Davis, and F. V. Richard, 1978, Ultrasonics Symposium, Proc. IEEE Cat. #78CH 1344-1SU, 60.
[58] G. Hatakoshi and S. Tanaka, Opt. Lett., 2, 142, 1978.
[59] S. K. Yao,and D. E. Thompson, Appl. Phys. Lett., 33, 635, 1978.
[60] T. Suhara, K. Kobayashi, H. Nishihara, and J. Koyama, Appl. Opt., 21,1966, 1982.
[61] T. Suhara, Y. Honda, H. Nishihara, and J. Koyama,Appl. Phys. Lett., 40, 120, 1982.
[62] T. Suhara, T. Shiona, H. Nishihara, and J. Koyama,J. Lightwave Technol., LT-1, 624, 1983.
[63] T. Suhara and H. Nishihara,IEEE J. Quantum Electron., QE-22, 845, 1986.
[64] H. Nishihara, M. Haruna, and T. Suhara, *Integrated Optics*, Ohm-Sha, 1985 (in Japanese).

Chapter 4
Electroabsorption Effect

4.1 Introduction

4.1.1 History

There have been three epochs in the short history of the research and development for electroabsorption modulators. The first was from the late 1950s, and the second was 1970s, and the third has extended through the late 80s to the present. In 1958, Franz [1] and Keldysh [2] reported ground breaking theoretical studies of the effect of an electric field on the absorption edge of a semiconductor. They predicted that, in the presence of an electric field, absorption would occur at photon energies lower than the forbidden energy gap (Franz-Keldysh effect). Their calculations of the shift of the optical absorption edge to lower energies under the influence of an electric field, and the sequent experimental observations of that shift by Williams [3,4], Moss [5], Boer et al.[6], and Vavilov and Britsyn [7] opened up a new branch of physics that has since grown and ramified[8,9].

Early research also indicated the Franz-Keldysh effect could potentially be used to modulate light intensity [10-12]. However, the nonlinearity of the effect coupled with material difficulties precluded any break throughs in this area. Another problem that limited performance was the large operating voltage and insertion loss of the bulk structure used in devices at that time.

In the 1970s, advances in double heterostructure growth by liquid-phase-epitaxy brought about renewed interest in applying the effect for light modulation, and there were many attempts at devising new electroabsorption light modulators, switches, and detectors[13-18]. In many of these devices (excluding modulators) the effect of nonlinearity is of little consequence. As far as modulators are concerned, several volts operation for an on/off ratio of more than 10 dB has been achieved. However, the rapid development of easy-to-use semiconductor laser diodes with direct modulation capacity in the early 80s left external modulators, at least up to now, with no part to play in optical fiber transmission systems.

Recent development in crystal growth and device processing

techniques and the limitations of directly modulated laser diodes have once again caused us to take a new look at this mechanism. In high-bit-rate (over a few Gbit/s), long-haul optical transmission systems, the frequency chirping associated with high-speed direct modulation of semiconductor laser diodes is a serious problem that limits transmission length and modulation speed. Effects to solve this problem resulted in the development of high-speed modulation [19], low-chirp [20] Franz-Keldysh modulators. Other promising advances in electroabsorption modulators have also been reported[21, 22]. In the mid-1980s, optical fiber transmission using such modulators [23] was accomplished [24, 25] and successful transmission was also demonstrated using modulators that were monolithically integrated with DFB laser diodes [26, 27]. At present, the record performance is a 3-dB bandwidth of more than 20 GHz [28].

The multiple quantum well structure, one of the newest areas of research, exhibits strong excitonic effects that modify the fundamental absorption edge of materials. The exciton effect results from the Coulomb interaction between the electron-hole pair in the crystal, and manifests itself as an increased steepness in the absorption coefficient. These effects are also present in the covalent semiconductors, but they can be observed only at very low temperatures due to the low excitonic binding energy (typically 3 meV for Ge and GaAs). In multiple quantum well structures, the binding energy can be as large as 10 meV and the excitonic structure can be observed at room temperature. When an electric field is applied perpendicularly to such structures, the absorption coefficient near the band edge decreases significantly. This is similar to Franz-Keldsh effect, but, because it is associated with quantum well structures, it is called the quantum-confined Stark effect (QCSE). Based on this effect, high-speed low-driving voltage electroabsorption modulators have been developed. The details are described in Section 4.3.

4.1.2 Theoretical description

A simple way of expressing the Franz-Keldysh effect is to say that at a given energy there is a greater probability of finding the electron (or hole) inside the energy gap (Fig. 4.1), or that the tunneling probability increases when an electric field is present. In the presence of an electric field, the valence and conduction band states of the

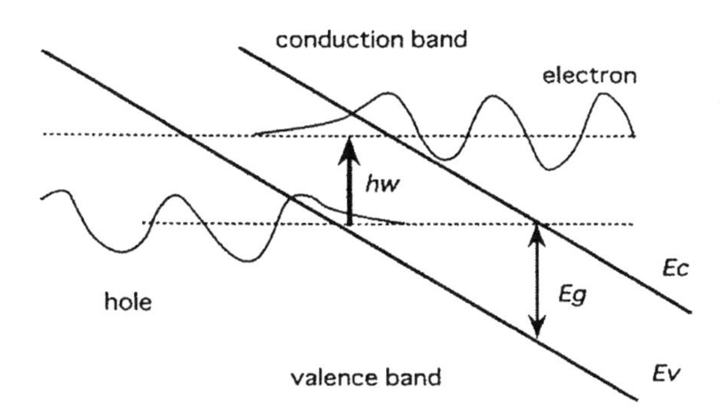

Fig. 4.1. Energy band diagram under high electric field.

crystal are perturbed by mixing states whose crystal momenta differ by a vector parallel to the electric field vector. This mixing yields a nonzero probability for the valence band electrons to absorb photons whose energies are less than the bandgap energy. Hence, the fundamental optical absorption edge appears to shift to smaller photon energies (or longer wavelengths).

The original theoretical treatments have been extended by Callaway and Tharmalingam who derived expressions based on approximations to the Airy integrals, applicable to low fields and photon energies moderately close to bandgap energy Eg. The simple approach introduced by Tharmalingam [29] will be followed here as it was outlined in the Landau's and Lifshitz's text [30].

In an external field F, the potential energy of a particle is U = -Fx + const. The constant is selected as U = 0 ; hence x = 0 then U = -Fx. The Schrodinger equation in this case is

$$d^2\Psi/dx^2 + (2m/\hbar^2)(\eta + Fx)\Psi = 0. \qquad (1)$$

Instead of a coordinate, the dimensionless variable

$$\xi=(x + \eta/F)(2mF/\hbar^2)^{1/3} \qquad (2)$$

is introduced, so Eq. (1) becomes

$$\Psi'' + \eta = 0. \qquad (3)$$

Equation (3) includes no energy parameter. Therefore, the eigen function for an optional energy value is obtained if a solution for the required condition is solved. The solution of Eq. (3) has the form

$$\Psi(\eta) = CAi(-\eta), \quad (4)$$

where

$$Ai(\eta)= \int \cos(u^3/3 + u\eta)du \quad (5)$$

is the Airy function and C is a normalization constant. The Aity function decays exponentially for η large and positive as

$$Ai(\eta) =C/2|\eta|^{1/4} \exp(-2|\eta|^{3/2}/3) \quad (6)$$

and oscillates for large negative values of its argument as

$$Ai(-\eta)=C/\eta^{1/4}\sin(2\eta^{3/2}/3 + \eta/4). \quad (7)$$

The normalization constant C is

$$C=(2m)^{1/3}/\eta^{1/2}F^{1/6}\hbar^{2/3}. \quad (8)$$

For direct-gap, homogeneous, and defect-free crystals, the absorption coefficient α in the presence of electric field F can be described by [27, 28,17]

$$\alpha(h\omega,F) = R\theta^{1/2}(|dAi(\eta)/d\eta|^2 - \eta|Ai(\eta)|^2) \quad (9)$$

$$= \xi F^{1/3}[Ai'(\eta) - \eta[Ai(\eta)]^2]/h\omega, \quad (10)$$

where $Ai(\eta)$ is the Airy function, F is in V/cm, $h\omega$ is in eV, and R, η, and θ are

$$R = 2e^2C_0^2(2\mu/\hbar)^{3/2}/(\hbar\omega cnm^*{}_e{}^2), \quad (11)$$

$$\eta = (E_g - \hbar\omega)/h\theta, \qquad (12)$$

$$\theta^3 = e^2F^2/(2\mu\hbar). \qquad (13)$$

Here m^*_e and e are the mass and charge of an electron, respectively; n is the refractive index, c is the speed of light, $\hbar$ is Planck's constant/ 2π, μ is the reduced mass $m^*_e m^*_h/(m^*_e + m^*_h)$, and C_0 is a constant of the material (dimensions of momentum) that theoretically represents the strength of the transition. C_0 will not be given explicitly, since no calculation of the absorption coefficient from the first principles is intended. C_0 and the gap energy, E_g, will rather be used as free parameters in order to fit the limiting case of Eq.(9) for F—> 0 to measured values of the absorption coefficient. For F—> 0, due to direct allowed transitions behind the edge ($h\omega > E_g$), Eq.(9) becomes the familiar expression for the absorption coefficient

$$\alpha(h\omega,0) = R\pi(h\omega - E_g)^{1/2}/\hbar^{1/2}\,, \qquad (14)$$

which gives α—> 0 as F—> 0 in front of the edge ($\hbar\omega < E_g$). C_0 and E_g are adjusted so that Eq.(13) reproduces as closely as possible the upper part of the experimental absorption edge.

Since with the electroabsorption (EA) effect, we are interested only in the spectral region $\hbar\omega < E_g$, the Airy bracket can be well approximated by the exponential term $0.067\exp(-2.5\eta^{3/2})$. Thus, Eq.(9) reduces to

$$\alpha(\hbar\omega,F)= \Delta\alpha(\hbar\omega,F) = AF^{1/3}\exp(-B\Delta E^{3/2}F^{-1})/\hbar\omega, \qquad (15)$$

where $\Delta E = E_g - \hbar\omega > 0$, and A and B are material parameters proportional to $(\mu m^*_e)^{4/3}$ and $(\mu m^*_e)^{1/2}$, respectively.

The energy range over which the fundamental absorption edge is shifted can be described by the field-broadening parameter, $\hbar\theta$, as

$$\hbar\theta = 4\text{x}10^6(m^*_e/\mu)^{1/3}F^{2/3} \qquad (16).$$

In Eqs.(14) and (15) $\hbar\omega$, $\hbar\theta$, and ΔE are in eV; F is in V/cm.

4.1.3 Device Design and Experimental

First, modulation experiments were done by using bulk type structures like those mentioned in the introduction. It is necessary to apply a large uniform field across the region of interest and a p-n semiconductor junction or metal-semiconductor junction (Schottky barrier) is used to obtain fields of this magnitude in semiconductors. Two configurations were used in the electrabsorption experiment: one parallel and the other perpendicular. Figures 4.2 (a) and (b) show the experimental arrangement.

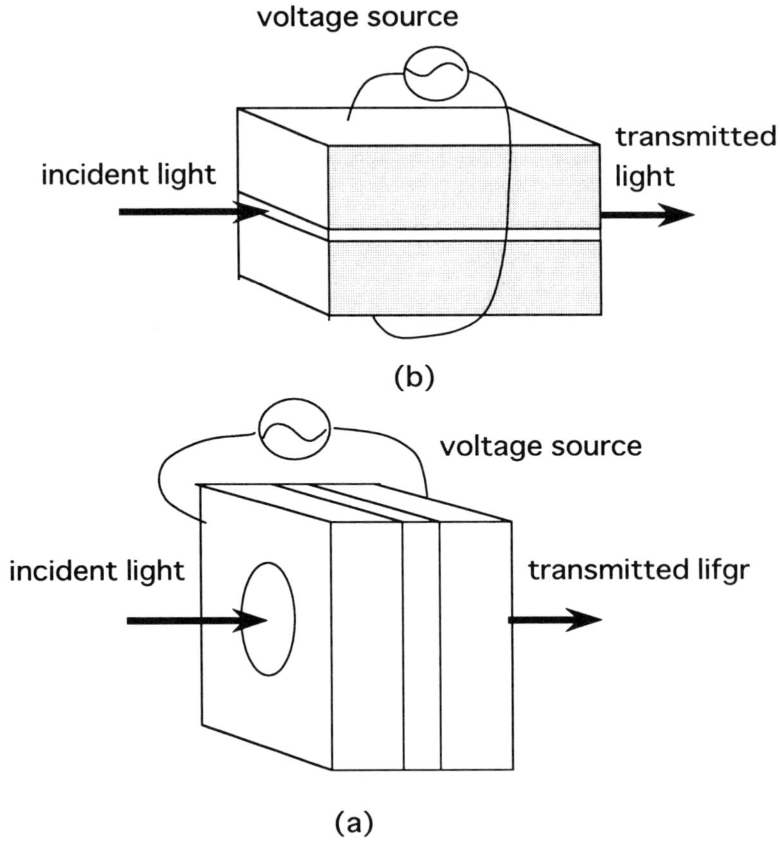

Fig. 4.2. A schematic configuration of Franz-Keldysh modulators.
(a) Incident light beam is parallel to the applied electric field.
(b) Incident light beam is perpendicular to the applied electric field.

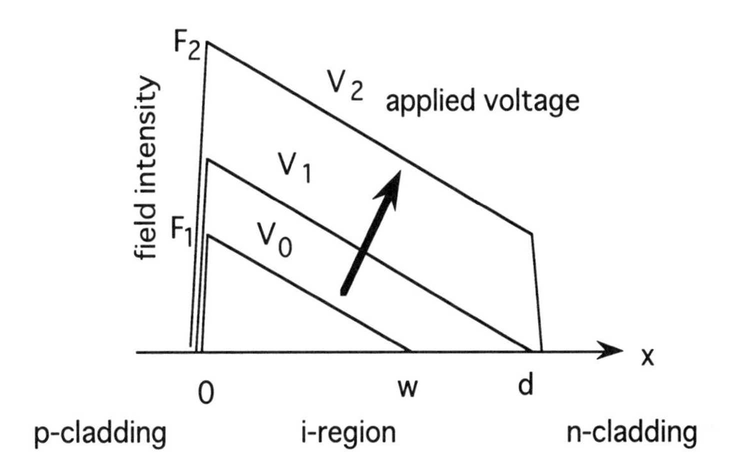

Fig. 4.3. Electric field profiles in the double-heterojunction waveguide of width d. One field (F_1) whose depletion depth W falls within the waveguide and F_2 with W outside the waveguide.

Transparent electrodes were used to measure electric field effects on optical properties of semiconductors where the applied electric field is parallel and perpendicular to the incident light, respectively.

In a reverse biased p-n junction where the p- and n-type regions are uniformly doped, the electric field is linear with distance across the depletion region, attaining its maximum value at the junction and falling to zero at the other side of that region. Increasing the voltage simply widens the depletion region and increases the maximum field in the junction. If one side of the junction has a higher carrier concentration, its depletion edge can be ignored. In this case, the depletion depth d is given by $d = cF/eN$, where N is the free carrier concentration (in the higher resistance region) and F is the peak value of the electric field[32]. If the depletion region has a uniform charge distribution, the field falls off linearly with depth.

The field distribution is shown in Fig.4.3. The maximum depth to which a field can be applied is determined from the field at which the material breaks down. In GaAs, the break down field has $F_b = 5x10^5 V/cm$. This means that the maximum depletion width is inversely proportional to the carrier concentration. Numerically, the relationship between the maximum depth d_m in micrometers and the carrier concentration N in one cubic centimeter is $d_m = 1.5x10^{16}/N$.

Modulation efficiency will be optimized when the depletion region extends across the guide. This could impose strong purity requirements for thicker waveguides or transverse geometry where the incident light is perpendicular to the semiconductor layers and parallel to the electric field. For the latter configuration, the optical interaction length will be limited and only a small on/off ratio will be possible. In addition, the field intensity will change along the perpendicular direction to semiconductor layers.

Neglecting reflection losses, the transmission of light in an absorbing medium follows Lambert's law:

$$I_t(z)=I_i\exp[-\int \alpha(y)dy], \qquad (17)$$

where I_i is the incident light intensity and z is the total optical path length traveled. The change in transmitted light intensity given by EA modulation is

$$\Delta I/I=(I_0-I_F)/I_0$$

$$=1-\exp\{-\int \alpha_F[F(y)]dy + \int \alpha_0(y)dy\}$$

$$=1-\exp(-\Omega), \qquad (18)$$

where the subscript F on α refers to the conditions when the electric field is applied and 0 when the field is off.

<u>A. Waveguide geometry[16,17]</u>. The waveguide structure can produce large optical interaction length and homogeneous field intensity. For light coupled into the waveguide (light propagating parallel to the junction), we assume the spatial dependence of the light intensity within a thick guide as shown in Fig. 4.4. Thus, Eq. (17) becomes

$$I_t / I_i = \Sigma[\int S_n(x)\exp(-\alpha(F(x))Ldx][\int S_n(x)dx]^{-1}, \qquad (19)$$

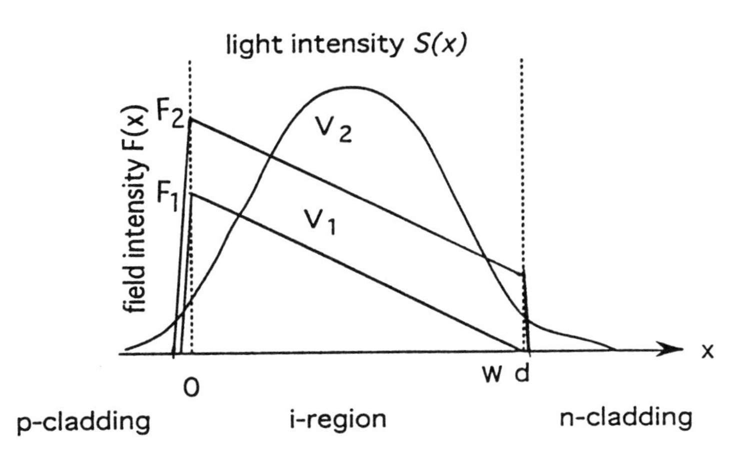

Fig. 4.4. Spatial dependence of the lowest mode light intensity.

where α is now constant along z but depends upon the electric field in the x direction. $S_n(x)$ represents the spatial shape of the light intensity of the n-th mode within the waveguide. For depletion conditions in the waveguide region, F is linear in x direction, as shown in Fig. 4.4.

F_1 in Fig. 4.4 is the maximum electric field within the double heterojunction (DHJ) when the depletion depth W is less than the thickness d of the DHJ. In this case,

$$F(x) = F_1(1 - x/W) \qquad (20)$$

$$F_1 = eN_D W/\kappa\varepsilon_0$$

$$= [2eN_D(V_D - V_A)/\kappa\varepsilon_0]^{1/2} \qquad (21)$$

where N_D is the net donor concentration, κ is the static dielectric constant of the junction material, ε_0 is the permittivity of free space, and $V = V_D - V_A$ is the difference between the applied and the flat-band voltage. When the applied voltage is such that the electric field reaches through the junction (W>d),

$$F(x) = F_2(1 - x/d) + V/d - F_2/2, \qquad (22)$$

where F_2 is now the maximum electric field corresponding to $W = d$ in Eq. (20).

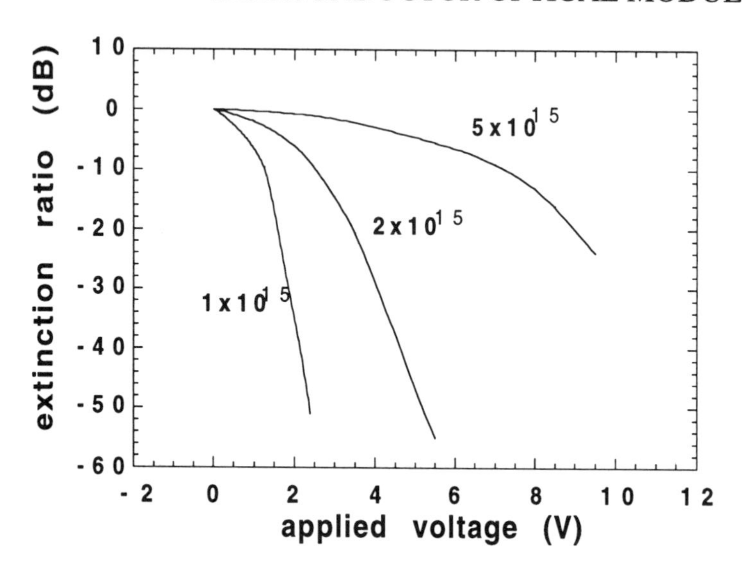

Fig. 4.5. Extinction ratio induced by electroabsorption in InGaAsP/InP double-heterostructure with N_D as a parameter. E_g-hω = 5 meV, d=1 μm, and L=500 μm.

Assuming $S(x)=\sin^2(\pi x/d)$ for the lowest propagation mode, equation (19) becomes

$$I_t/I_i=(2/d)\int \sin^2(\pi x/d)\exp\{-\xi L[\text{Airy}]F(x)^{1/3}/h\omega\}dx, \qquad (23)$$

where L is the waveguide length and [Airy] is the Airy in the parentheses in (10). The upper limit of integration (whether d or W) and the corresponding expression for F(x) [whether Eq. (19) or (21)] are determined from the condition d>W or d<W.

I_t/I_i has been calculated from Eq. (23) as a function of $|V_D-V_A|$ bias under the following conditions: InGaAsP E_g=0.8 eV at 300 K; incident unpolarized monochromatic light; d=1 μm, L=500 μm. Light attenuation in dB for various values of net donor concentrations N_D are shown in Fig. 4.5.

It should be noted that for these conditions optimum light attenuation occurs for $N_D=10^{15}cm^{-3}$ or less. However, no substantial increase in light attenuation is obtained by reducing the net donor concentration of the active region below $10^{15}cm^{-3}$. The reason for the sudden increase in light attenuation below $5x10^{16}cm^{-3}$ can be explained by considering the reach-through voltage V_1 in the DHJ.

When the bias voltage is less than V_1 only parts of the light intensity envelope S(x) within the DHJ are affected by the electric field, i.e., the electric field induces light attenuation only at points x<W, and the over-all light attenuation within the junction is relatively small. This is the case for $N_D = 5x10^{16}cm^{-3}$ in Fig. 4.5, where V_1 is about 4.1 V. When V_1 is reached, the entire envelope S(x) is affected by the electric field and there will be a drastic increase in light attenuation within the junction.

For practical device concentration, DHJ's having $N_D > 5x10^{16}cm^{-3}$ in the active region would require large voltages to achieve reach-through. This in turn, would lead to a high electric field near the junction and thus eventual breakdown of the device. This can be avoided by using a thin active layer for small reach-through voltage along with a suitable optimum detuning energy. The thin active layer in turn will result in small optical confinement and small light attenuation. Therefore, the maximum light attenuation is made as small as possible under the practical use requirement. In recent bulk EA modulators, using a 0.3 μm thick active layer and 250-350 μm sample have resulted in a more than 20 dB on/off ratio and a 3-dB bandwidth of 20 GHz.

Light attenuation as a function of $|V_D - V_A|$ bias for various photon

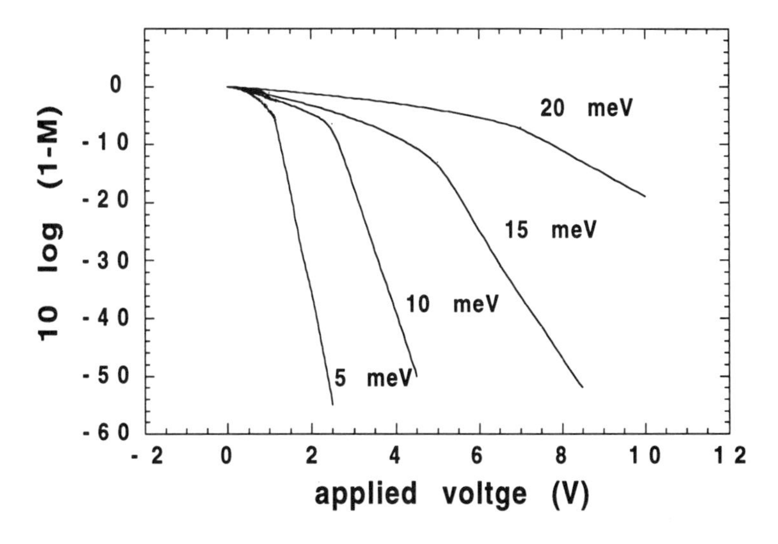

Fig. 4.6. Extinction ratio induced by electroabsorption in InGaAsP/InP double-heterostructure with the detuning energy as a parameter. $N_D = 10^{15}cm^{-3}$, d=1 μm, and L=500 μm.

energies hω is shown in Fig. 4.6. Here the value of N_D was $10^{15}cm^{-3}$. The small detuning energy gives a small bias but a large transmission loss due to residual absorption from the band tail. For practical use, the detuning energy should be about 20 meV.

B. Tranverse geometry. For light incident perpendicular to the junction equation (18) will take the form

$$I_t/I_i=\exp\{-\xi\int[\text{Airy}]F(x)^{1/3}dx\}$$

$$=\exp(-\alpha\Omega), \qquad (24)$$

where α is some average absorption coefficient within the depletion region and where now the light propagates along x-direction.

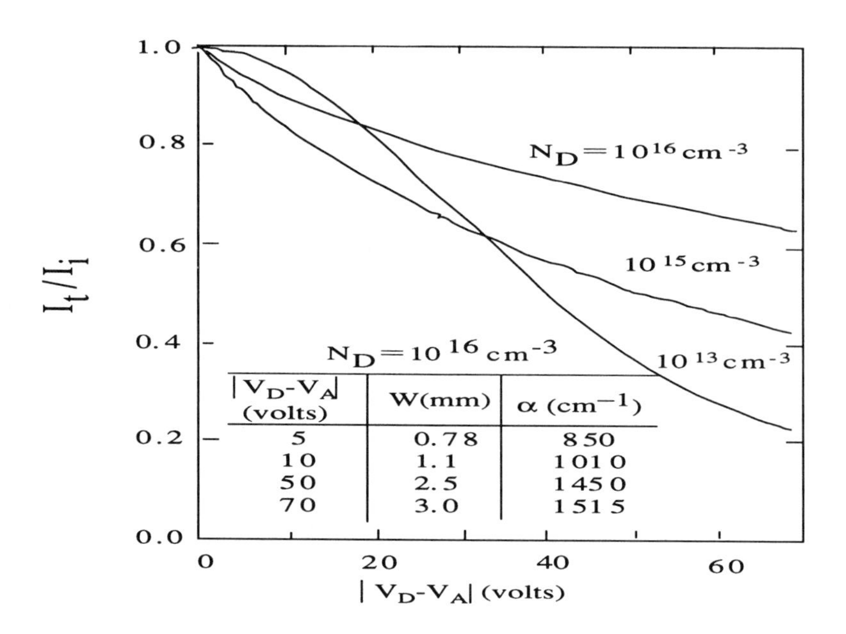

$\|V_D-V_A\|$ (volts)	W(mm)	α (cm^{-1})
5	0.78	850
10	1.1	1010
50	2.5	1450
70	3.0	1515

Fig. 4.7. The extinction ratio as a function of reverse bias and donor concentration when the light is transmitted perpendicular to the junction. The absorbed GaAs layer thickness is 50 μm and ℏω =5 meV.

Figure 4.7 shows I_t/I_i in a DHJ with d of 50 μm calculated using Eq. (15) for various values of N_D and $|V_D-V_A|$. Here the detuning energy E_g-hω was 5 meV.

In contrast to the parallel geometry, maximum attenuation of light is achieved not only with low N_D, but also with considerably larger voltages. Best results would be achieved with a thichness of d 10-50 μm; film that was any thicker would yield larger attenuation, but also larger insertion losses. For a moderate concentration, say $N_D = 10^{15}$-10^{16}cm^{-3}, the device would have to be driven into avalanche mode. To minimize insertion losses, the thickness of the active layer should not exceed the maximum voltage depletion depth. The electrode contact could be either transparent or have a ring configuration.

As will be described in Chapter 9, the normal mode of the surface, such as in transverse geometry, inherenly has a low extinction ratio and large driving voltage. To overcome these disadvatages, optical resonance using a Fabry-Perot interferometer is used, and the effective optical length is increased.

In this chapter we have discussed the absorption coefficient change (electroabsorption effect) associated with applied electric field. Refractive index change (electrorefraction effect), which is based on the Kramers-Kronig relation, is also interested because of its relatively large effect compared with electrooptic (Pockels) effect. This is discussed in the Chapter 3.

Prior work has been confined to InP and GaAs at low fields, but with the development of ultra-low-loss and low-dispersion long-wave optical fibers for telecommunications, GaSb, InAs, and InSb and their mixtures of these crystals are also important [31]. Moreover, II-VI and other III-V semiconductor materials such as ZnS, ZnSe, GaN, and InGaN that operate at visible light regions including green and blue light are now strongly being investigated from the viewpoint of application to optical signal processings as well as short-haul optical fiber transmissions. However, the Franz-Keldysh effect is universal. Only parameters such as effective mass and energy band gap of each materials are different.

4.2 Quantum Confined Franz-Keldysh Effect

It is for fields perpendicular to the layers that quantum well electroabsorption is qualitatively different from bulk electroabsorption, and most of the MQW modulators presented so far are of this configuration. Before we begin the discussion of the quantum-confined Stark effect (QCSE), it is useful to start by neglecting excitonic effects. This is because we have already discussed Franz-Keldysh effect, which has been used for a long time and is still being applied in practical applications, and this approach is very useful for studing the excitonic effect. Moreover, recent research on and applications of the QCSE indicate that a thick quantum well produces highly efficient electroabsorption, though the quantum size effect decreases as the well thickness increases and the excitonic effect also decreases. This model is known as the quantum-confined Franz-Keldysh effect [33] (QCFK). When excitonic effects are neglected, the electron and hole wavefunctions in the plane become plane waves, and for the case without an electric field, the absorption in the quantum well is a series of steps, one step for each allowed transition. Calculations illustrating this are shown in Fig. 4. 8 when the potential barrier is infinite. The applied electric field is 10^5 V/cm for a 15-nm-thick GaAs quantum well.

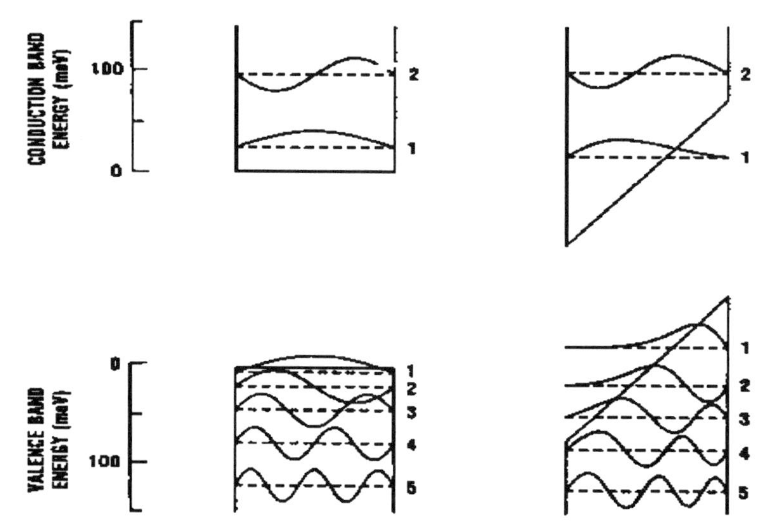

Fig. 4. 8. Calculated wavefunctions and energy levels for a 15-nm-thick two-band GaAs quantum well with infinite barriers at 0 and 10^5 V/cm [33].

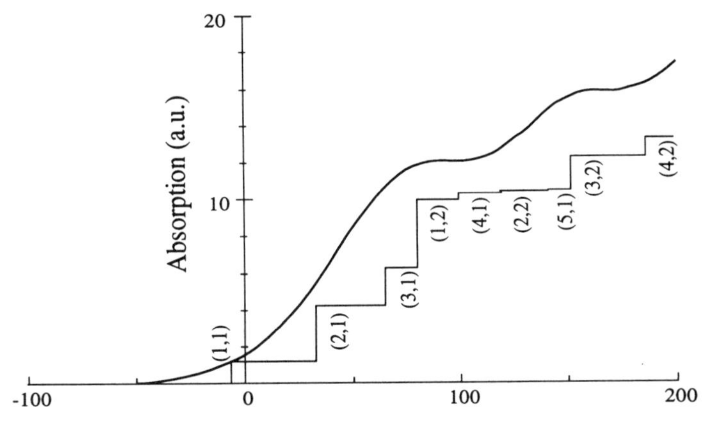

Fig. 4. 9. Calculated absorption of a 15-nm-thick two-band GaAs QW at 10^5 V/cm neglecting excitonic effects. The individual transitions are labeled (n_V, n_C), where n_V (n_C) is the valence (conduction) subband number. The smooth line is the calculated Franz-Keldysh effect for bulk material [33].

Figure 4. 9, which shows the absorption calculated with a field, illustrates several features of this electroabsorption. For instance, the lowest transition energy is reduced by the applicating field, and the electron and hole in these states are pulled to opposite sides of the well by the field. In addition, the overlap integral between electron and hole wavefunctions decreases, reducing the step height for this transition. When a strong field is applied, the forbidden transitions are in some cases stronger than the allowed ones.

The relation of this quantum well electroabsorption model to the Franz-Keldysh effect is interesting; it has been shown analytically that in the limit as the well thickness becomes infinite, this model becomes formally identical to the Franz-Keldysh effect [33, 34] correctly predicting bulk electroabsorption when excitons are neglected. For a 30-nm-thick layer, the electroabsorption in this model is practically almost identical to the Franz-Keldysh effect at 10^5 V/cm. For the thickness within the Bohr radius, the electroabsorption is apparently very different in character from the Franz-Keldysh effect.

Recent MQW modulators with InGaAsP/InGaAsP structures operating at long wavelength regions have quantum well over 10-nm thick and highly efficient electroabsorption has been reported. In such structures, the valence band offset is made small in order to overcome the optical saturation at high input optical intensities due to the hole-pile up, resulting in less electron confinement. This situation is similar to the above QCFK effect except for the exciton transitions. The reason these structures exhibit such high electroabsorption efficiency will be described in the last section.

4.3 Quantum Confined Stark Effect

It is well known that an electric field affects the optical properties of bulk semiconductors near the optical absorption edge. As described in the previous section, the Franz-Keldysh effect predicts a weak absorption tail at photon energies below the nominal bandgap with applied field. Quantum well electroabsorption differs from that in bulk in several ways. Most importantly, the mechanisms differ greatly depending on whether the electric field is parallel or perpendicular to the quantum wells, with qualitatively different behavior appearing for the perpenducular field case (the quantum-confined Stark effect (QCSE)[35-36]. The QCSE has received a lot of attention for devices because it is very large even at room temperature, allowing low-driving-energy optical modulators and switches.

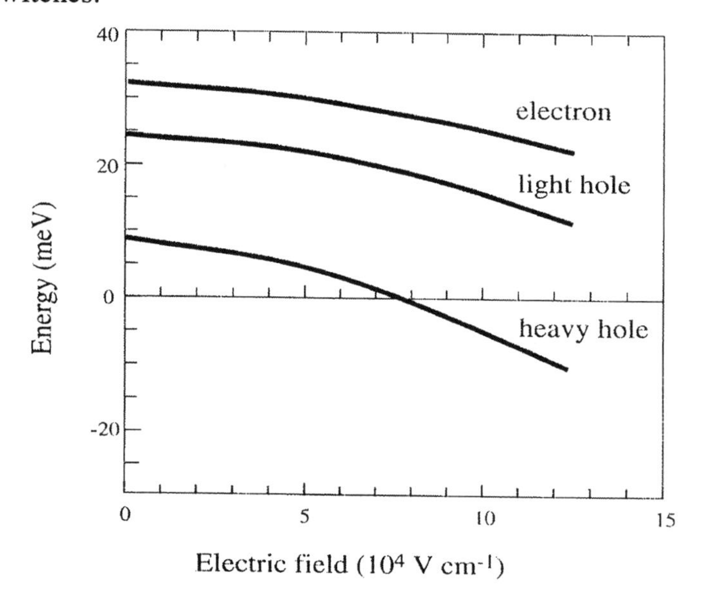

Fig. 4. 10. Electroabsorption for fields perpendicular to the layers.

Various QCSE models have been reported [35-40] and based on theoretical considerations, the lowest exciton states have been calculated in the presence of perpendicular field as shown in Fig. 4. 10. Figure 4. 10 shows the first electron, heavy-hole, and light-hole states for a 9.5-nm GaAs quantum well with AlGaAs barriers. The peak energy shifts as the applied field increases and this shift is approximately quadratic at low fields [37].

Figure 4. 11 shows the calculated heavy-hole exciton binding energy for 5-, 10-, 15- and 20-nm-thick GaAs quantum wells as a function of field applied perpendicular to the layers [41]. Exciton becomes significantly broader as the field is applied. This is because the Coulomb attraction is weaker when the electron and hole are pulled apart by the field. There is a significant decrease in the magnitude of the binding energy as the field is applied. The absolute magnitude of the energy shift from the binding energy is not very significant compared with the shift of the single-particle states, but the increase in size can result in a significant change in the electron-hole overlap, reducing the absorption strength of the exciton line.

The energy shift is approximately proportional to the four power of the quantum well at low field, according to the variational calculation [37]. Therefore, there is an optimum quantum well thickness at which a highly efficient electroabsorption effect will arise. Details on this issue will be discussed in a later section along with the figure of merit for modulators.

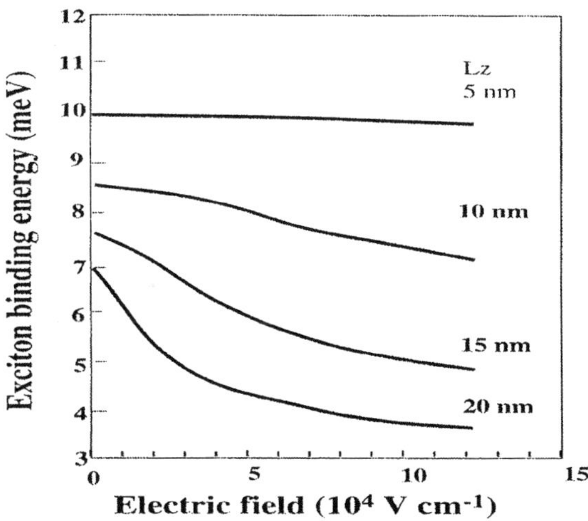

Fig. 4. 11. Calculated heavy-hole exciton binding energy for GaAs QWs for fields perpendicular to the layers [41].

4.3.1 Anisotropy of Electroabsorption

An interesting property of quantum well structures is that, due to the two-dimensional confinement, it appears the band-to-band transitions are anisotropic for light propagating parallel to the plane of the layers. In particular, there should be no heavy-hole absorption, or at least it should be significantly reduced, for light polarization perpendicular to the layers. This has been proposed as an explanation for the polarization-dependent gain exhibited by quantum well lasers [42, 43]. The anisotropy of absorption for light propagating along the plane of the layers has been reported for GaAs/AlGaAs and InGaAs/InAlAs MQWs [44, 45]. The selection rules for absorption depend on the quantum well band structure. The quantum well size effect has the symmetry of a uniaxial perturbation, which splits the heavy hole (hh) and light hole (lh) valence bands at the zone center. From these symmetry considerations, the predicted oscillator strengths of the hh and lh band-to-band transitions are 3/4 and 1/4 respectively for electric field vector **e** parallel to the plane, and 0 and 1 respectively for **e** perpendicular to the plane [46].

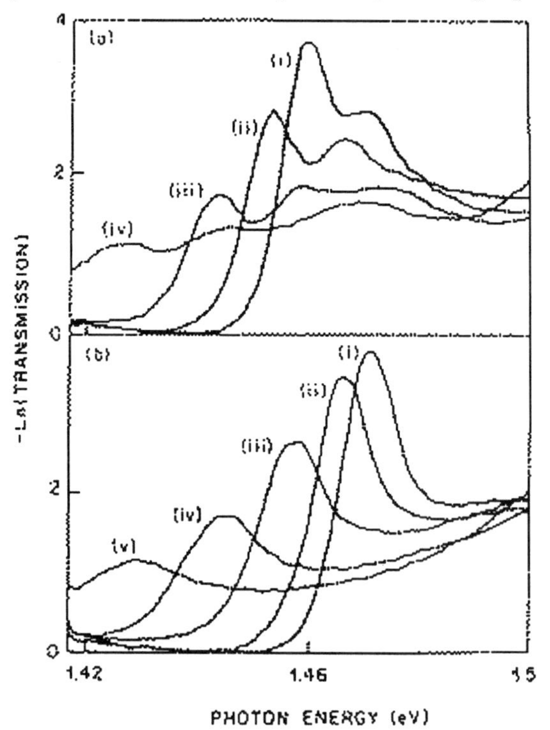

Fig. 4. 12. Experimantal spectra for electric fields perpendicular to the quantum well layers with GaAs/AlGaAs quantum wells.

(a) Incident light polarization parallel to the plane of the layers.

(b) Incident light polarization perpendicular to the plane of the layers [47].

Figure 4. 12 shows the experimental spectra for electric fields perpendicular to the quantum well layers with 9.4-nm GaAs quantum wells [47]. The spectra clearly show that the excitons do not broaden very much with field, and the peaks shift. Polarzation dependence is also clear. As shown in Fig. 4. 13, the results are similar for InGaAs/InAlAs MQWs with well thickness of 7.4 nm. The difference between GaAs/AlGaAs and InGaAs/InAlAs is the presence of a small bump for TM polarization in the spectra for the latter , which is considered to be due to the valence band mixing.

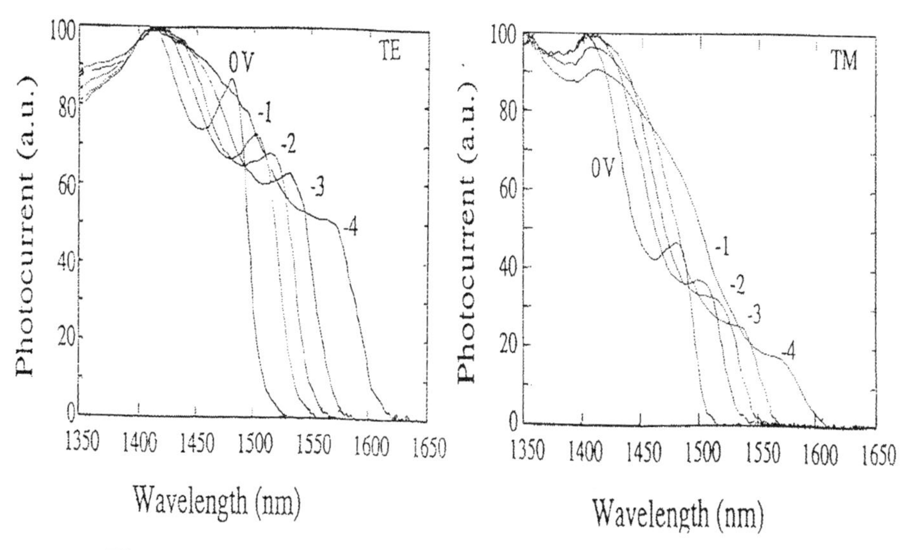

Fig. 4. 13. Experimantal spectra for electric fields perpendicular to the quantum well layers with InGaAs/InAlAs quantum wells. (a) Incident light polarization parallel to the plane of the layers.(b) Incident light polarization perpendicular to the plane of the layers

On the other hand, as can be seen from the absorption spectra for InGaAsP/InGaAsP MQWs operating at 1.55-μm wavelength in Fig. 4. 14, the behavior of the excitons in these MQWs is very different from those in InGaAs/InAlAs MQWs. This is due to the weak electron confinement in the wells, which will be discussed in the following section. The spectral broadening associated with the applied electric fields is due solely to the tunneling of carriers. In systems other than InGaAsP/InGaAsP MQWs, such as GaAs/AlGaAs and InGaAs/InAlAs, the potential height is relatively large and theoretical estimation under

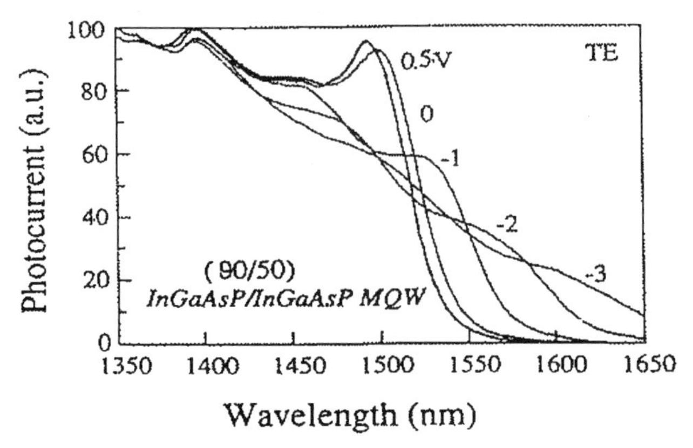

Fig. 4. 14. Experimantal spectra for electric fields prpendicular to the quantum well layers with InGaAsP/ InGaAsP quantum wells.

applied field is obtained under the assumption that potential profiles of the barrier regions remain flat, independent of the applied fields. On the other hand, in actual QW structures, when the confinement weakens due to the tilted potential profiles of the barrier regions, QW eigenstates change with the applied field from true bound states to quasi-bound states [48].

4.3.2 Strong and Weak Quantum Confinement

We overlook the theoretical calculations on the QCSE and neglect the effects of the finite barrier heights, both in the values of the energy levels and in the possibility of tunnelling out of the wells. To obtain a quantitative comparison with experimental results, some preliminary comments are necessary. From the viewpoint of practical applications to modulators, quantum wells operating at long wavelength region, where optical fibers are transparent and their dispersion is low, are important. The InGaAsP/InGaAsP quantum well usually used for operation in that region has less electron confinement. This is because in an attempt to overcome optical saturation at higher optical intensities due to hole-pile up, the QWs are designed so that the valence band offset is small. In such situations, the discrete energy level for electron becomes quasi-bound states with a finite energy width due to the weakened confinement.

Figure 4. 15 shows the calculated shapes of the absorption spectra

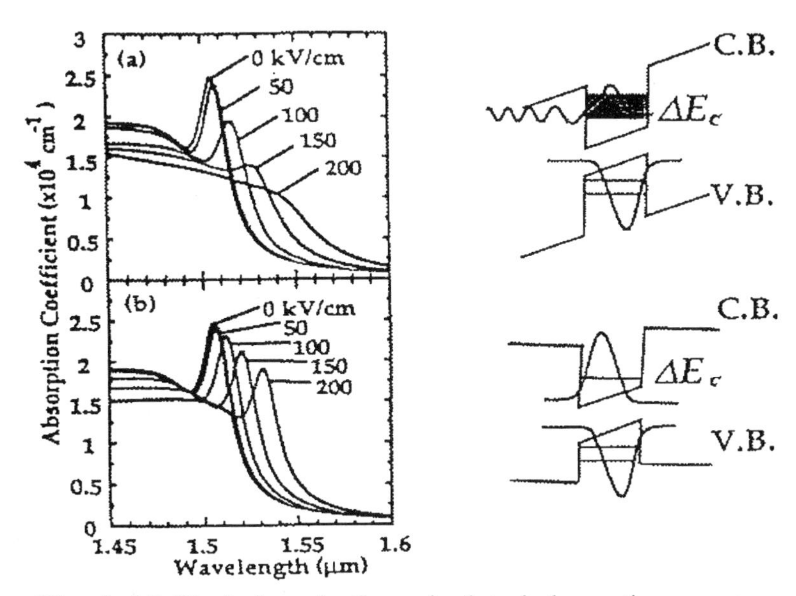

Fig. 4. 15. Variations in the calculated absorption spectra under 5 different electric field strengths with an increment of 50 kV/cm. Upper lines show the spectra of QW structure with tilt barriers; i. e., weak confinement, and lower lines show those with flat barrier ; i. e., strong confinement.

in a QW structure under an electric field strength of zero and 150 kV/cm, respectively, together with the wave functions [49]. In the presence of an electric field, the wave functions have an oscillatory tail that extends out of the well, as shown in the inset of Fig. 4. 15. As a result, the field-induced broadening is remarkable, i.e. the weakened confinement makes the discrete level being quasi-bound with a finite width.

As mentioned above, the behavior of the excitons is very different in quantum wells with different barrier components even if the field is applied perpendicularly to the layers. Experimetal results and calculations are shown in Figs. 4. 16 and 4. 17 for a 6-nm-thick InGaAsP QW with InGaAsP barriers whose photoluminescent (PL) wavelength is 1.3-μm and a 10-nm-thick InGaAsP QW whose PL wavelength is 1.2 μm. In the calculation, conduction band offsets of $0.5\Delta E_g$ for the 6-nm QW structure and $0.55\Delta E_g$ for the 10-nm QW structure were assumed, where ΔE_g means the band gap energy difference. Hence, the estimated barrier heights from the ground level of the conduction band are 40 and 130 meV, respectively. As can be seen from the figures, the barrier height difference has an important effect on the field-induced broadening and carrier tunnelling dominatly contributes to the spectral broadening. Other broadening mechnisms such as inhomogeneous field distribution and thickness and composition fluctuations play a minor role.

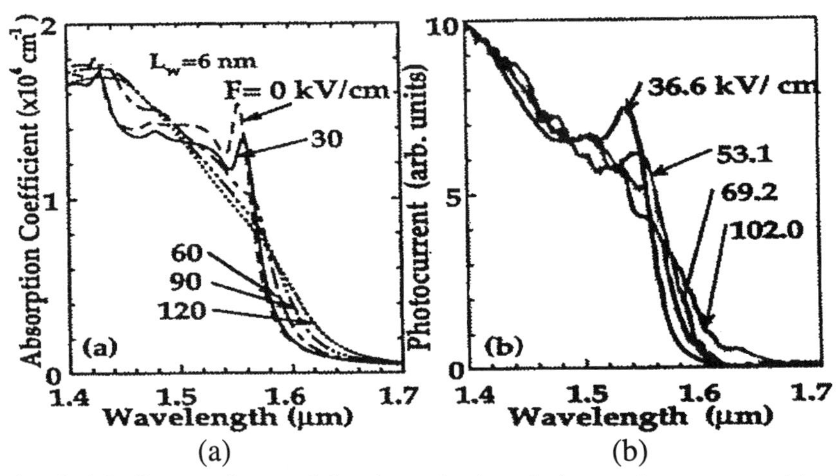

(a) (b)

Fig. 4. 16. Comparison of the the calculated absorption spectra (a) and the measured photocurrent spectra (b) for a 6-nm InGaAsP/InGaAsP single QW structure.

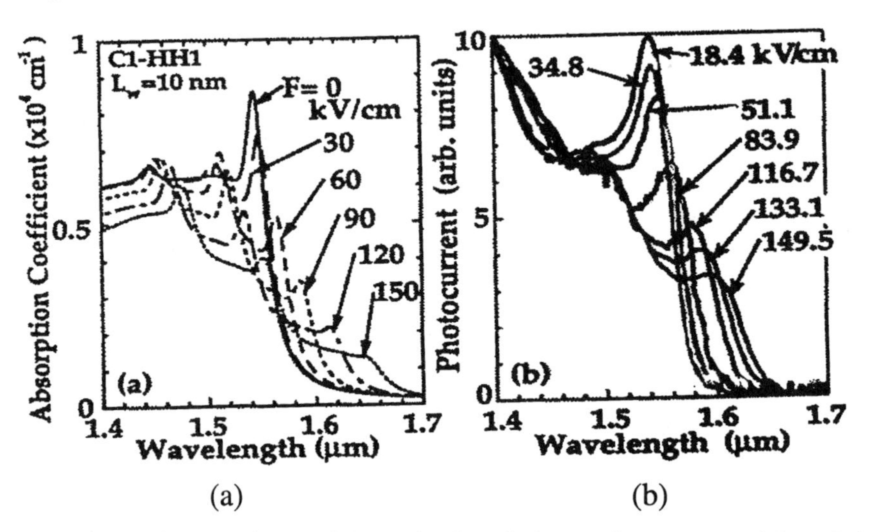

(a) (b)

Fig. 4. 17. Comparison of the calculated absorption spectra (a) and the measured photocurrent spectra (b) for a 10-nm InGaAsP/InGaAsP single QW structure.

In contrast to InGaAsP/InGaAsP MQWs, InGaAs/InAlAs MQWs have a large ΔE_C and show strong electron confinement. Therefore, when electric field is applied perpendicular to the layers, the exciton resonances and their shifts can be observed clearly.

This quantum confinement difference affects device characteristics. In particular, when tensile strain is introduced to obtain polarization insensitive modulation, both MQWs can operate as polarization insensitive modulators, but the power saturation level is very different between them. This is because tensile strain introduction brings about the increase of valence band offset between wells and barriers, resulting in hole-pile up under high input power. The details of this will be described in Chapter 7.

4.4 Wannnier-Stark Localization

When quantum wells are separated by very thin barriers in a periodic structure, the discrete energy levels broaden into minibands because of the resonant tunneling effect. Carriers are delocalized and the superlattice (SL) structure exhibits to some extent three-dimensional (3D) behavior. The application of a low electric field to such a structure breaks the resonance because the energy levels of adjacent quantum wells are misaligned by eFd, where e is the elementary electron charge, F is the applied electric field, and d is the SL period.

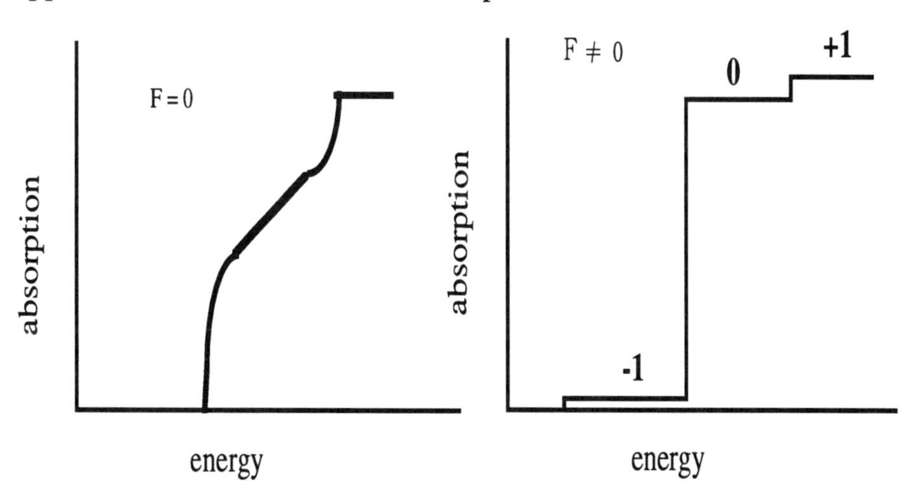

Fig. 4. 18. Absorption spectra of an ideal superlattice without applied electric field and in the high-field limit.

Carriers tend to localize and the structure recovers its two-dimensional (2D) behavior. This is called as Wannier-Stark localization [50-52] and could be used to achieve very-low-voltage optical-waveguide modulation in the range of 1.5 μm wavelength [53]. Bleuse et al. [50] have shown that the SL absorption spectrum could be viewed as the sum of absorption steps corresponding to transitions connecting holes and electrons localized in wells separated by p periods and occurring at energies E_{QW} + peFd (p = 0, +/-1, +/-2,-------) where E_{QW} is the fundamental transition energy associated with the isolated quantum well. When eFd approaches the SL total miniband width D, only the (p = 0) and (p = +1) transitions have significant oscillator strengths and the absorption spectra of a SL at zero field have the shape shown in Fig. 4. 18.

Two interesting electroabsorption configurations can be distinguished [53].

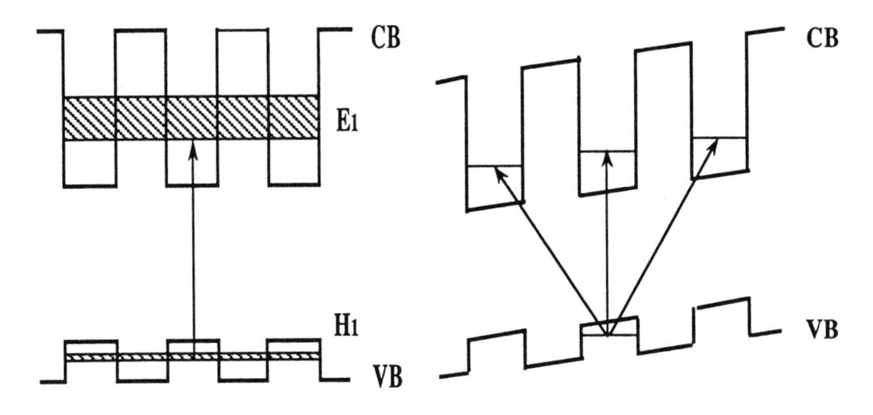

Fig. 4. 19. Sketch of the heterostructure energy band without and with applied field for the Wannier-Stark effect .

1) Incident photon energy is higher than the zero-field SL bandgap and lower than the single quantum well bandgap (i.e., $E_{QW}- \Delta/2 < h\nu_0 < E_{QW}$): There is a negative absorption variation commonly referred to as a "blue-shift" [54] that has been successfully used to realize a self electrooptical bistable device [55], electroabsorptive modulator arrays [56], as well as a normally-off asymmetric Fabry-Perot reflection modulator [57,58], all in the GaAs/AlGaAs material system.

2) Incident photon energy is higher than the zero-field SL bandgap

and lower than the single quantum well bandgap (i.e., $h\nu_0 < E_{QW} - \Delta/2$): There is a positive absorption variation from the (p = -1) oblique transition, which is similar to the conventional "red-shift" electroabsorption effects except that it can be obtained with much smaller electric fields. This (p = -1) oblique transition has been used to achieve efficient very-low-drive-voltage optical-waveguide modulation [59], and both high extinction ratio and low on-state attenuation have been demonstrated by using a InGaAs/InAlAs superlattice [60]. Figure 4. 19 shows schematics of Wannier-Stark localization when an electric field is applied.

A p-i-n heterostructure grown by MBE with a 0.26-μm-thick intrinsic region made of a 20-period, 6.5-nm InGaAs/2nm InAlAs SL (absorption edge 1.5 μm) and surrounded by two 8-period, 2.5-nm InGaAs/3 nm InAlAs confining SLs has been fabricated [61]. In this heterostructure, a high-mesa, 3.5-μm-wide ridge was formed by IBE (ion-beam-etching) and a low-capacitance contact pad was patterned with polyimide. In the low field domain, the extinction ratio was 10 dB with a 0.9 V drive voltage and a 18-GHz bandwidth, while in the high field domain, it was 17 dB with 2.5 V and a 20 GHz bandwidth [61]. Much higher efficiency of 20 GHz/V was also obtained, but power saturation was observed under relatively low levels (0.3 to 1.2 mW). The reason for this power saturation is not well understood yet, but it may be due to the weak field intensity that hardly sweeps the generated carriers.

Thin-barrier superlattices are promising systems for elecrtroabsorptive devices for several reasons. First, the thin barrier eliminates any problems associated with the time scale [62, 63] of photoexcited carrier transport inside the structure. High-speed optical switching with a switching time of 33 ps has been demonstrated using a SL structure with 3.5-nm-thick $Al_{0.3}Ga_{0.7}As$ barriers [64]. Structures based on Wannier-Stark localization are expected to show good transport properties since device operation is based on the efficient interwell coupling. Second, for a given number of wells, reducing barrier width also decreases the overall thickness of the active layer. The external voltage required to obtain the desired electric field can thus be minimized. Third, the Al compounds usually used for barriers, such as AlAs, AlGaAs, and InAlAs, are chemically active and easily oxidized, resulting in the formation of deep level traps that increase the background impurity levels. The thinner the barriers including Al are, the better the background impurity levels become.

4.5 Electroabsorption for Coupled QWs

In optical modulators, two crucial parameters determining their potential applications are drive voltage and achievable contrast ratios. Both are affected by device structure, as discussed previoussly. The use of coupled wells and superlattices to obtain changes in the optical properties at low electric fieds and consequently low voltage modulation has been tried. In this section, we will discuss the coupled wells and superlattices.

Theoretical treatments of coupled wells have tended to concentrate on the evaluation of absorption peak shifts in terms of energy and the calculation of carrier wavefunctions and their overlaps, with few full calculations of absorption spectrum [65, 66] and little consideration of using coupled wells as optical modulators.

Experimental results have been reported that indicate significant differences between uncoupled multiple quantum well structures and coupled quantum well structures [67-71]. Large shifts in wavelength of absorption peaks with applied voltage [67-72] offer the possibility of optical modulation in devices at lower drive voltages than with uncoupled wells.

Figure 4. 20 shows schematic energy diagrams of a coupled quantum well without and with applied electric field. The superlattice is a series of quantum wells coupled by the resonant tunneling effect. This coupling results in the broadening of the energy levels into subbands of widths ΔE_1 and ΔH_1. The SL band gap is smaller than that of the isolated

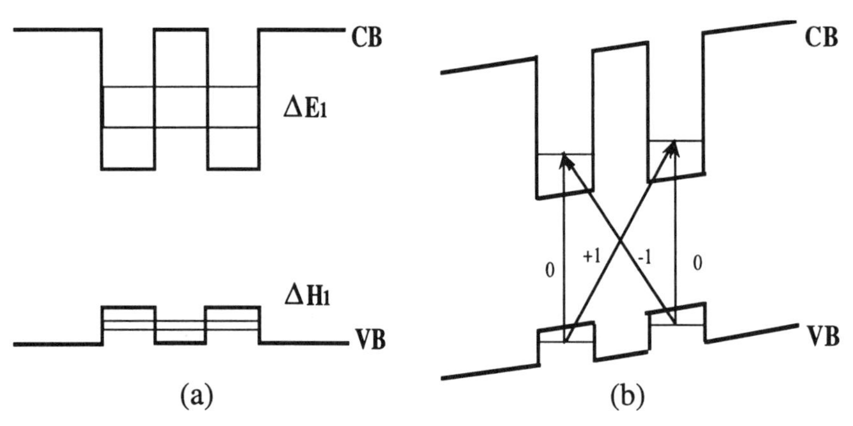

Fig. 4. 20. Coupled quantum well potential profiles in (a) the absence and (b) presence of an electric field F.

quantum well by $(\Delta E_1 + \Delta H_1)/2$. When an electric field is applied along the growth axis, the resonance condition is absent, as the genuine energy levels in the consecutive quantum wells become misaligned by eFd, where d is the superlattice period and F is electric field. The tunneling probability should hence decrease drastically, which in turn means that the eigenstates would tend to localize over a few adjacent quantum wells. The absorption coefficient should then tend towards the step function corresponding to a series of uncoupled quantum wells, thus, producing a blue shift of the absorption edge on the order of $(\Delta E_1 + \Delta H_1)/2$.

Asymmetric coupled quantum well systems have received a great deal of attention. In such a system, the real-space indirect transition between wide and narrow wells contributes to the electroabsorptive and electrorefractive properties of the system. As shown in Fig. 4. 21, this interwell transition is the lowest energy transition in the system at electric fields higher than the resonance field and can be made to dominate the system's below-band-gap optical properties. The absorption characteristics of this transition depend on the geometry of the system. As the electric field increases beyond resonance, the strength of the interwell transition decreases at a rate determined by the thickness of the tunnel barrier.

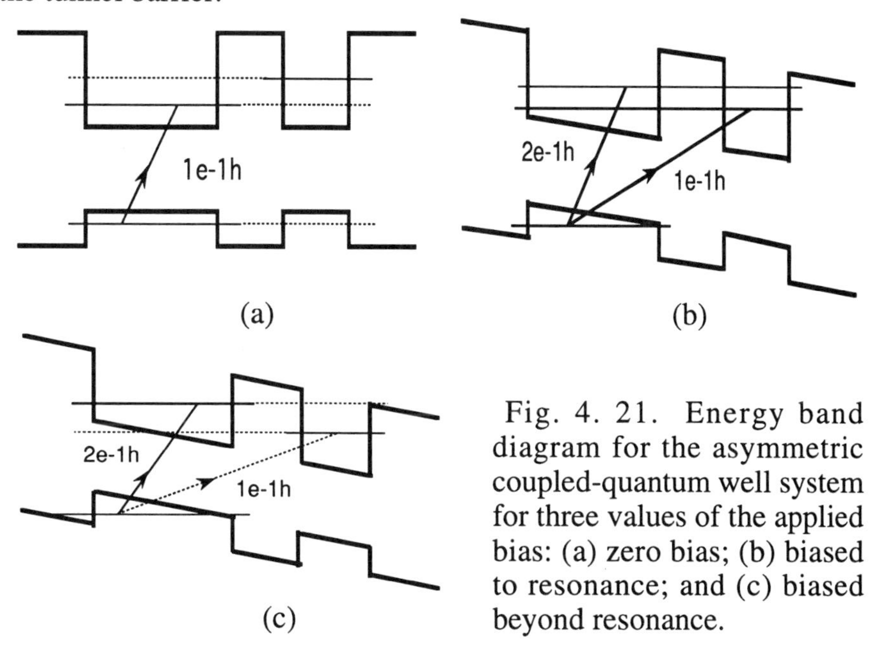

Fig. 4. 21. Energy band diagram for the asymmetric coupled-quantum well system for three values of the applied bias: (a) zero bias; (b) biased to resonance; and (c) biased beyond resonance.

The phase shifts measured in InGaAs/InAlAs waveguides containing asymmetric coupled well samples are several times larger than those measured in uncoupled well samples [73] and for amplitude-modulation the former are much more effective. The devices contain 7.5- and 5.0-nm InGaAs quantum wells with a 1.6-nm InAlAs barrier and operate at 150 deg/ 6V and 13 dB/6V as a phase modulator and electroabsorption modulator, respectively. This device has been improved to operate at 13 dB/ 4V with a 3 dB bandwidth of 4 GHz [74] by another group.

4.6 Miscellaneous Effects

Most work has concentrated on square quantum wells, but nonsquare ones have been investigated because square quantum wells may not be the best structure for devices. Many works on nonsquare wells are only calculations and the superiority of such wells has not been well confirmed experimentally. None the less, they have some potential for improving device characteristics. Following the axiom "simple is best", these artificial structures will probably be selected sooner or later, but the recent rapid progress in crystal growth and processing technologies permit no prediction of the timetable. Recent developments in MBE and MOVPE technology have made it possible to grow non-square QWs with high precision.

Figure 4. 22 shows some proposed structures of nonsquare wells [75]. A Stark shift has seen observed in photocurrent measurements of parabolic [76], step-like [77, 78], graded-gap [79], symmetric and asymmetric triangular [80, 81], and inverse parabolic QWs [82]. In addition, some preliminary calculations [75-86] have been made, and an enhancememt of the QCSE was found in some structures [67, 71, 73]. In general [75], wider wells have a larger Stark shift, independent of shape and symmetry. The symmetric and asymmetric triangular wells have the largest Stark shift, but they also exhibit a drop of the wavefunction overlap. In contrast, the square well has a small shift as well as a small overlap reduction.

On the whole, each of these calculations has certain limitations and the effectiveness or reality of the nonsquare structures depends on to what extent the calculation is based on practical conditions.

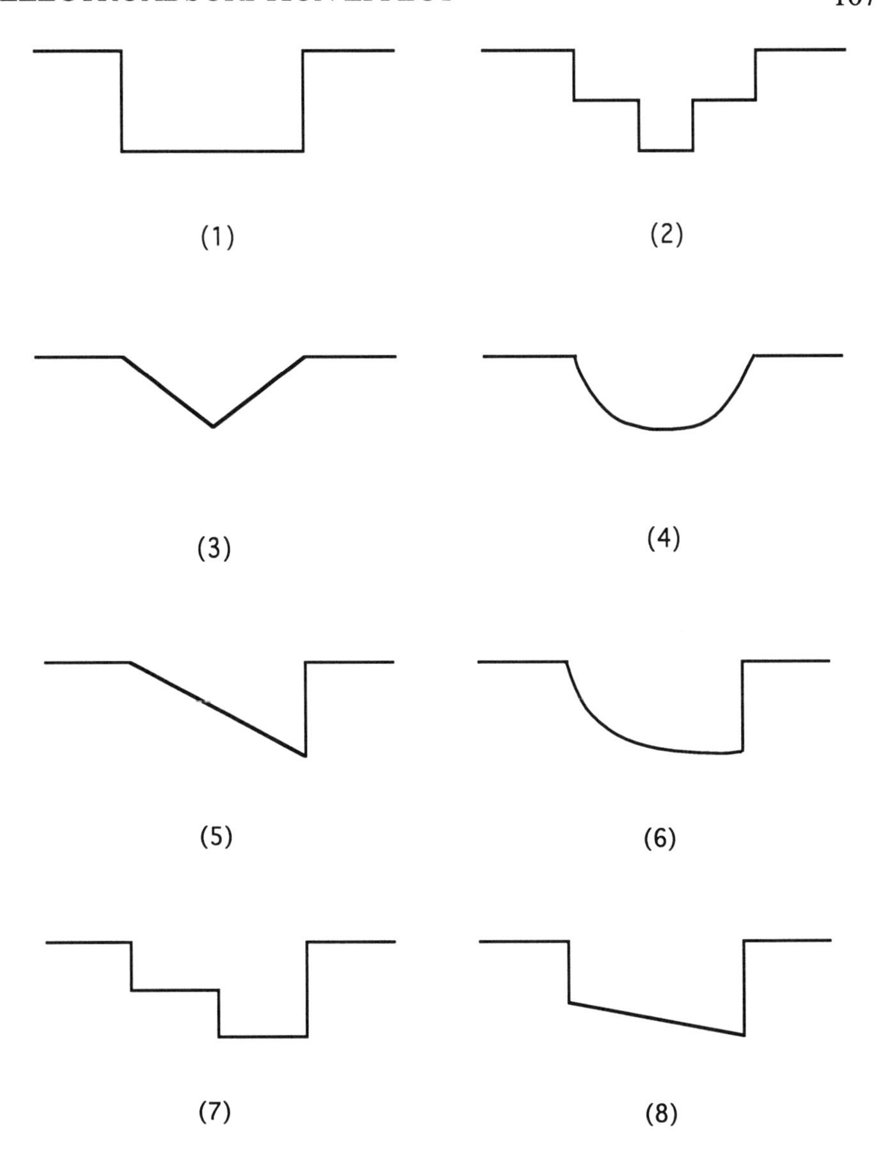

Fig. 4. 22. Schematic conduction band diagram of eight well structures with different shapes (symmetric wells (1)-(5) and asymmetric wells (6)-(8)), composed of $Al_{0.8}Ga_{0.2}As$ in the barriers $Al_xGa_{1-x}As$ in the well layer with varying composition from $x = 0$ (bottom) to $x = 0.4$ (top).

4.7 References

[1] W. Franz, Z. Naturforsch., 13a, 484, 1958.
[2] L. V. Keldysh, Zh. Eksp. Teor. Fiz., 34, 1138, 1958 [Sov. Phys.-JETP., 7, 788, 1958].
[3] R. Williams, Phys. Rev., 117, 1487, 1960.
[4] R. Williams, Phys. Rev., 126, 442, 1962.
[5] T. S. Moss, J. Appl. Phys., 32, 2136, 1961.
[6] K. W. Boer, H. J. Hasche, and U. Kummel, Z. Phys., 155, 170, 1959.
[7] V. S. Valilov, and K. I. Britsyn, Fiz. Tverd. Tela., 2, 1937, 1960 [Sov. Phys.-Solid State, 2, 1746, 1961].
[8] M. Cardona, Solid State Physics, Suppl. 11, edited by F. Seitz, D. Turnbull, and H. Ehrenreich (Academic, New York, 1969).
[9] "*Semiconductors and Semimetals,*" edited by R. K. Willardson and A. C. Beer (Academic, New York, 1972), Vol. 9.
[10] G. Racette, Proc. IEEE, 52, 716, 1964.
[11] P. Handler, Phys. Rev., 137A, 1862, 1965.
[12] F. K. Reinhart, Appl. Phys. Lett., 22, 372, 1973.
[13] G. E. Stillman, C. M. Wolfe, and I. Melngailis, Appl. Phys. Lett., 25, 36, 1974.
[14] J. C. Dyment, F. P. Kapron, and A. J. Spring Thorpe, Inst. Phys. Conf., Ser. No. 24 (Institute of Physics, London,1975), p. 200.
[15] G. E. Stillman, C. M. Wolfe, C.O. Bozler, and J. A. Rossi, Appl. Phys. Lett., 28, 544, 1976.
[16] N. Bottka and L. D. Hucheson, J. Appl. Phys., 46, 2645, 1975.
[17] N. Bottka, Opt. Eng., 17, 530, 1978.
[18] J. C. Cambell, J. C. DeWinter, M. A. Pollack, and R. E. Nahory, Appl. Phys. Lett., 32, 471, 1978.
[19] N. K. Dutta, and N. A. Olessen, Electron. Lett., 20, 634, 1984.
[20] Y. Noda, M. Suzuki, Y. Kushiro, and S. Akiba, Electron. Lett., 21, 1182, 1985.
[21] Y. Noda, M. Suzuki, Y. Kushiro, and S. Akiba, J. Lightwave Technol., LT-4, 1445, 1986.
[22] H. Soda, K. Nakai, H. Ishikawa, and H. Imai, Electron. Lett, 21, 1232, 1987.
[23] H. Tanaka, M. Suzuki, and S. Akiba, OEC'88, 3B2-1, Oct. 1988, Tokyo.
[24] M. Suzuki, Y. Noda, H. Tanaka, S. Akiba, Y. Kushiro, and H. Isshiki, J. Lightwave Technol., LT-5, 1277, 1987.

[25] H. Soda, K. Nakai, and H. Ishikawa, European Confence on Optical Communication, ECOC'88, 1988, p.227.
[26] T. Okiyama, I. Yokota, H. Nishimoto, K. Hironishi, T. Hori matsu, T. Touge, and H. Soda, ECOC'89, MoA1-3, 1989.
[27] N. Henmi, S. Fujita, T. Saito, M. Yamaguchi, M. Shikada, and J. Namiki, in Digest of Conference on Optical Fiber Communication, OFC'90, 1990, Paper No. PD8-1.
[28] G. Mak, C. Rolland, K. E. Fox, and C. Baauw, Photon. Technol. Lett., 2, 10, 1990.
[29] K. Tharmalingam, Phys. Rev., 130, 2204, 1963.
[30] L. D. Landau and E. M. Lifshitz, "*Quantum Mechanics,*" 2nd ed. Addison Wesley, Reading, Massachusetts, 1965.
[31] B. R. Bennett and R. A. Soref, IEEE Quantum Electron., QE-23, 2159, 1987. Detailed references are therein.
[32] S. M. Sze, "*Physics of semiconductor devices*", John Wiley & Sons, Inc., 1969.
[33] D. A. B. Miller, and D. S. Chemla, and S. Schmitt-Rink, Phys. Rev., B33, 6976, 1986.
[34] S. Schmitt-Rink, D. A. B. Miller, and D. S. Chemla, Phys. Rev., B35, 8113, 1987. and ibid. Adv. Phys., 38, 89, 1989.
[35] D. A. B. Miller, D. S. Chemla, T. C. Damen, A. C. Gossard, W. Wiegmann, T. H. Wood, and C. A. Burrus, Phys. Rev. Lett., 53, 2174, 1984.
[36] D. A. B. Miller, D. S. Chemla, T. C. Damen, A. C. Gossard, W. Wiegmann, T. H. Wood, and C. A. Burrus, Phys. Rev., B32, 1043, 1985.
[37] G. Bastard, Phys. Rev., B30, 3547, 1985.
[38] J. A. Brum and G. Bastard, Phys. Rev., B31, 3893, 1986.
[39] G. D. Sanders and K. K. Bajaj, Phys. Rev., B35, 2308, 1987.
[40] S. Hong and J. Singh. J. Appl. Phys.. 62. 1994. 1987.
[41] S. Nojima, Phys. Rev., B37, 9087, 1988.
[42] M. Yamanishi and I. Suemune, Jpn. J. Appl. Phys., 23, L35, 1984.
[43] M. Yamada, S. Ogita, M. Yamagishi, K. Tabata, N. Nakaya, M. Asada, and Y. Suematsu, Appl. Phys. Lett., 45, 324, 1984.
[44] J. S. Weiner, D.S. Chemla, D. A. B. Miller, H. A. Haus, A. C. Gossard, W. Wiegmann, and. C. A. Burrus, Appl. Phys. Lett., 47, 664, 1985.
[45] K. Wakita, Y. Kawamura, M. Nakao, and H. Asahi, IEEE J. Quantum Electron., 23, 2210, 1987.
[46] D. D. Sell, S. E. Stokowski, R. Dingle, and J.V.DiLodrenzo,

Phys. Rev., B7, 4568, 1973.
[47] J. S. Weiner, D. A. B. Miller, D. S. Chemla, T. C. Damen, C. A. Burrus, T. H. Wood, A. C. Gossard, and W. Wiegmann, Appl. Phys. Lett., 47, 1148, 1985.
[48] A. K. Ghatak, IEEE J. Ouantum Electron., 24, 1524, 1988.
[49] T. Yamanaka, K. Wakita, and K. Yokoyama, Appl. Phys. Lett., 65, 1540, 1995.
[50] J. Bleuse, G. Bastard, and P. Voisin, Phys. Rev. Lett., 60, 220, 1988.
[51] E. E. Mendez, F. Agullo-Rueda, and J. M. Hong, ibid, 60, 2426, 1988.
[52] P. Voison, J. Bleuse, C. Bouche, S. Gaillard, C. Alibert, and A. Regreny, ibid, 61, 1369, 1988.
[53] E. Bigan, M. Allovon, M. Carre, C. Braud, A. Carenco, and P. Voisin, J. Quantum Electron., 28, 214, 1992.
[54] J. Bleuse, P. Voisin, M. Allovon, and M. Quillec, Appl. Phys. Lett., 53, 2632, 1988.
[55] I. Bar-Joseph, K. W. Goosen, J. M. Kuo, R. F. Kopf, D. A. B. Miller, and D. S. Chemla, ibid, 55, 340, 1989.
[56] G. R. Olbright, T. E. Zipperian, J. Klem, and G. R. Hadley, J. Opt. Soc. Amer. B, 8, 346, 1991.
[57] K. -K. Law, R. H. Yan, J. Merz, and L. A. Coldren, Appl. Phys. Lett., 56, 1886, 1990.
[58] K. -K. Law, R. H. Yan, L. A. Coldren, and J. Merz, ibid, 57, 1345, 1990.
[59] E. Bigan, M. Allovon, M. Carre, and P. Voisin, ibid, 57, 327, 1990.
[60] E. Bigan, M. Allovon, M. Carre, C. Brud, A. Carenco, and P. Voisin, Electron. Lett., 28, 48, 1992.
[61] F. Devaux, E. Bigan, M. Allovon, J. C. Harmand, and M. Carre, ECOC'92, TuB6.4, P.217, 1992.
[62] H. Schneider, K.v. Klitzing, and K. Ploog, Superlatt. Microstruct. 5, 383, 1989.
[63] G. Livescu, D. A. B. Miller, T. Scizer, D. J. Burrows, J. E. Cunningham, A. C. Gossard, and J. H. English, Appl. Phys. Lett., 54, 748, 1989.
[64] G. D. Boyd, A. M. Fox, D. A. B. Miller, L. M. F. Chirovsky, L. A. D'Asaro, J. M. Kuo, and R. F. Kopf, Appl. Physs. Lett., 57, 1843 1990.
[65] J. Lee, M. O. Vassel, E. S. Kotels, and B. Elman, phys. Rev., B38,

10057, 1989.
[66] D. Atkinson, G. Parry, and E. J. Austin, Semicon. Sci. Technol., 5, 516, 1990.
[67] Y. J. Chen, E. S. Kotels, E. S. Elman, and C. A. Armienco, Phys. Rev, B36, 4562, 1987.
[68] H. Q. Le, J. J. Zayhowski, and W. D. Goodhue, Appl. Phys. Lett., 50, 1518, 1987.
[69] M. N. Islam, R. L. Hillman, D. A. B. Miller, D. S. Chemla, A. C. Gossard, and J. H. English, ibid, 50, 1098, 1987.
[70] J. W. Little, J. K. Whisnant, R. P. Leavitt, and R. A. Willson, ibid, 51, 1786, 1987.
[71] Y. Tokuda, K. Kanamoto, N. Tsukada, and T. Nakayama, ibid, 54, 1232, 1989.
[72] N. Debbar, S. Hong, J. Singh, P. Bhattacharya, and R. Sahai, J. Appl. Phys., 65, 383, 1989.
[73] R. P. Leavitt, K. J. Ritter, J. W. Little, S. C. Horst, and K. W. Steijn, QELS'89, MBB1, 1989.
[74] K. J. Ritter, J. K. Whinant, S. C. Horst, and J. W. Little, CLEO'90, CTHM2.
[75] W. Chen, and T. G. Andersson, Semicon. Sci. Technol., 7, 828, 1992.
[76] T. Ishikawa, S. Nishimura, and K. Tada, Jpn. J. Appl. Phys., 29, 1466, 1990.
[77] M. Morita, K. Goto, and T. Suzuki, ibid, 29, L1663, 1990.
[78] Y. Suzaki, S. Arai, S. Baba, and M. Koutoku, Phon. technol. Lett., 3, 1110, 1991.
[79] T. Ishikawa and K. Tada, Jpn. J. Appl. Phys., 28, L1982, 1989.
[80] K. -K. Law, R. H. Yan, A. C. Gossard, and J. L. Merz, J. Appl. Phys., 67, 6461, 1990.
[81] P. W. Yu, D. C. Reynolds, G. D. Sanders, K. K. Bajaj, C. E. Stutz, and K. R. Evans, Phys. Rev., B43, 4344, 1991.
[82] W. Q. Chen, S. M. Wang, T. G. Andersson, and J. Thordson, J. Appl. Phys., 74, 6247, 1993.
[83] G. D. Sanders, and K. K. Bajaj, J. Vac. Sci. Technol., B5, 1295, 1987.
[84] T. Hiroshima and K. Nishi, J. Appl. Phys., 62, 3360, 1987.
[85] G. D. Sanders, and K. K. Bajaj, J. Appl. Phys., 67, 6461, 1990.
[86] Y. Susa and T. Nakahara, Appl. Phys. Lett., 60, 2324, 1992.

Chapter 5
Various Modulation

5.1 Direct Modulation

5.1.1 Introduction

In the previous sections we introduce electrooptic effect and electroabsorption modulation and some devices using these phenomena. In this chapter we describe other modulation mechanisms and their applications. At first we introduce direct modulation using semiconductor laser diodes, then we discuss the effects associated with injected carriers; laser diode gate switches, carrier injection modulators or switches, BRAQWESTs or barrier resevoir and quantum well electron transfer structure modulators. And then other mechanisms used for semiconductor modulators or switches are described.

For a long time, laser diodes have been used as light sources and modulators because of their simplicity, small-size, relatively broad bandwidth and high modulation efficiency. This book is concerned with external modulators, not with laser diodes. However, direct modulation is very useful and popular and is important for applications such as external modulators, so we will briefly summarize direct modulation in this section, giving special attention to frequency modulation [1,2].

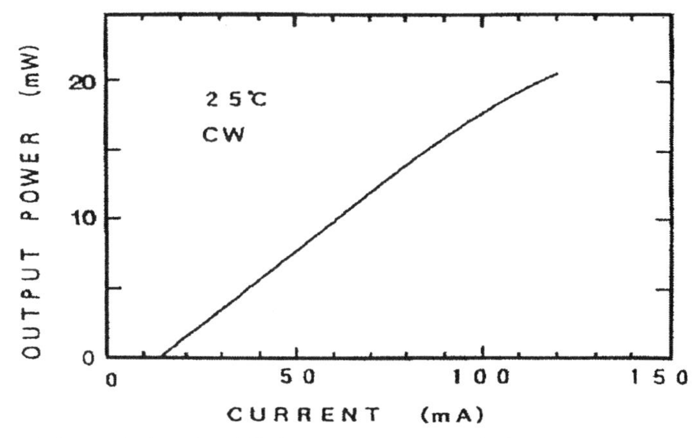

Fig. 5.1. Typical light output versus injection current for a semiconductor laser diode.

The laser-diode modulation is based mainly on changes in the power of light emitted associated with change in input current. The relationship between light output and input current for laser diodes is shown in Fig. 5.1. When the modulation frequency is not high and the laser diode structure is single-mode, the relationship between the input current I and the light output L is linear. That is, L is proportional to $[I-I_t]$, where I_t is the threshold current for lasing operation. This linearity is usually nearly perfect and in some cases this characteristic is used in analogue light sources, though in most usual applications laser diodes are used as digital light sources. The slope efficiency of the L-I curve (or differential quantum efficiency) is over 40 percent.

5.1.2 Relaxation Frequency

When the modulation frequency increases with input-current amplitude hold constant, a phase delay between the output light and the input current is brought about, producing relaxation oscillations in the output light, as shown in Fig. 5.2. The amplitude of the light shows a resonance-like peak in frequency and after the resonance it abruptly decreases. This resonance frequency is expressed as follows [3].

$$f_r = [\{(\tau/\tau_p)(1 + A)(I_b/I_t - 1)\}^{1/2}]/2\pi\tau,$$

where τ is the carrier lifetime due to spontaneous emission and non-radiative recombination, τ_p is the photon lifetime, A is a parameter determined by the material and structure of the laser's active layer, and I_b is the bias current. The resonance frequency f_r increases as the input current increases but it usually saturates at a few GHz, even when the bias current I_b increases.

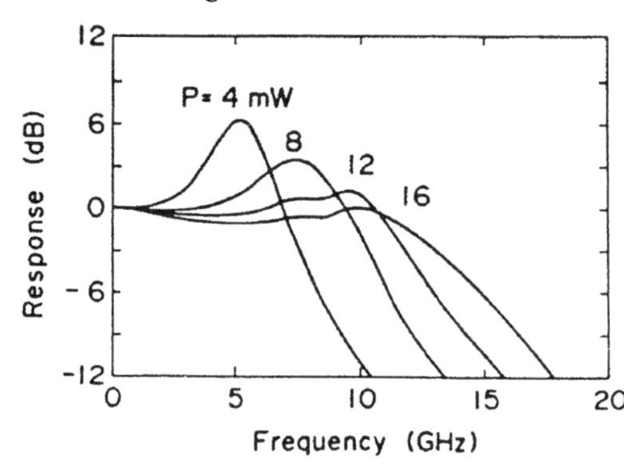

Fig. 5.2. Frequency response of the small-signal modulation for a semiconductor laser diode.

The direct modulation frequency is limited by the relaxation frequency and therefore high frequency modulation is necessary to obtain a high relaxation frequency. There are several devices that can be used to improve the relaxation frequency: shortening of photon lifetime τ_p by shortening the laser cavity length, strengthening the differential gain by using suitable detuning energy and introducing quantum well structures into active layers, facet-coating for optimum reflectivity, etc. However, standard Fabry-Perot type (necessary-cavity mirrors) laser diodes operate with multi-mode oscillation when they are modulated at high frequencies (Fig. 5.3) [4]. This is because the modal losses in Fabry-Perot lasers are primarily due to the mirror loss, which is independent of frequency and equal for all modes; net gain (gain minus loss) differences between various longitudinal modes are very small. As a result, many modes can reach lasing threshold, leading to multi-mode oscillations.

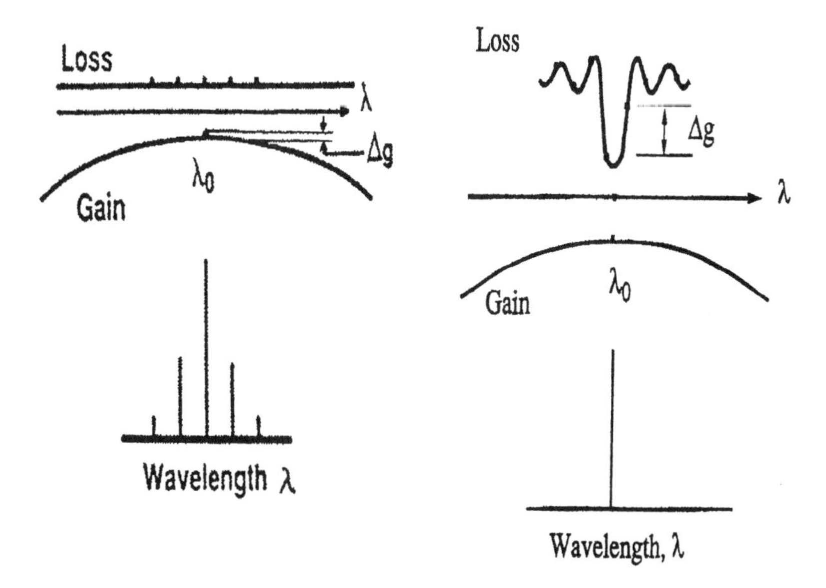

Fig. 5.3. Emitted light spectrum from a Fabry-Perot type semiconductor laser diode drived with a large-signal modulation.

Fig. 5.4. Operation principle of a DFB laser diode.

Distributed feedback (DFB) or distributed Bragg reflector laser diodes were devised to overcome this drawback [5]. Periodic perturbations in the laser or waveguide (other than mesa mirror facets for lasing) produce distributed feedback due to the stop-band effect. A corrugated guide reflects the light in it if the grating has a period which is a multiple of half of the wavelength of that light. Only the mode near the Bragg wavelength is reflected constructively (Bragg reflection). Figure 5.4 shows the operation principle.

Based on these principles, DFB or DBR laser diodes can operate in dynamic single-mode operation. The relaxation oscillation frequencies up to a few tens of GHz have been developed (the top frequency is over 20 GHz for a 3-dB bandwidth)[6]. Nowadays, the relaxation frequency limitation has been overcome and magnitude does not pose a severe problem in the future improvement of lasers, rather it is frequency chirping that appears as the next problem awaiting solution in field of direct modulation.

5.1.3 Chirping

In the direct modulation of the laser diodes commonly used in transmission systems, the refractive index of the laser cavity changes with the changes in carrier density associated with current modulation and temperature changes in the active layer. This results in changes in oscillating wavelength. This phenomenon induces an increase in the wavelength of the emitted light. Chirping is defined as the ratio of the laser gain change to current change. This ratio affects the lasers emitting linewidth [7]. In ordinary optical fiber the transmission-loss minimum is at 1.55 μm and the dispersion minimum is at 1.3 μm. The linewidth broadening produces a group delay in the optical fiber after long distance transmission, due to the dispersion. Figure 5.5 is a schematic of operating principles related to chirping [8]. Such chirping limits the transmission distance.

In large-signal modulation such as pulse modulation, optical power changes in a complicated problem due to the relaxation oscillation. Carrier variation is also large and this aggravates chirping. The spectrum broadens about 30 GHz due to the chirping and has an intricate structure with many peaks. The relaxation oscillation changes with the modulation bias condition. Therefore, the chirping also changes as with the bias condition. Generally, the relaxation oscillation is suppressed as the bias current increases to become larger than the threshold current. Therefore, in order to suppress the chirping associated with the direct modulation, drive conditions must be optimized.

5.1.4 Frequency Modulation

In coherent transmission systems and optical-wave transmission systems, unlike direct modulation, the chirping phenomenon is used positively for frequency modulation. Frequency modulation and chirping are intrinsically similar phenomena.

Laser diodes can operate as frequency modulators using the mechanism discussed above. The emitted light's wavelength is usually constant in the DC bias condition. When the diode is biased under AC conditions, the wavelength changes with input current intensity. There are three regions of frequency domains in laser diode response [9,10], as shown in Figs. 5.5 and 5.6:

(1) A lower frequency region (below 10 MHz); the frequency modulation index increases monotonously with increase in the modulation frequency. The emission wavelength shifts longer (red shift) as the input current increases.

(2) A medium frequency region (between about 10 MHz and 1 GHz); response is flat irrespective of modulation frequency. The frequency deviation is a blue shift (emitting wavelength becomes shorter) as the input current increases.

(3) A higher frequency region (above about 1 GHz); response increases abruptly in the vicinity of 1 GHz and shows a resonance-like peak, then decreases abruptly.

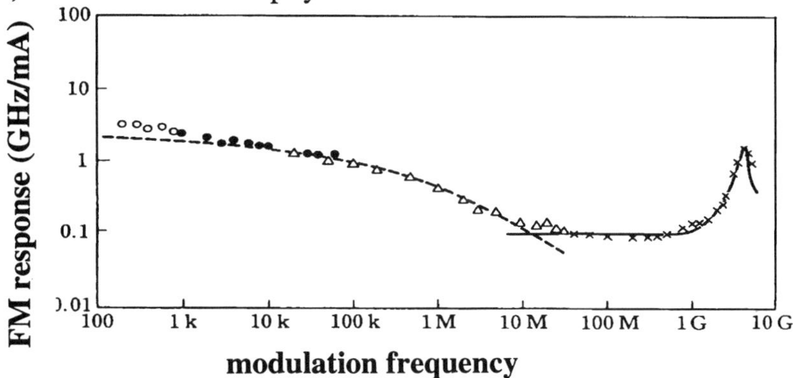

Fig. 5.5. Frequency response of a conventional semiconductor laser diode.

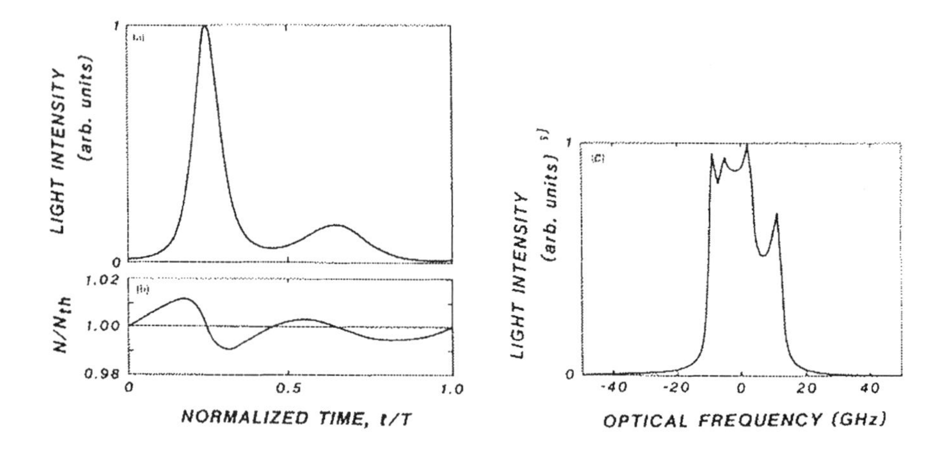

Fig. 5.6. Spectrum broadening of a laser diode due to chirping. (a) Calculated light waveform and (b) Calculated carrier density waveform. Carrier density N is normarized by the density N_{th} at threshold. (c) Time-averaged emission spectrum derived from (a) and (b) [8].

The interface frequencies where these fequency regions meet depends on device structures and operating conditions. In region (1), the frequency modulation is brought about by the rise in temperature due to current injection, and this thermal effect cannot respond to increases in modulation frequency. The frequency modulation due to temperature rise is caused by the cavity-length expansion and by increase in the refractive index due to energy-band-gap shrinkage, both of which make the emission wavelength increase.

In region (3), the frequency modulation is caused by variation in carrier density, and the phase of the frequency modulation before and after the resonance frequency changes by 180 degrees.

Semiconductor laser diodes show very intricate frequency-modulation characteristics, which differ according to the magnitude of the modulation frequency. Especially between region (1) and region (2), frequency deviation changes completely, from a red shift to a blue shift; this poses a serious problem from a practical point of view.

Some experiments are proposed to solve this problem; one possibility is that even in the lower frequency region, frequency deviation could be kept in the blue-shift direction by enhancing the modulation efficiency in the medium frequency region. A second possible method is to make the phase in frequency region (2) the same as that of region (1) by red-shifting the frequency deviation of region (2). A multi-electrode contact structure, which can apply the injection current into many sections independently to delibarately induce spatial hole-burning, has been proposed and demonstrated [8] (Fig. 5.7). Good frequency modulation efficiency has been achieved, resulting in a flat frequency response without a deep-drop in the low frequency region (as shown in Fig. 5.8). This structure has been used widely for frequency modulation.

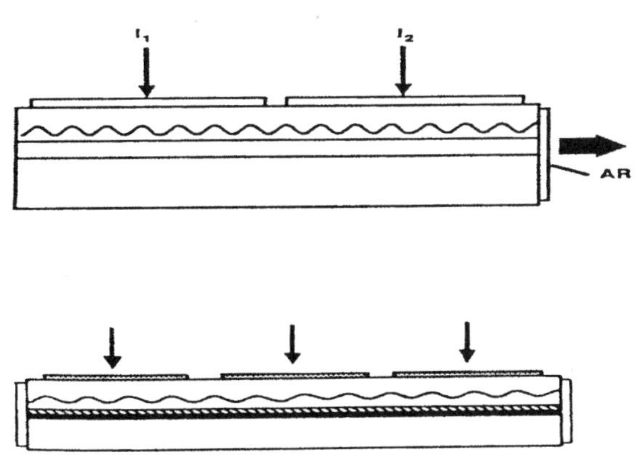

Fig. 5.7. Schematic structure of a multi-electrode DFB laser.

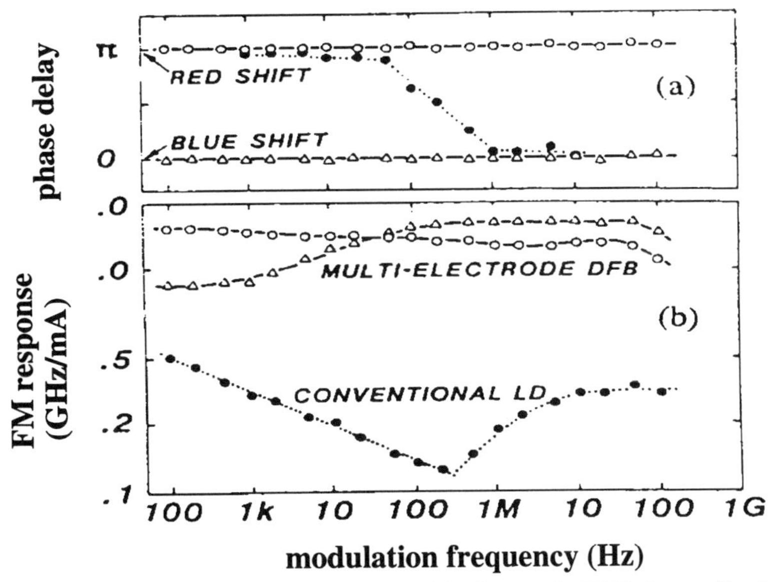

Fig. 5.8. Frequency response of a multi-electrode DFB laser diode.

5.2 Laser Diode Switch

A laser diode generally consists of an optical amplifier and an optical feedback section to lase. We can obtain an optical amplifier by eliminating the feedback from lasers. For conventional Fabry-Perot laser diodes, this is done by coating both facets of the laser mirrors with an anti-reflection material. Semiconductor waveguides generally have large insertion losses compared with those made of other materials such as $LiNbO_3$ and the SiO_2 used in planar lightwave circuits (PLC). Semiconductor waveguide switches are made with a p-n junction, and this laser diode structure is used as both an optical gate and an optical amplifier. This enables lossless optical switching operation and the integration of many elements into a small area, as well as low operating power. The operation principle is similar to the carrier injection modulation/switching described in the next subsection and modulation covered previously, therefore we describe the laser diode switch in this section.

Figure 5.9 is a schematic diagram of a semiconductor diode gate switch. When current is not injected to an optical amplifier, it only absorbs the incident light and the out put light is off, whereas the incident light is guided and output when a current is injected. Figure 5.10 shows the operation principle. The output optical power strongly depends on the injection current and at a certain current level the guided light intensity changes: it no longer decreases and the light is amplified.

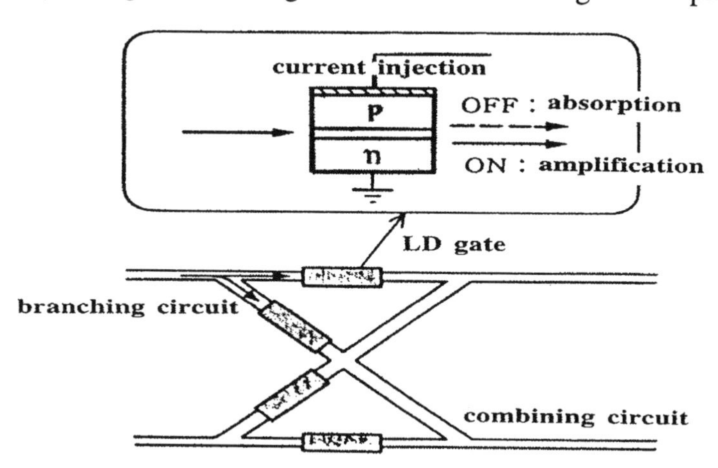

Fig. 5.9. Fundamental configuration of a laser diode gate switch.

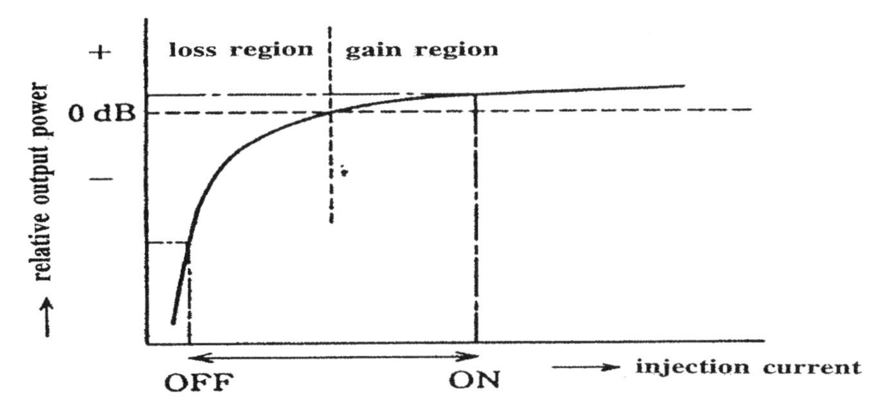

Fig. 5.10. Operation principle of a LD gate.

Since the first laser diode gate switch was proposed by Ikeda in 1981 [11], several types of semiconductor gate switches have been reported, the most recent ones being 4x4 gate amplifier switches [12-15]. Recently, lossless switching operation of the 4x4 laser gate switch was demonstrated with the injected gate current of about 16 mA for each gate [16]. The time response (switching speed) was less than 0.5 ns with a crosstalk level of less than -20 dB. A photograph of this device is shown in Fig. 5.11. Compactness is achieved through the use of blanching circuits with large splitting angles obtained by using the total reflection at the semiconductor-air interface [17-21]. A Y-branch circuit whose splitting angle is a small 4° ~ 6°, which is typical for semiconductor waveguides, increases switch to several centimeters. Such devices have low insertion loss because of their excellent optical gain and are suitable for large scale integration.

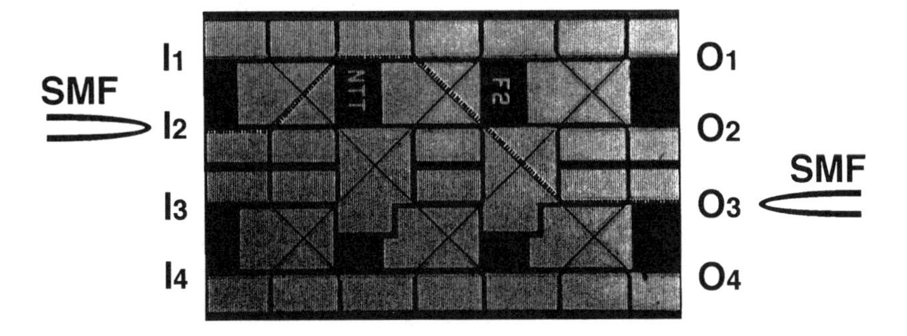

Fig. 5.11. Photograph of a 4x4 laser diode gate switch.

5.3 Carrier Injection Effect

The carrier-induced change in the refractive index associated with plasma dispersion is about two orders of magnitude larger than the change induced by the electrooptic effect [22-24]. Moreover, it is not dependent on polarization, unlike the case of the electrooptic effect. Thus polarization independent switches are thought to be feasible. This effect has the advantage of presenting wavelength indpendent operation possibilities. Moreover, the driving method is similar to that of laser diodes and optical amplifiers because of forward bias operation, which makes the easily monolithic integration of semiconductor amplifiers easy. Monolithic integration enables us to overcome the large propagation losses and insertion losses inherent in semiconductor waveguides. The drawbacks are propagation loss due to free carrier absorption and the switching speed is limited by carrier recombination.

The first trial using carrier-induced refractive index change in an optical switch was conducted in 1984 [25, 26]. Since then many groups have tried to fabricate large scale optical switch arrays for space division switching using carrier injection [27-35]. At present, 4x4 optical switch arrays have been built on III-V semiconductors comprising GaAs/GaAlAs alloys [29], where the fabrication technology is more mature but less interesting than in the 1.55 μm wavelength region, or comprising InP material with average currents per coupler in the range of 10 mA [34] or lossless and low-crosstalk (40 dB on/off ratio) optical switches [35].

Below, we briefly describe the carrier injection optical switches.

5.3.1 Operation Principle

When the carriers are injected from the p-n junction into the waveguide layer, the refractive index in this layer decreases. The carrier-induced change in the refractive index is dependent on the carrier density as a result of both the band-to-band anomalous dispersion and the free carrier plasma dispersion [23, 24]. The former index change is several times greater than the change due to plasma dispersion in InGaAsP/InP systems [24].

In order to efficiently take advantage of the anomalous dispersion, the input laser wavelength should be fine tuned to be very near, but slightly longer than, the bandgap wavelength of the InGaAsP layer. Unless this condition is satisfied there will be a large loss due to band-edge absorption when there is no current injection. Therefore, free carrier plasma dispersion is used to avoid band-edge absorption in the InGaAsP layer. The index change is given in Section 3.3. The estimated induced index change is about -5×10^{-3} at 1×10^{18} cm^{-3} injection current density for a 1.55-μm wavelength.

The induced change in the refractive index is about two orders of magnitude greater than that for the electrooptic effect. Although the increase in propagation loss due to absorption by injected free carriers must be considered, the loss is still low for a carrier density of 10^{18} cm^{-3}. However, in designing a photonic space division switching network on a large scale, it is nescessary to reduce the loss.

5.3.2 Digital Optical Switch

In modal-evaluation switches the propagation path of the optical signal through two intersecting waveguides is selected by inducing a small index asymmetry between the waveguides as shown in Fig. 5.12. The key attraction is their digital nature. This asymmetry is achieved by injecting current through an electrode into the waveguide or by reverse biasing the electrode to deplete charge carriers in the waveguide. Once sufficient index asymmetry is induced, they switch completely, and the control signal does not have to be held at a precise value. Due to this digital nature they are referred to as digital optical switches. These switches are fast (a few nanoseconds for carrier injection and tens of picoseconds for the carrier depletion mode [36]) and have an advantage over gain switches in their lower power requirements and reduced noise. Compared to the interferometric switches, they are more fabrication tolerant and can be polarization and wavelength insensitive [37].

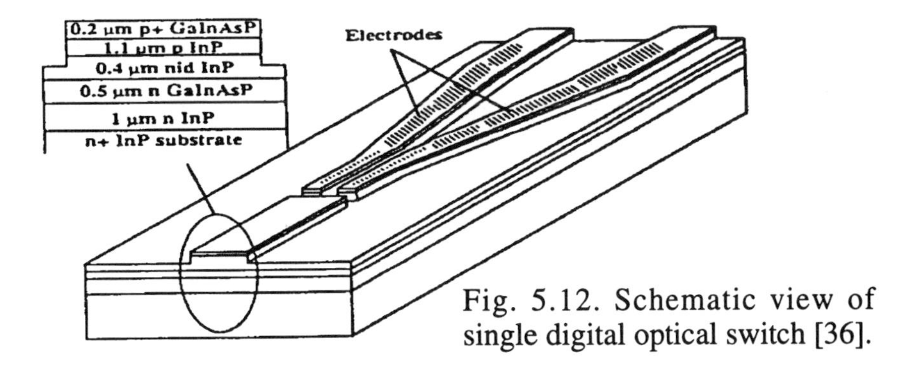

Fig. 5.12. Schematic view of single digital optical switch [36].

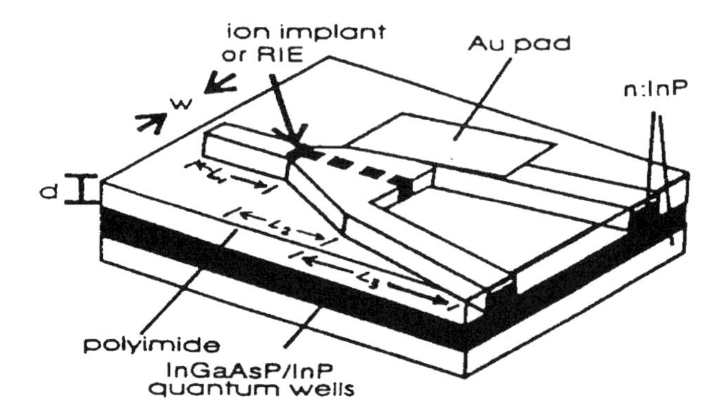

Fig. 5.13. Schematic of tapered Y-branch digital switch [45].

Digital optical switches have been demonstrated using $LiNbO_3$ for small opening angles [38], first by Alcatel Alsthom in InP [39] and followed by GTE [37, 40-42]. III-V semiconductor-based switches are expected to have potential for integration with other active devices such as optical amplifiers. 1x4 and 4x4 space switching matrices based on digital optical switches have already been reported [39] and used in optical crossconnect system experiments up to 10 Gbit/s [43]. Larger opening angles (0.5° to 2.0°) to minimize the switch size have been reported with only 1-mm-long electrodes [44, 45]. Figure 5.13 shows a tapered Y-branch switch with a wider taper angle followed by a narrow angle to maintain adabaticity [45]. This tapering provides digital switching characteritics with a crosstalk of -25 dB for an active length of only 800 µm and a measured bandwidth of 10 GHz that required a switching voltage of -4V via the quantum confined Stark effect.

5.3.3 Monolithic Integration with Amplifiers

One of the best ways to reduce the propagation loss and to obtain a loss-less optical switch is to use monolithically integrated traveling-wave amplifiers. Such amplifiers are the most desirable unit for large-scale photonic switching networks because they are small and enable the network to be enlarged without worrying about the loss budget.

A carrier-injection-type optical single-slip structure (S^3) switch (COSTA) with a traveling-wave amplifier (TWA) has been proposed [35, 46]. This switch consists of four unit cells, as shown in Fig. 5.14. Each cell consists of an X-crossing waveguide, two Y-branch switches, and an amplifier. The X-crossing angle is ten degrees and the Y-branch angle is 5 degrees. A butt-coupling structure was adopted for coupling between the passive and amplifier waveguides. This achieves about 80% coupling efficiency at the interface. A 4x4 COSTA for 1.3 μm wavelength has been developed. It comprises 16 TWA's and 32 Y-branch optical switches monolithically integrated on an InP substrate using MOVPE and dry etching techniques [29]. For switched path lossless operation including coupling loss between a single mode fiber and the device has been achieved with injection currents into the amplifier and switching section of 200 and 50 mA, respectively. The estimated internal gain in an amplifier was 25 dB.

This device is considered to be composed of a carrier injection optical switch and a laser diode gate switch described in Section 5.2.

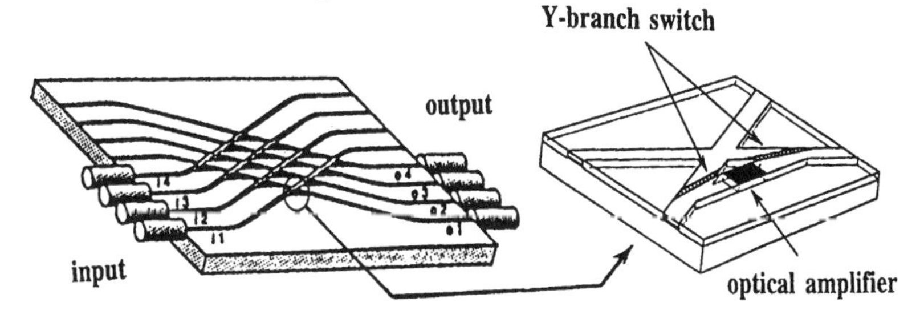

Fig. 5.14. Fundamental configuration of a carrier-injection type optical single-slip structure (S3) switch with traveling-wave amplifier (COSTA) [46].

5.4 Tunable-Electron-Density Modulators

5.4.1 Introduction

In MQW structures the quantum-confined Stark effect is usually used for electroabsorption and electrorefraction modulators. In this effect, the absorption coefficient and refractive index change by the electric field associated with the shift of the exciton resonance peak to a longer wavelength (red shift). On the other hand, bandfilling by carrier injection also provides the large index changes required for optical switching devices. However, the power dissipation is large and the speed is limited by the minority carrier recombination times. The multiple period barrier resevoir and quantum well electron transfer structure (BRAQWETS) modulators are another promising structure, particularly for electrorefractive modulators invented by the AT&T group [47]. In this device, electroabsorption and electrorefraction are provided by a synchronous, voltage-controlled transfer of electrons into multiple quantum wells. The principle electrooptic effect arises from the quenching of absorption upon correlated filling of single-particle and pair states in the phase space.

A modulation-doped field effect transistor (FET) has previously been employed in reflectivity studies of band filling in a single quantum well [48-50], but the close proximity of the gate electrode to the well precludes normal-incidence transmission measurements and has unacceptable losses for waveguide applications. A reservoir of charges for the quantum well as well as an isolating barrier for suppressing current flow is introduced and any number of multiple quantum wells can be stacked without density inhomogeneities, resulting in large optical confinement and small losses [47].

Unlike the QCSE, the increase in refractive index change, Δn, is not accompanied by an increasing $\Delta\alpha$ because of the blue shift of the absorption edge. This means that a smaller detuning of the photon energy from the zero-field absorption edge can be used to take advantage of the larger electrorefractive effect available there. Furthermore, the dependence of Δn on the applied voltage is linear in contrast to being predominantly quadratic for QCSE. The linearity of phase shift with voltage makes it valuable to the balanced "push pull" operation of Mach-Zehnder interferometers for chirp-free intensity modulation. Compact Mach-Zehnder interferometers [51, 52] with a low voltage-length product of 2.2 Vmm for π phase shift at 1.5 μm and an intersecting waveguide optical switch [53] based on InGaAs/InGaAlAs/InAlAs BRAQWETS have been demonstrated. Recently, another groups reported polarization independent operation by using InGaAs/InP tensile strained quantum well layers [54] and high speed operation by using a graded-gap spacer layer [55]. Detailed modeling and design considerations for high performance waveguide phase modulators were reported [56]. In this section, we briefly discuss the BRAQWETS modulators and optical switches.

5.4.2 Operation Principles

Figure 5.15 shows the absorption coefficient of the MQW structure as a function of carrier density [50]. As the carrier density increases, the exciton oscillator strength decreases. That is, electron transfer in from the electron reservoir to fill additional conduction subband states in quantum wells results in quenching of some excitonic and interband absorption, effectively shifting the absorption edge to a higher energy. Current injection will also affect the absorption coefficient but the speed is limited by the minority carrier recombination lifetime. BRAQWETS is distinguished by being a majority-carrier unipolar device. It involves only electrons in a quantum well whose density can be modulated by

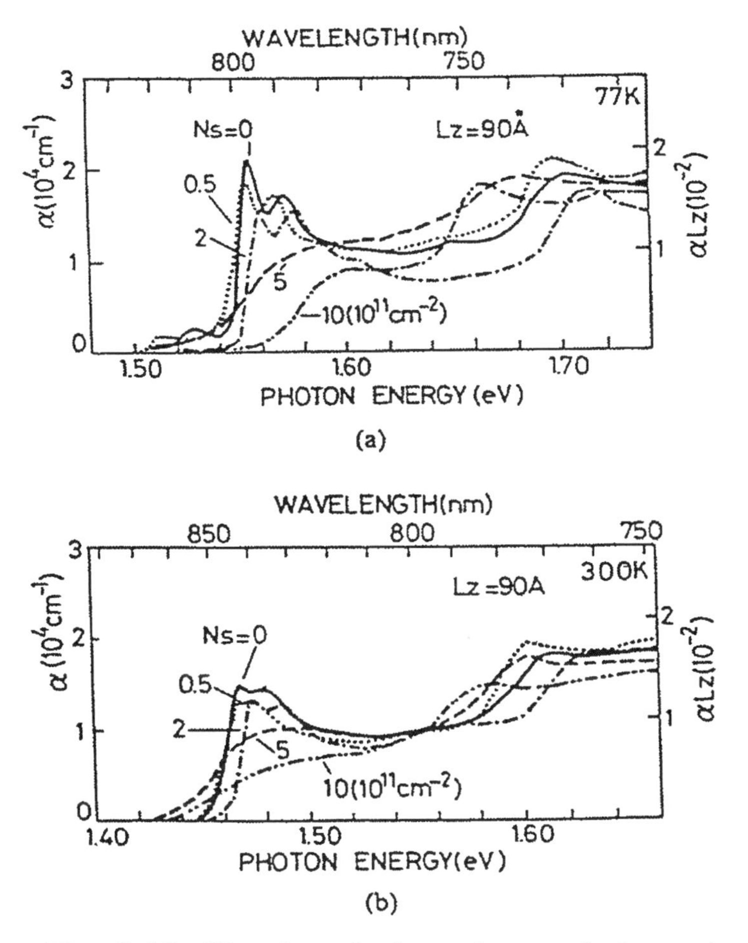

Fig. 5.15 Wavelength dependence of absorption coefficient of MQW structures as a function of carrier density [50].

an electric field.

Figure 5.16 shows the fundamental period of the device [47], which has been repeated five times to form the core of a waveguide. This block forms the active core of an n-i-n InAlAs/InGaAlAs waveguide heterostructure. Bottom and top have Ohmic metal contacts. A 50-nm n-type doped InGaAlAs layer provides the reservoir of electrons. It is isolated from a 9-nm InGaAs quantum well by an undoped InGaAlAs spacer layer that ensures that the quantum well is free from impurities. The third constituent is the 55.5-nm InAlAs region that is p-doped in the middle. A12 nm layer undoped InGaAlAs provides the transition to the next building block.

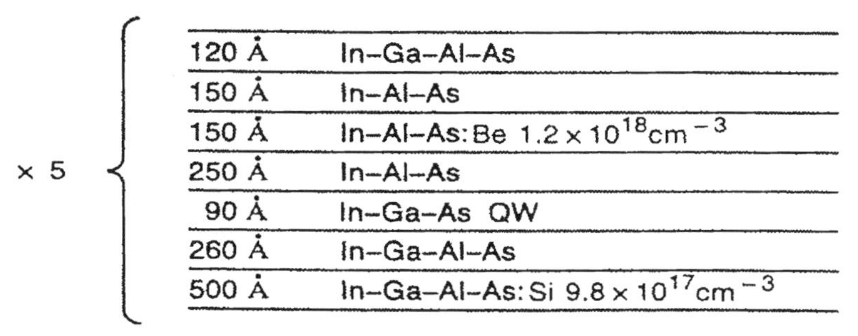

Fig. 5.16 Fundamental period of BRAQWETS device repeated five times to form the core of a waveguide.

Figure 5.17 shows the band diagram of one period without bias and with bias (at room temperature) and the associated absorption spectra[53]. The quantum well rides on the left slope of the potential hill formed by the n-i-p-i-n layer sequence. The p-type doped region is always completely depleted. The quantum well is clearly above the Fermi energy (broken line in Fig. 5.17) and thus empty. The difference in energy between the ground state of the well and the Fermi energy (and thus the electron density in the quantum well at zero bias) can be controlled by sliding the quantum well up and down the potential slope, i.e., changing the quantum well position between the barriers.

Since the bandfilling leads to a blue shift of the absorption edge, very large $\Delta n/\Delta\alpha$ values are obtained, thus, virtually eliminating the cross-talk degradation at high voltages due to the residual electroabsorption commonly found in quantum well switching devices.

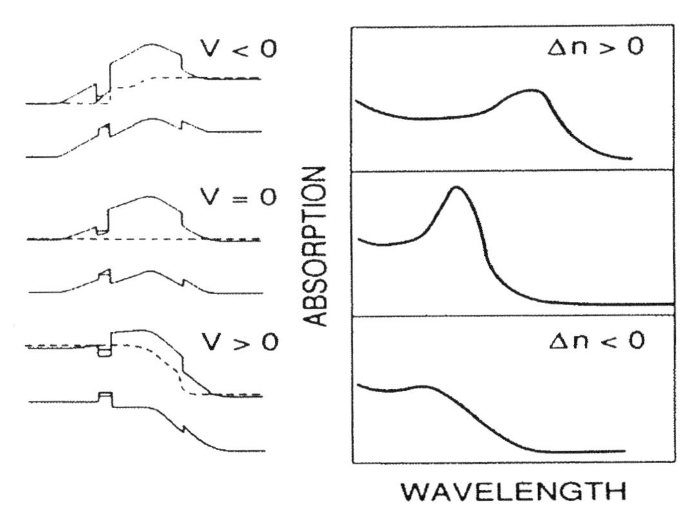

Fig. 5.17 Schematic of band structure and associated absorption spectrum of one BRAQWETS period as a function of applied bias. Electron transfer into the well, causing a blue-shift of the absorption via bandfilling (after [53]).

5.4.3 Device Performances

As described in the introduction, 1x1 interferometric modulators using BRAQWETS having the smallest voltage-length product of 2.2 Vmm were reported[52]. To maximize the overlap between the waveguide mode and the electro-optically active quantum wells and thus increase the voltage sensitivity of the device, the thickness of the InGaAlAs reservoir and the InAlAs barrier layers was reduced without any change in the basic operation of the BRAQWETS. The total optical confinement factor Γ_{SQW} is 0.073 for the eight InGaAs quantum wells. In the 700 μm long active length devices at wavelength, π phase shift is produced by a $V_\pi = 3.2$ V single arm drive. The corresponding extinction ratio (power on/power off) is 9.4 dB.

The modulation bandwidth with a 3 dB electrical bandwidth of 5.7 GHz has been reported and the fundamental response time is determined by the voltage-dependent speed of carrier escape from the well, which is indicated by using pump-probe measurments [57]. A reduction of device parasitics has resulted in the high bandwidth of 15 GHz. Reduced modulator capacitance is achieved by deep etching of the waveguide with straight sidewalls provided by reactive ion etching and planarization by low dielectric constant polyimide. For example, with an etching depth of 4 μm, and waveguide length L = 286 μm, and polyimide with dielectric constant $\varepsilon = 3$, modulator capacitance of less than 0.2 pF is obtained.

Recently it has been reported that the intrinsic speed of the BRAQWETS, which was limited due to a slow escape of electrons (τ = 20-50 ps) from the quantum well back to the reservoir [57], is reduced to $\tau = 1$ ps by using a graded-gap spacer layer and reducing the triangular barrier for escape of electrons [55]. Figure 5.18 shows the band diagram of fast BRAQWETS with a graded gap spacer layer for different bias voltages [58].

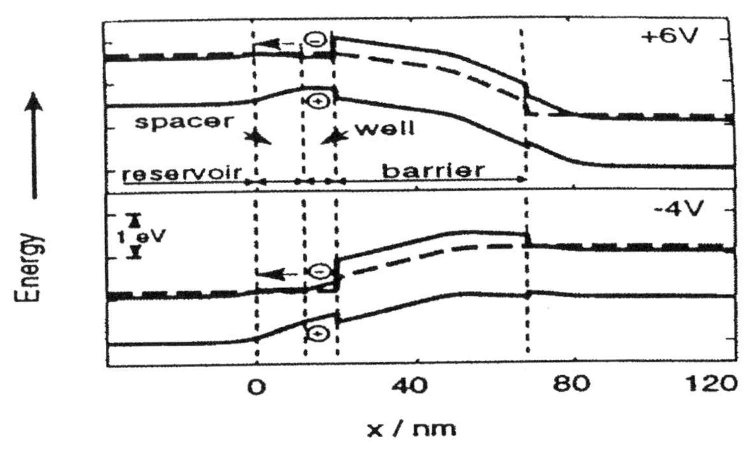

Fig. 5.18 Band diagram of fast BRAQWETS with graded gap spacer layer for different bias voltages [58].

Intersecting waveguide geometry which provides very compact crosspoints but requires large refractive index changes for effective switching is demonstrated [53]. The measured switching voltages are 13 V.

Y-branch electrooptic waveguide switches using BRAQWETS are demonstrated. At 0 V, a 50:50 split is obtained as designed; for positive applied voltage, the output light is directed away from the active port and for negative applied voltage, light exits from the active port [59].

Although $\Delta n/F$ in the BRAQWETS is significantly larger than that of the QCSE in normal incidence, the enhancement is not as large in the waveguide geometry. This is due to the extra layer thickness required for the barrier, reservoir, and spacers in addition to the electrooptically active quantum well [59]. For example, a five-period BRAQWETS with a total thickness of 760 nm contains only 45 nm of quantum well. When this five-period structure is enclosed by n-doped InAslAs cladding layers to form a n-i-n single-mode slab waveguide, the optical confinement of the mode to the quantum wells is only 5 percent. The applied voltage drops across a total undoped region of 510 nm, yielding a phase shift coefficient for this structure of 50 degrees/Vmm at wavelength 1.58 μm. The devices for highly efficient, high-speed applications using such bandfilling effects need further development.

5.5 Wavelength Conversion/Frequency Modulation

5.5.1 Introduction

Functional optical devices with bistability, optical logic, or optical wavelength conversion can be realized by semiconductor laser diodes and semiconductor amplifiers. These devices have many advantages such as gain, the ability to operate with small input power and small size, and the potential to provide the capability for monolithic integration [61].

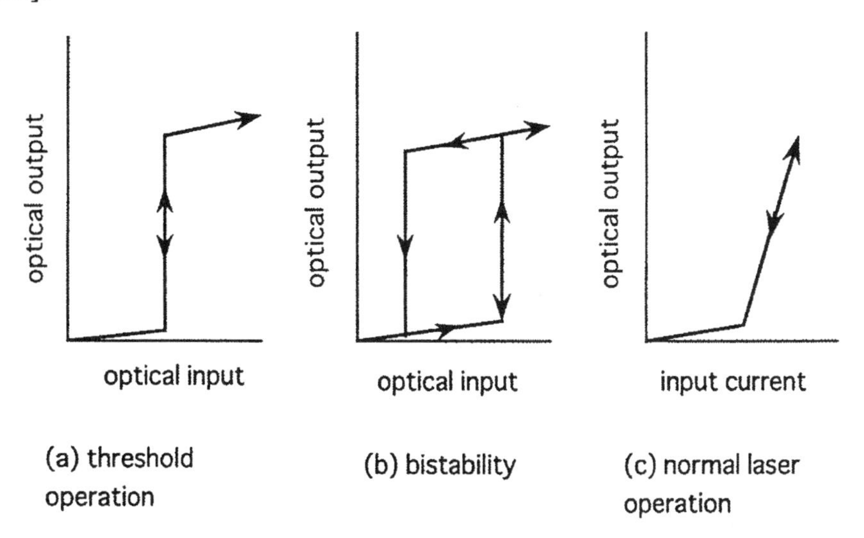

Fig. 5.19. Various characteristics of a multi-electrode semiconductor laser. (a) indicates threshold operation, (b) optical bistability, (c) normal laser operation.

Frequency modulation has been tried for a long time using laser diodes as described in Section 5.1. In these devices, the operating speed has been considered to be limited by carrier lifetime and reaches 10 GHz at most. When the input power of optical bistable devices increases, the on-switch time is shortened and off-time is limited by carrier lifetime. When we use switching between TE and TM polarizations and operate these devices only the on-switch state, we can operate this switch with 50-100 GHz [62].

Recently, a pump-probe method has clarified ultra-high speed change in gain and refractive index due to intra-band electron relaxation in semiconductor lasers and amplifiers as a gain medium. This method efficiently generates four-wave mixing based on the nonlinear effect, and wavelength conversion corresponding to over THz can be applied for signal processing with modulation speed of 100 Gbit/s.

Optical wavelength is generally converted by using the following mechanisms:

(1) Change in the absorption coefficient by using a saturable absorber in a DFB or DBR laser cavity irradiated with optical inputs [63-65].

(2) Gain quenching in semiconductor lasers [66] or semiconductor amplifiers [67].

(3) Four-wave mixing in semiconductor lasers with gain [68, 69].

In this section 5.5 we briefly discuss wavelength conversion based on the all-optical method [70].

5.5.2 Saturable Absorber

A nonlinear phenomenon that the absorption coefficient saturates with input light intensity is called saturable absorption. When materials indicating saturable absorption are set in a semiconductor laser cavity, functional optical devices such as optical bistability, threshold operation, and wavelength conversion are achieved. A semiconductor laser whose electrode is divided into multiple sections is fabricated for this purpose. One of the sections operates as a saturable absorber which is biased with zero or smaller injected current. In the saturable absorber the absorption is large when the input optical power is small, whereas the absorption saturates and abruptly becomes transparent when the input power increases. At the same time semiconductor lasers lase. Figure 5.19 shows the shematic examples of multi-electrode semiconductor laser characteristics.

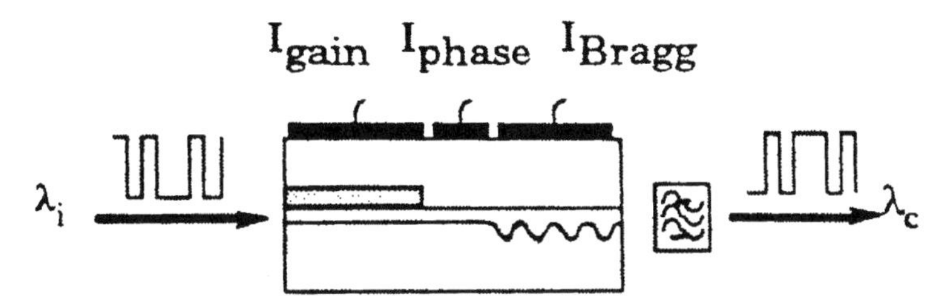

Fig. 5.20. A principles of wavelength converters based on DBR lasers.

A multi-electrode laser with DFB/DBR structures, which can operate a single-mode oscillation, can also operate as a wavelength converter [63-65]. The input signal is launched into the gain-section where it causes gain saturation that controls the oscillation of the laser. Figure 5.20 shows the principles of wavelength converters based on semiconductor lasers. The output wavelength is determined by the bias currents and the output wavelength range by the tuning properties of the laser. Compared with semiconductor optical amplifier converters, laser converters have the advantage that no external CW sources are needed, as discussed in the following section. The necessary input power levels are generally higher than for optical amplifier converter. The output power characteristic becomes very steep for the input wavelength because of laser narrow gain width. The maxmum data rate is determined by the laser resonance frequency, and conversion of 10 Gbit/s signals is possible [71,72].

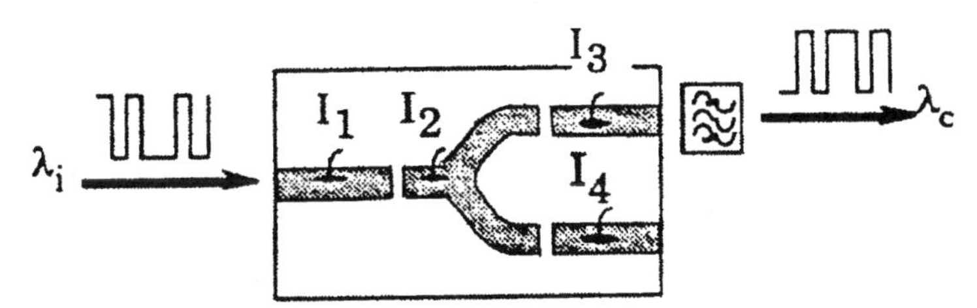

Fig. 5.21. A principle of wavelength converters based on Y-lasers.

The Y-laser converter (Fig. 5.21) is another wavelength converter that has been used with success. The principle is to injection lock the laser to the input signal at marks I_1, I_2, I_3 and I_4 and to let it operate at its own lasing wavelength during the spaces in the figure[73]. Conversion at 5 Gbit/s and over wavelength spans of 40 nm has been reported [74].

5.5.3 Semiconductor Amplifiers

Cross gain modulation in semiconductor optical amplifiers is a simple scheme for wavelength conversion [70]. The input signal (at l_i) modulates the gain by saturation and thereby modulates a CW input with desired output wavelength (l_c) [75, 76], as shown in Fig. 5.22. The conversion is independent of the input signal polarization since the gain of the semiconductor optical amplifier is polarization insensitive. The cross gain modulation scheme inverts the converted signal as seen from Fig. 5.22.

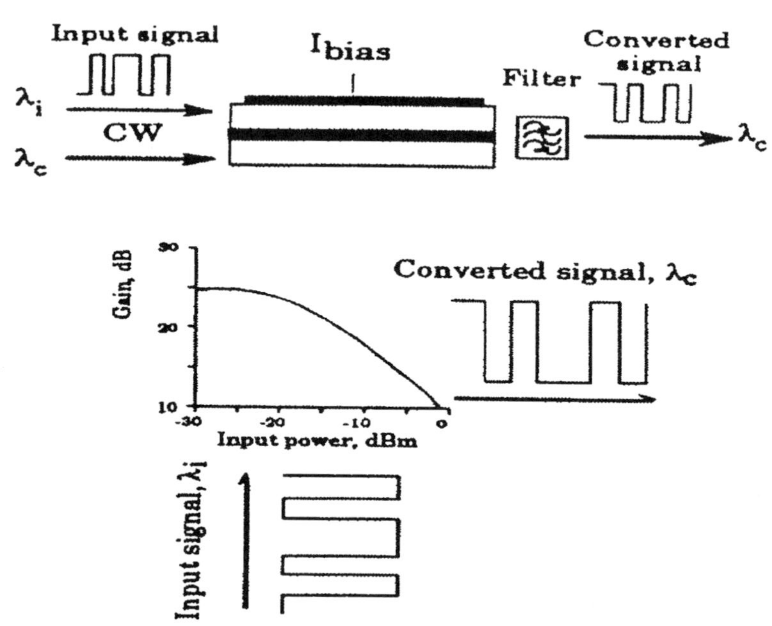

Fig. 5.22. Schematic of wavelength converters based on cross gain modulation in semiconductor optical amplifier (SOA).

The conversion can be performed over the entire gain bandwidth of the amplifier (i.e., 40-50 nm), but the best results are obtained going from longer to shorter wavelengths because the differential gain is largest at the short wavelength end. With a high injection current as well as a high optical power level in the semiconductor optical amplifier, conversion at data rates as high as 20 Gbit/s has been demonstrated [77, 78]. This converter is attractive because of its simplicity, however, it gives extinction ratio degradation which results in excess penalties for conversion to longer wavelengths. This problem can be avoided using cross phase modulation where the amplifier is placed in an interferometric configuration.

Figure 5.23 shows the schematic of interferometric semiconductor optical amplifier converters; Mach-Zehnder and Michelson configuration [75]. The two amplifiers are operated asymmetrically (either different bias or different input power levels), thus the CW light is modulated by the phase difference that is introduced between the two interferometer arms when the input signal depletes the injected carrier concentration in the amplifiers. Only small input signals are needed to introduce a π phase difference between the two arms, thus with the interferometer high extinction ratios can be obtained for conversion to both longer and shorter wavelengths. A 20 Gbit/s conversion experiment with cross phase modulation in a single semiconductor optical modulator has been reported [79].

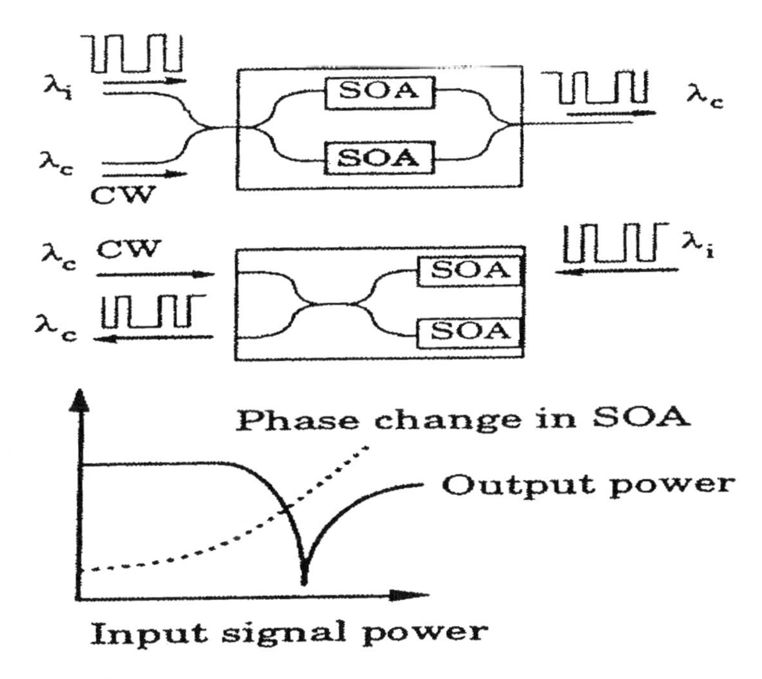

Fig. 5.23. Schematic of interferometric semiconductor optical amplifier converters with Mach-Zehnder and Michelson configurations.

5.5.4 Side-Injection-Light-Controlled Bistable Laser-Diode

The conventional bistable laser diode consists of a laser waveguide and a saturable absorption region which is controlled by optical triggering. It is applicable to optical logic processing because of its distinct threshold and its memory and gain characteristics [80, 81]. Optical multiplex/ demultiplex and time-division multiplex (TDM) switching have been demonstrated with bistable laser arrays and lithium niobate optical modulators [82]. The input wavelength tolerance, however, and the signal isolation are inadequate. A side-injection-light-controlled bistable laser diode (SILC-BLD) has been developed [83], and Fig. 5.24 shows the structure of this device. It consists of a main DFB laser for output and an orthogonally crossed subwaveguide for input. The subamplifier picks up input light and conducts it to the saturable absorption region in the main laser cavity. The input signal and output laser cavity are completely isolated, so there is no back reflection or signal leakage. This results in wide input wavelength sensitivity by reducing the influence of the laser cavity.

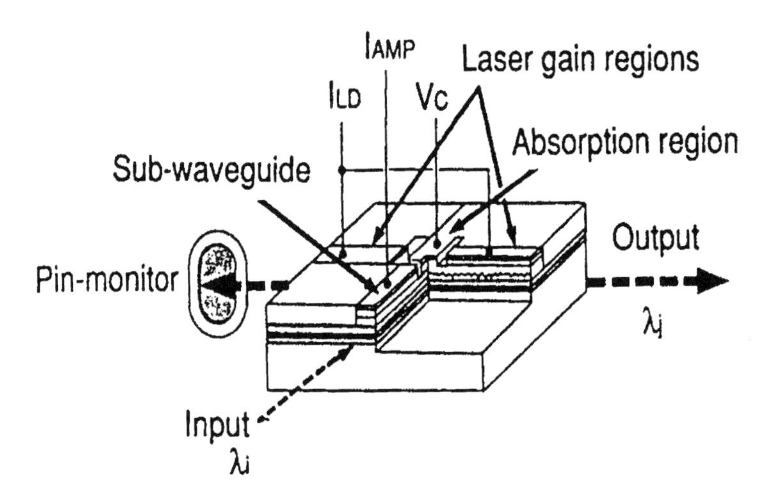

Fig. 5.24. Structure of side-injection-light controlled bistable laser diode.

As shown in Fig. 5.25 [84], direct optical demultiplexing of 1-Gbits NRZ signals has been accomplished with a module which has the functions of digital regeneration, bit-length conversion, wavelength conversion, and optical gating on the chip. This device can be used to make a wavelength convertor for TDM/WDM systems. Moreover, all-optical memory set and reset functions have been demonstrated with this device [85].

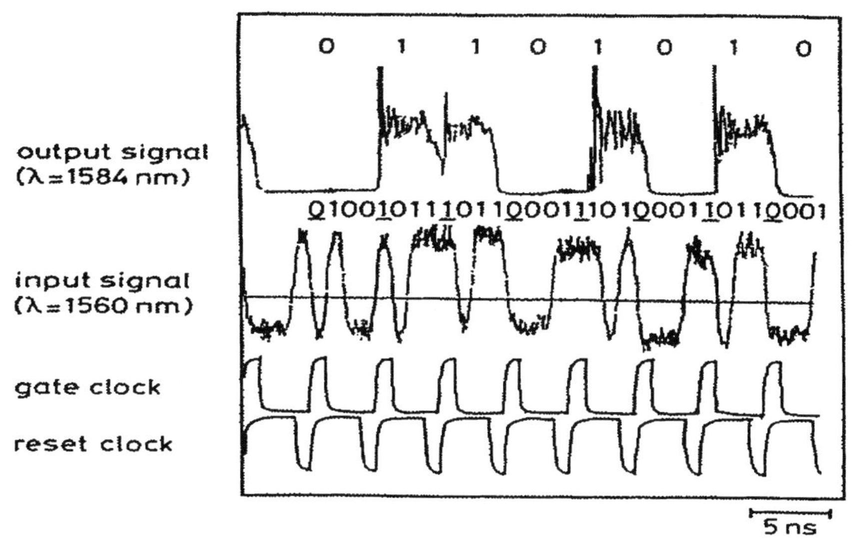

Fig. 5.25. Optical demultiplexing NRZ signals from 1 Gbit/s to 250 Mbit/s.

5.5.5 Four-Wave Mixing

Wavelength conversion is necessary in frequency switching to transform incident light wavelength into other wavelengths. A semiconductor laser can achieve this function through the third order nonlinear effect.

In general, the polarization in materials is expressed by the power series of electric field intensity as the following:

$$P(F) = \varepsilon_0\chi^{(1)}F + \varepsilon_0\chi^{(2)}F^2 + \varepsilon_0\chi^{(3)}F^3 + \text{-------},$$

where $\chi^{(2)}$ expresses second harmonic generation and linear electrooptic (Pockels) effect, and $\chi^{(3)}$ expresses third harmonic generation and quadratic electrooptic (Kerr) effect. Using $\chi^{(3)}$, we can realize optical frequency conversion. For example, when the incident light angular frequencies consist of ω_1, ω_2, and ω_3, the sum or difference of them such as $\omega_1+\omega_2-\omega_3$ that never exist in incident light can be generated. In the case of semiconductor lasers the gain bandwidth is as wide as a few tens of nanometers. When two different wavelengths consisting of a strong pump light (angular frequency ω_p) and a weak probe light (ω_q) are injected, a new light (signal light ω_s) is output in a symmetric position to the probe light as the pimp light is centerized.This situation is shown in Figs. 5.26 and 5.27. The relation between ω_p, ω_q, and ω_s exists as $\omega_s = 2\omega_p-\omega_q$.This effect is called four-wave mixing because four waves take part in the phenomenon.

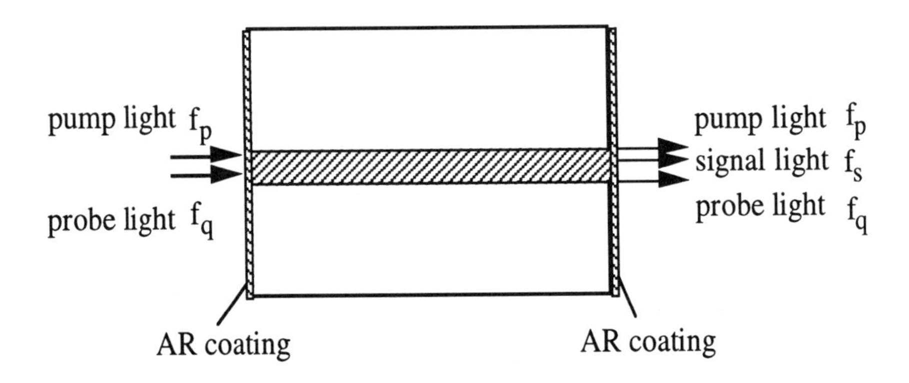

Fig. 5.26. Semiconductor optical amplifier and beam configuration for four wave mixing.

To obtain four-wave mixing in nonlinear optics, two or three light beams are irradiated in a nonlinear material with a certain incident angle. When each incident light has the same frequency, it is called degenerate four-wave mixing; when the frequency is different, it is called nondegenerate four-wave mixing. When a semiconductor otical amplifier is used for nonlinear material, nondegenerate four-wave mixing is used. This is because optical amplifiers have a single mode waveguide and a pump light and a probe light are in the same mode propagating in the same direction, i.e., collinear configuration. This results in an unresolved angle, making it necessary to resolve frequency domain. In order to obtain angular resolved four-wave mixing, a special structure providing multi-mode propagation is necessary.

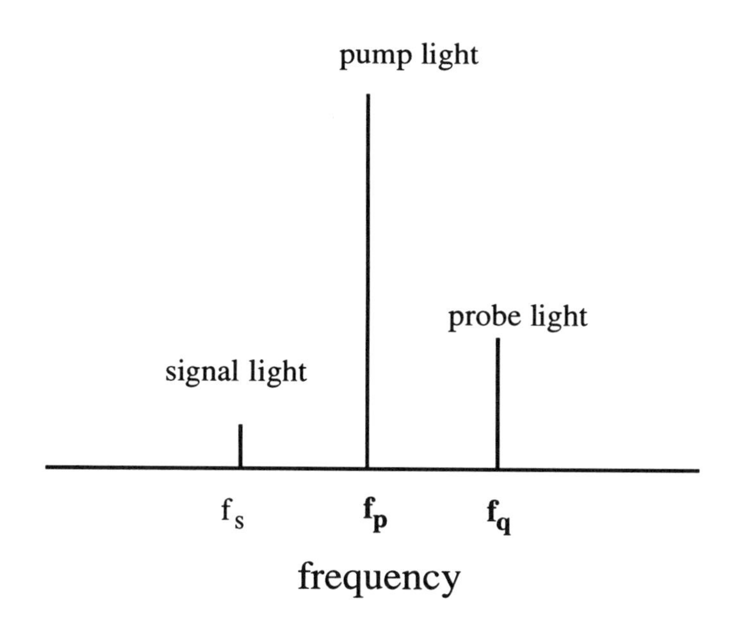

Fig. 5.27. Spectrum of four wave mixing.

In order to generate four-wave mixing, the nonlinear effect must respond to the different frequency f_d of the two incident light frequencies f_d-f_p-f_q. When f_d is below a few GHz, the main nonlinear effect is caused by the change in carrier density due to inter-band transition. This frequency range is called as nearly quasi four-wave mixing.
When f_d increases more than a few GHz, the change in carrier number cannot respond to it and the nonlinear effect due to intra-band relaxation becomes dominant. The frequency range between a few 100 GHz and a few THz is called highly nondegenerate four-wave mixing.

Schnabel et al. [68] reported 3.8 THs wavelength conversion of optical pulses with a width of 30 ps using a strained MQW laser amplifier operating at 1.55 µm. They also succeeded in the polarisation-insensitive wavelength conversion [69] of signals with 275 GHz (= 2nm) bandwidths, which were frequency-multiplexed by 10 channel signals.

Four-wave mixing has the advantage of being extremely fast, but convertion efficiency is not very high and decreaes with the separation between the pump and input signal. Therefore, conversion over long wavelength spans is difficult and requires relatively large optical power levels.

5.6 References

[1] G. P. Agrawal and N. K. Dutta, *Semiconductor Lasers*, second ed.,Van Nostrand Reinhold, 1993.
[2] G. H. B. Thompson, *Physics Semiconductor Laser Diodes*, Chi chester:John Wiley & Sons, 1980.
[3] T. Ikegami and Y. Suematsu, Proc. IEEE 55, 122, 1967; Electron. Comm. Jpn, 51-B, 51, 1968; 53-B, 69, 1970.
[4] T. Ikegami, ECOC'75, Tech. Dig., 111, 1975.
[5] Y. Suematsu, S. Arai, and K. Kishino, IEEE J. Lightwave Technol., LT-1, 161, 1983.
[6] J. E. Bowers, B. R. Hemenway, A. H. Gnauck, and D. P. Wilt, IEEE J. Quantum Electron., QE-22, 833, 1986.
[7] C. H. Henry, IEEE J. Quantum Electron., QE-18, 1982, 259, 1982.
[8] Y. Yoshikuni, T. Matsuoka, G. Motosugi, and N. Yamanaka, Appl. Phys. Lett., 45, 820, 1984.
[9] Y. Yoshikuni and G. Motosugi, IEEE J. Lightwave Technol., LT-5, 516, 1987.
[10] K. Kobatashi ans I. Mito, IEEE J. Lightwave Technol., 6, 1623, 1988.
[11] M. Ikeda, Electron. Lett., 17, 899, 1981.
[12] D. A. O. Davies, P. S. Mudhar, M. A. Fisher, A. A. H. Mace, and M. J. Adams, Electron. Lett., 28, 1521, 1992.
[13] M. Gustavsson, B. Lagerstrom, L. Thylen, M. Janson, L. Lungren, A. -C. Morner, M. Rask, and B. Stoltz, Electron. Lett., 28, 2223, 1992.
[14] J. F. Vinchant, M. Renavd, M. Erman, J. L. Peyer, P. Parry, and P. Pagnod-Rossiaux, IEE Proc. Optoelectron., 140, 301, 1993.
[15] T. Kirihara, M. Ogawa, H. Inoue, H. Kodera, and K. Ishida, IEEE Photon. Technol. Lett., 6, 218, 1994.
[16] S. Oku, Y. Shibata, T. Takeshita, and M. Ikeda, IEE Proc.-Optoelectron., 143, 1996.
[17] P. Albrecht, W. Doldissen, U. Niggebrugge, H. P. Nolting, and H. Schmid, 14th ECOC'88, Technical digest Part 1, 235.
[18] S. Oku, K. Yoshino, M. Ikeda, M. Okamoto, and T. Kawakami, Intn'l Topical Meeting on Photon. Switching PS90, Kobe, Japan, p. 98.
[19] I. H. White, J. J. S. Watts, J. E. Carroll, C. J. Armistead, D. J. Moule, and J. A. Champelovier, Electron. Lett., 26, 617, 1990.

[20] J. D. Burton, P. J. Fiddyment, M. J. Robertson, and P. Sully, 13th IEEE Int'l Semiconductor Laser Conf., Takamatsu, Japan, 1992, p.128.
[21] D. K. Probst, L. G. Perrymore, B. C. Johnson, R. J. Blackwell, J. A. Priest, and C. L. Balestra, IEEE Photon. Technol. Lett., 4, 1139, 1992.
[22] F. R. Nash, J. Appl. Phys., 44, 4696, 1973.
[23] C. H. B.Thompson, *"Physics of Semiconductor Laser Devices"*, Wiley, Chichester, 1980.
[24] J. Manning, R. Olshansky, and C. B.Su, IEEE J. Quantum Elec tron., QE-19, 1525, 1983.
[25] O. Mikami and H. Nakagome, Electron. Lett., 20, 228, 1984.
[26] O. Mikami and H. Nakagome, Opt.and Quantum Electron., 17, 449, 1985.
[27] K. Ishida, H. Nakamura,T. Kadoi, and H. Inoue, Appl. Phys. Lett., 50, 141, 1987.
[28] K. Wakao, N. Nakai,M. Kuno, and S. Yamakoshi, IEEE J. on Selected Areas in Comm., 6,1199, 1988.
[29] H. Inoue, H. Nakamura, K.Morosawa, Y. Sasaki, T. Katsuyama, and N. Chinone, IEEE J. on Selected Area in Comm., 6, 1262, 1988.
[30] F. Ito, M. Matsuura, and T.Tanifuji, IEEE J. Quantum Electron. 25, 1677, 1989.
[31] F. Ito and T. Tanifuji, Appl. Phys. Lett., 54, 134, 1989.
[32] H. Inoue, T. Kato,Y. Takahashi, E. Amada, and K. Ishida, Optical Engineering, 29, 191, 1990.
[33] J. F. Vinchant, J. A. Cavailles, M.Erman, P. Jarry, M. Renaud, J. Lightwave Technol., 10, 63, 1992.
[34] E. Lallier, A. Enard, D. Rondi, G. Glastre, R. Blondeau, M. Papuchon, N. Vodjdani, ECOC/IOOC'91, Postdeadline Paper, P.41.
[35] T. Kirihara, M. Ogawa, H. Inoue, and K. Ishida, Dig. of 4th Top. Meet. on Opt. Am. and Their Appl.(OA'93), Yokohama, Kanaga wa, Japan, Jul.4-6, SuB4(1993).
[36] J. F. Vinchant, M. Renaud, Ph. Jarry, A. Goutelle, J. L. Peyre, and M. Erman, OFC'93, Technical Digest, San Jose, Ca, February 21-26, pp. 30-31.
[37] W. H. Nelson, CLEO/PACIFIC RIM'95, Technical Digest, Makuhari Messe, Chiba, July 10-14, pp. 81-82.
[38] Y. Silberberg, P. Peerlmutter, J. E. Baron, Appl.Phys. Lett.,51, 1230, 1987.

[39] J. F. Vinchant, J. A. Cavailles, M. Erman, Ph. Jarry, and M. Renaud, J. Lightwave Technol., 10, pp. 63-70, 1992.
[40] W. H. Nelson, A. N. M. Masum Choudhury, M. Abdalla, R. Bryant, W. Niland, E. Vaughn, W. Powazinik, LEOS'92 Conference Proceedings, Bostin, Ma, November 16-19, 1992, pp. 608-609.
[41] W. H. Nelson, A. N. M. Masum Choudhury, M. Abdalla, R. Bryant, E. Meland, W. Niland, W. Powazinik, OFC'94 Technical Digest, San Jose, Ca, February 20-25, 1994, pp. 53-54.
[42] W. H. Nelson, A. N. M. Masum Choudhury, M. Abdalla, R. Bryant, E. Meland, W. Niland, IEEE Photon. Technol. Lett., 6, 1332-1334, 1994.
[43] A. Jourdan, G. Soulage, G. Da Loura, B. Clesca, P. Doussiere, C. Duchet, D. Leclerc, J. F. Vinvhant, and M. Sotom, OFC'95 Technical Digest, San Diego, Ca, February 28-March1, 1994, pp. 277-278.
[44] M. Renaud, J. F. Vinchant, A. Goutelle, B. Martin, G. Ripoche, M. Bachmann, P. Pagnod, and F. Gaborit, ECOC'95 Technical Digest, Brussels, September 17-21, 1995, pp. 99-102.
[45] M. N. Khan, J. E. Zucker, L. L. Buhl, B. I. Miller, and C. A. Burrus, ECOC'95 Technical Digest, Brussels, September 17-21, 1995, pp. 103-106.
[46] H. Inoue, T. Kirihara, Y. Sasaki, and K. Ishida, IEEE Photon. Technol. Lett., 2, 214-215, 1990.
[47] M. Wegener, J. E. Zucker, T. Y. Chang, N. J. Sauer, K. L. Jones, and D. S. Chemla, Phys. Rev., B41, 3097, 1990.
[48] D. S. Chemla, I. Bar-Joseph, C. Klingshirn, D. A. Miller, J. M. Kuo, and T. Y. Chang, Appl. Phys. Lett., 50, 585, 1987.
[49] A. Kastalsky, J. H. Abeles, and R. F. Lehney, Appl. Phys. Lett., 50, 708, 1987.
[50] H. Sakaki, H. Yoshimura, and T. Matsusue, Jpn. J. Appl. Phys., 26, L1104, 1987.
[51] J. E. Zucker, M. Wegener, K. L. Jones, T. Y. Chang, N. J. Sauer, and D. S. Chemla, Appl. Phys. Lett., 56, 1951, 1990.
[52] J. E. Zucker, K. L. Jones, T. Y. Chang, N. J. Sauer, B. Tell, K. Brown-Goebeler, M. Wegener, and D. S. Chemla, Electron. Lett., 26, 2029, 1990.
[53] J. E. Zucker, K. L. Jones, G. R. Jacobovitz, B. Tell, K. Brown-Goebeler, T. Y. Chang, N. J. Sauer, M. D. Divino, M. Wegener, and D. S. Chemla, IEEE Photon. Technol. Lett., 2, 804, 1990.
[54] N. Agrawal et al., CLEO'93, CThS68, 1993.

[55] N. Agrawal et al., Conf. on InP and Related Materials, MB5, 1994.
[56] M. K. Chin, T. Y. Chang, W. S. C. Chang, IEEE J. Quantum Electron., 28, 2596, 1992.
[57] J. E. Zucker, K. L. Jones, M. Wegener, T. Y. Chang, N. J. Sauer, M. D. Divino, and D. S. Chemla, Appl. Phys. Lett., 59, 201, 1991.
[58] N. Agrawal, C. Bornholdt, H. -J. Ehrke, D. Franke, D. Hoffmann, F. Kappe, R. Langenhorst, G. G. Mekonnen, F. W. Reier, and C. M. Weinert, in OEC'94 Techn. Dig., July 1994, 14A2-2, P. 114. or N. Agrawal and M. Wegener, Appl. Phys. Lett., 65, 685, 1994.
[59] Y. Chen, J. E. Zucker, T. Y. Chang, N. J. Sauer, B. Tell, and K. Brown-Goebeler, Integrated Photonics Technical Digest, 1993, P. 176.
[60] J. E. Zucker, SPIE vol. 1283, Quantum-Well and Superlattice Physcs III, Mar. 1990, San Diego, California.
[61] H. Kawaguchi, "*Bistabilities and Nonlinearities in Laser Diodes*", ARTECH HOUSE, 1944.
[62] H. Kawaguchi, I. H. White, M. J. Offside, and J. E. Carroll, Opt. Lett., 17, 130, 1992.
[63] H. Kawaguchi, K. Magari, H. Yasaka, M. Fukuda, and K. Oe, Electron.Lett., 23, 1088, 1987.
[64] H. Kawaguchi, K. Magari, H. Yasaka, M. Fukuda, and K. Oe, IEEE J. Quantum Electron., QE-24, 2153, 1988.
[65] S. Yamakoshi , in Techn. Dig. of OFC'88, Jan. 1988, New Orleans, post dead line paper PD 10.
[66] T. Durhuus, R. J. S. Pedersen, B. Mikkelsen, K. E. Stubkjaer, M. Oberg, and S. Nilsson, IEEE Photn. Technol. Lett., 5, 86, 1993.
[67] B. Glance, J. M. Wiesenfeld, U. Koren, A. H. Gnauck, H. M. Presby, and A. Jourdan, Electron. Lett., 28, 1714, 1992.
[68] R. Schnabel, W. Pieper, R. Ludwig, and H. G. Weber, Electron. Lett., 29, 821, 1993.
[69] R. Schnabel, U. Hilbk, Th. Hermes, P. MeiBner, CV. Helmolt, K. Magari, F. Raub, W. Pieper, F. J. Westphal, R. Ludwig, L. Kuller, and H. G. Weber, IEEE Photn. Technol. Lett., 6, 56, 1994.
[70] K. E. Stubkjaer, T. Durhuus, B. Mikkelsen, C. Joergensen, R. J. Pedersen, C. Braagaard, M. Vaa, S. L. Danielsen, P. Doussiere, G. Garabedian, C. Graver, A. Jourdan, J. Jacquet, D. Leclerc, M. Erman, M. Klenk, Techn. Dig. of ECOC'94, P. 635.
[71] R. J. S. Pedersen, B. Mikkelsen, T. Durhuus, C. Braagaard, C. Joergensen, K. E. Stubkjaer, M. Oberg, and S. Nilsson, OFC'94, ThQ3, San Jose, California, Feb. 1994.

[72] H. Yasaka, H. Ishii, K. Takahata, K. Oe, Y. Yoshikuni, and H. Tsuchiya, Electron. Lett., 30, 133, 1994.
[73] W. Idler, M. Schilling, D. Baums, K. Dutting, K. Wunstel, and O. Hildebrand, Proc. of ECOC'92, Berlin, Sept. 1992, P. 449.
[74] E. Lach, D. Baums,K. Daub, T. Feeser, W. Idler, G. Laube, G. Luz, M. Schilling,, and K. Wunstel, Proc. of ECOC'93, vol.2, P. 137, Montreux, Switzerland, Sept. 1993.
[75] T. Durhuus, B. Fernier, P. Garabedian, F. Leblond, J. L. Lafragette, B. Mikkelsen, C. G. Joergensen, and K. E. Stabkjaer, Proc. of CLEO'92, Anaheim, May 1992, CThS4.
[76] B. Glance, J. M. Wiesenfeld, U. Koren, A. H. Gnauck, H. M. Presby, and A. Jourdan, Techn. Dig. of CLEO'92, Anaheim, May 1992, post deadline paper CPD27.
[77] B. Mikkelsen, M. Vaa, R. J. Pedersen, T. Durhuus, C. G. Joergensen, C. Braagaard, N. Storkfelt, K. E. Stabkjaer, P. Doussiere, G. Garabedian, C. Graver, E. Derouin, T. Fillon, and M. Klenk, Proc. of ECOC'93, vol.3, P.73, Montreux, Switzerland, Sept. 1993.
[78] J. M. Wiesenfeld, Electron. Lett., 30, 720, 1994.
[79] D. M. Patrick and R. J. Manning, Electron. Lett., 30, 252, 1994.
[80] T. Odagawa,T. Machida, K. Tanaka, T. Sanada, and K. Wakao, Trans. IEICE Jpn., J74-C-1, No. 11, pp. 465-469, 1991.
[81] H. Uenohara, H. Iwamura, and M. Naganuma, J. J. Appl. Phys., 29, No. 12, pp. L2442-2444, 1990.
[82] S. Suzuki, T. Terakado, K. Komatsu, K. Nakashima, A. Suzuki, and M. Kondo, IEEE J. Lightwave Technol., LT-4, pp. 894, 1986.
[83] K. Nonaka, Y. Noguchi, H. Tsuda, and T. Kurokawa, IEEE Photon. Technol. Lett., 7, No. 1, pp. 29-31, 1995.
[84] K. Nonaka, and T. Kurokawa, Electron. Lett., 31, No. 21, pp. 1865-1866, 1995.
[85] K. Nonaka, Y. Noguchi, and T. Kurokawa, CLEO-Pacific Rim'95, WJ4, 1995.

Chapter 6
Photonic Switching Devices

6.1 Introduction

Ultrafast and wideband signal transmission and processing technologies are required for constructing future communications networks, and photonic switching technology is one of the most important research and development issues in the development of fully-optical communication networks where information will be carried by optical signals through nodes as well as through links. The past ten years have seen great progress in trunk transmission techniques using optical fiber cables and semiconductor lasers, and optical- fiber transmission systems having a capacity of 10 Gbit/s has been in practical use [1].

On the other hand, networks that are entirely optical also require node systems that handle optical signals. That is, these networks require switches and cross-connects that manipulate optical signals directly, without O/E conversion. The research on a photonic node system, however, is still in the basic stages because a practical switching device has not yet been developed. Most optical components have been developed for point-to-point transmissions with a single light wave, but photonic switching devices are required to operate with many input and with output signals and many frequencies. Photonic switching devices are therefore hard to make because they need many input and output fiber connections, have extremely rigorous polarization conditions, must perform complicated operations, and must be implemented on large chips.

6.2 Classification of Photonics Switching Devices

Photonics switches are classified, depending on the structure of the path of the lightwave, into guided-wave types and free-space types. Free-space switches control the optical path normally incident to the device active layers. These surface-normal switches are described in Chapter 9. This chapter discusses the guided-wave types.

Guided-wave type switches can use the long interaction length between electric and photonic systems. The physical mechanisms used for switching are similar to those in used in modulators: changes of refractive index or absorption coefficient in the active regions. That is, the switching mechanism is based on the Pockels (electrooptic, EO) effect, Kerr effect, Franz-Keldysh effect, quantum confined Stark effect, or plasma effect. The switching speed of a current-injection-controlled switch is limited to a few gigaherz because the lifetime of injected carriers is only a few nanoseconds, and the speed of an electric-field-controlled switch is usually limited by the device capacitance and is expected to be at most 100 GHz.

The following three types of operation modes are used for switching devices: space division, time division, and wavelength (frequency) division. Wavelength division devices, where optical amplifiers and bistability laser diodes are used for wavelength conversion, are discussed in Chapter 5. This chapter discusses space-division switches in detail.

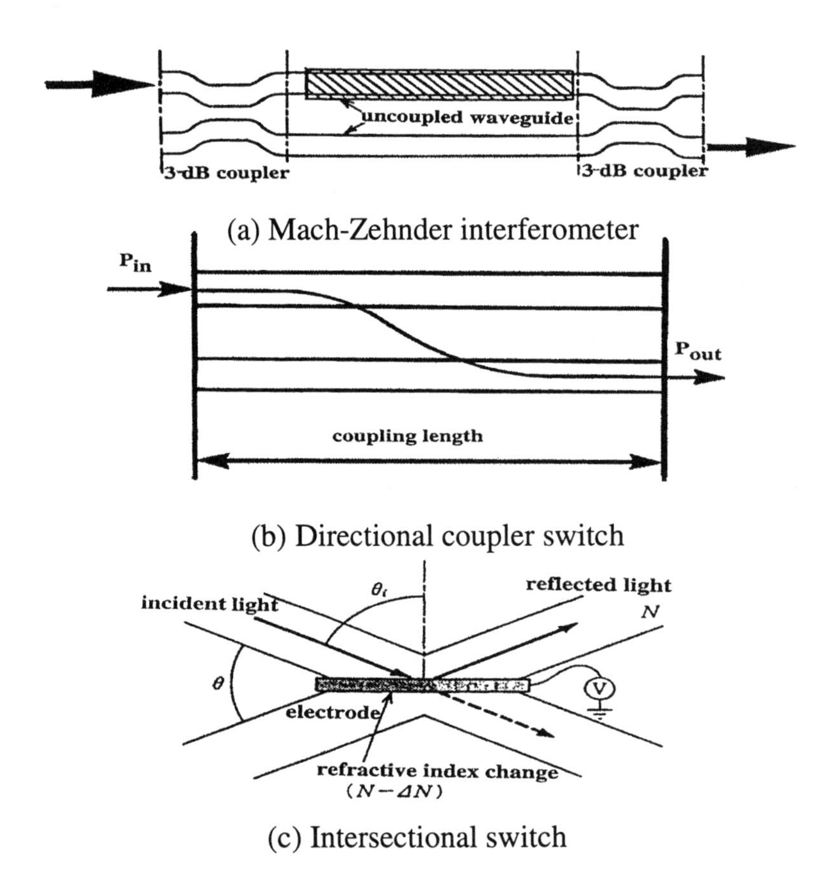

Fig. 6.1. Various switch structures.

This is the most important type of guided-wave switch, and three types of waveguide structures have been reported: the Mach-Zehnder interferometer, the directional coupler, and the intersectional-type structure (Fig. 6.1).

6.3 Mach-Zehnder Interferometer

In classical optics, Mach-Zehnder interferometers are used for measuring the refractive index of a material. Nowadays they are used as intensity modulators, which are attractive for use in high-bit-rate and long-haul optical-fiber communication systems. When the bite rate is more than 10 Gbit/s, it is advantageous to use chirpless modulators such as the Mach-Zehnder type instead of electroabsorption modulators.

The Mach-Zehnder interferometer was proposed independently by M. Mach and L. A. Zehnder in 1891, and guided-wave interferometers capable of modulating a laser beam by exploring the electrooptic effect were fabricated using LiNbO3 in 1980 [2-4]. After the first demonstration of Mach-Zehnder devices using semiconductors was reported by Donnelly et al. in 1984 [5], GaAs Mach-Zehnder modulators were developed [6-8]. Since then many devices have been developed using bulk GaAs/AlGaAs [9,10], and InGaAsP/InP [11-13], multiple quantum well (MQW) GaAs/AlGaAs[14], InGaAsP/InP [15-17] and InGaAs/InAlAs[18], and InGaAlAs/InAlAs barrier, reservoir, and quantum well electron transfer structures (BRAQWETS) [19,20].

Interferometers made from group III-V materials have the potential of being integrated with lasers, detectors, and high-speed electric devices to create a truly monolithic integrated optical circuit, and strong research and development efforts in many laboratories are being devoted to these devices.

6.3.1 Principle

The Mach-Zehnder(MZ) type modulator consists of two waveguides parallel to each other and two couplers that combine the parallel waveguides (Fig. 6.2). The change in refractive index has been divided into a linear electrooptic (EO) effect and a quadratic electrooptic (QEO) effect and carrier effects. These are discussed in detail in Chapters 3 and 4.

The optical radiation is split into two branches; one is manipulated and then the two branches are recombined. In the process of

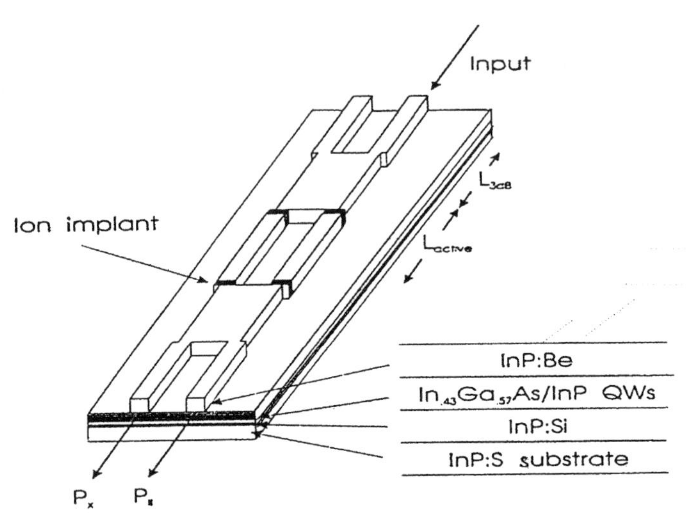

Fig. 6.2. Schematic of a Mach-Zehnder switch [61].

recombination, the electric fields or carrier injections produce variable optical interference, which is then analyzed.

In the analysis below, the complex mode-amplitude for the propagation-direction z is described by the expression $A_0 \exp i(\omega t\text{-}kz)$. If we omit the time-dependence of the electric fields and assume a 3-dB coupler, we can calculate the optical powers P_3 and P_4 at the output of the interferometer:

$$P_3 = kE_0^2\{1-\cos(\beta\Delta L)\}$$

$$P_4 = kE_0^2\{1+\cos(\beta\Delta L)\}$$

$$P_3 - P_4 = -2kE_0^2 \cos(\beta\Delta L),$$

where E_O, β, k, ΔL, and P are respectively the electric field at the input, the propagation constant of the sample defined as $\beta = 2\pi n/\lambda$, a proportional constant factor, the difference length, and optical power. The quadrature condition is fulfilled, when the $\cos(\beta\Delta L)$ is zero, that is

$$\beta\Delta L = 2\pi n\Delta L/\lambda,$$

where n is refractive index of the sample and λ is the optical wavelength in air. The output intensity then changes as $\beta\Delta L$ changes.

6.3.2 Characteristics

A. Drive Voltage. For MZ modulators, a phase shift π appears between two arms. If the output waveguides of the modulator are single-mode, the total optical field at the output can be regarded to a first approximation as a weighted sum of the optical field from the two arms. The weight factors depend on the power splitting ratio of the two Y-junctions and the photoabsorption in the modulator section. In particular, since the quantum confined Stark effect leads to nonlinear refractive index change with changes in voltage, in a push-pull configuration the larger phase shift change occurs in the more deeply dc-biased arm. As discussed in section 3.5, the usual drive voltage for a π phase shift is several volts.

B. Chirping and Transmission Experiments. Since the MZ modulator has a cyclic transfer function, red and blue shifts can be easily chosen by using different slopes of the transfer function. The ability to control the frequency chirp of MZ intensity modulators has made them well suited for multigigabit, long-haul optical-fiber transmission experiments [22]. A completely chirp-free device, however, is not always desirable as shown in Fig. 1.14, where the best chirp parameter for long distance optical fiber transmission is -0.6 ~ -0.8 for 10 Gbit/s. 10 Gbit/s modulation using InP/InGaAsP [17] and InAlAs/InGaAs [23] MZ modulators has been reported. These modulators have a 3-dB bandwidth in excess of 15 GHz and are operated in a push-pull drive configuration with only 2 V peak to peak. 10 Gbit/s nonreturn to zero format eye diagrams with an extinction ratio over 10 dB are demonstrated.

Recenly, a monolithic integration of MZ modulator and a laser diode has been reported and a transmission experiment has been demonstrated [24]. The details of this report are discussed in section 3.3 of Chapter 8.

6.4 Directional Couplers

6.4.1 Introduction

Semiconductor optical waveguide switches are critical components in photonic circuits for optical signal processing. $LiNbO_3$ has been the material of choice for optical space switches, and single-chip 16x16 switching matrices have been demonstrated [25]. A second choice is

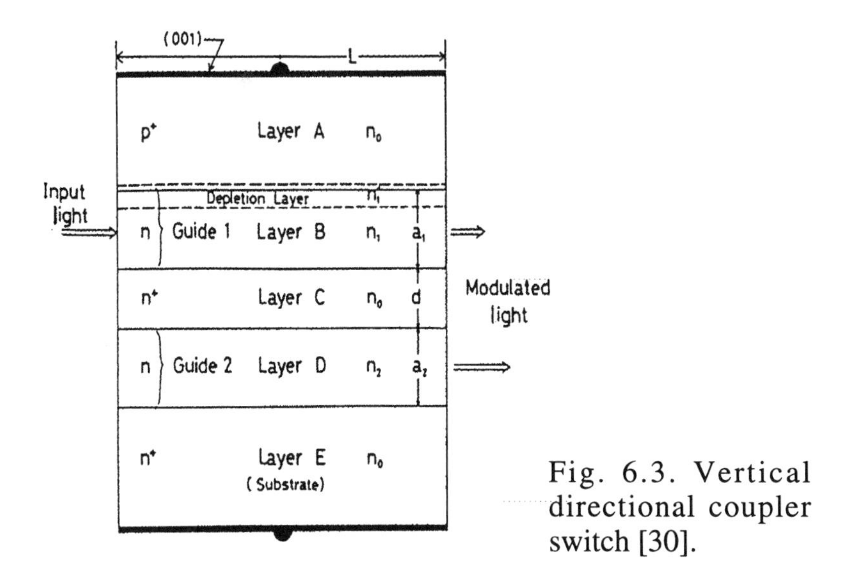

Fig. 6.3. Vertical directional coupler switch [30].

semiconductor materials, and although the semiconductor devices are less advanced than the $LiNbO_3$ devices, important progress has been achieved over the past few years. In large-scale monolithic applications where many devices must be fabricated on a single wafer, it is essential to minimize the length of each switch. Semiconductor switches appear a more suitable solution over the longer term because of their lower power consumption, smaller size, and their possibility of regenerating the signal through optical amplification.

An optical directional coupler can be composed of two or more parallel optical waveguides set so close to each other that the optical power in one guide can transfer to the other [26, 27]. By perturbing the coupling or the synchronism between two coupled guides distributionally, one can obtain a new kind of modulator whose basic theory was proposed by several groups [27-29].

The first structure demonstrated for a semiconductor directional coupler consists of p^+, n, n^+, n, and n^+ type semiconductor layers. Figure 6.3 shows this vertical-type directional coupler [30]. Two n-type layers operate as coupled optical waveguides. The exchange of light power between them can be controlled by a reverse voltage applied to the p^+-n junction. This type has advantages of precisely controlling the layer thickness with atomically smooth interfaces because crystal growth technique can be used, compared with the following lateral type that uses processing techniques for the control. But because the diameter of the focused spot at the entrance facet (~2 μm) is much larger than that of the guides (a few tenths of a micrometer), one normally excites an unknown linear combination of the guided modes in the structure, and this makes the output patterns unpredictable. Therefore, nothing about this type has been reported since the first report [30]. Up to now, these vertical-type devices have been reported using electric-field-induced refractive index and absorption changes in the MQW [31, 32].

Figure 6.4 shows the rib-waveguide structure [33, 34]. This laterally coupled type has so far been demonstrated, as a metal-gap optical stripline type [35-37], a channel-stop strip type [38], and a MOS type [39]. One problem, however, is that the modulating electric field and the optical field weakly overlap each other in the former two types of switch, and another is that the MOS devices have a cuttoff frequency of only 3 kHz. The device shown in Fig. 6.4 was composed of a pair of adjacent single-mode rib waveguides with a Au Schottky barrier. The effective refractive index of the rib region is larger than that of the surround, and light is confined to the rib region. A disadvanage of this type is the need to use a precise fabrication technique for etching the rib structure.

Many kinds of electrooptic directional coupler devices have been

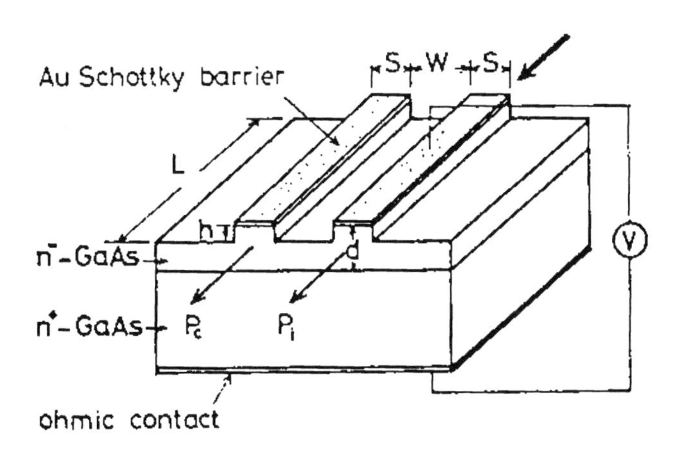

Fig. 6.4. Lateral directional coupler switch [33].

demonstrated using GaAs stepped Δβ junction [40], GaAs/AlGaAs [34] and InGaAsP/InP [41] double heterostructues (DH). These structures were made by liquid phase epitaxy and wet chemical etching and they are fairly large. Shorter coupling lengths and lower switching voltages are neccessary. Dry etching, which is an excellent tool for microfabrication of semiconductor devices, and molecular beam epitaxy have been used in order to improve the device characteristics [42].

Figure 6.5 shows the schematic diagram of an AlGaAs/GaAs DH directional coupler switch with a pn junction [42] and its output power variation with changes in the dc reverse bias voltage. The device length is 0.98 mm, which was chosen to coincide with the coupling length of the dirctional coupler switch calculated by Marcatili [28]. The outputs P_i and P_c are respectively those from the input-side and coupled-side waveguides. Output power exchange occurs at a reverse bias voltage of about 5 V, and extinction ratios for the crossover state (0 V) and for the straight-through state (5 V) are respectively 17 dB and 14 dB. Reactive ion etching provides precise control of the small lateral dimensions.

Excitonic electrorefraction in reverse-biased InGaAsP/InP MQW structures has been reported to enable the fabricating of directional-coupler switches that have active lengths under 600 μm and that operate at 1.3 μm and 1.55 μm [43]. These are the first directional-coupler switches using the MQW structure. The key difference from the bulk InGaAsP at the same detuning from the band gap is that a quadratic electro-optic coefficient is 13 times larger, while a linear electro-optic coefficient is of the same order. At -20 V most of the light emerges from one channel to the other.

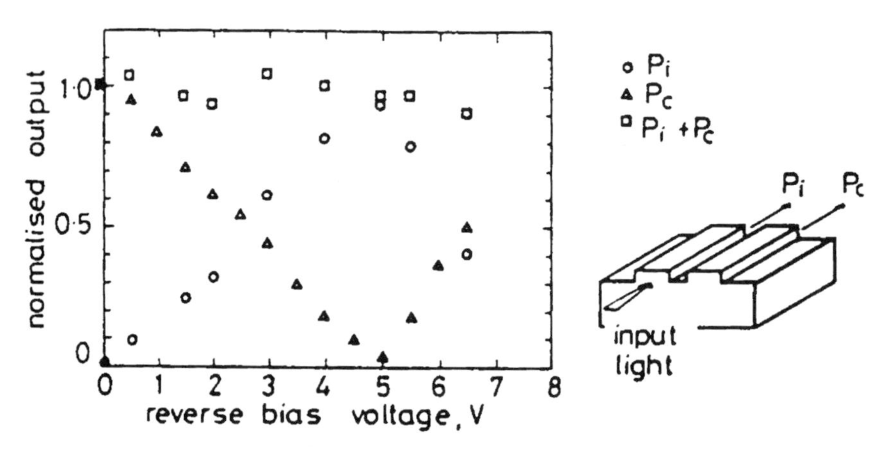

Fig. 6.5. Schematic diagram of a AlGaAs/GaAs DH directional coupler switch with a pn junction and switching characteristics [42].

6.4.2 Directional-Coupler Switch

Several types of space switch matrix with a scale greater than 2x2 have been developed, using directional couplers [42-45], a laser amplifier gate [46, 47], and the Y-junction [48]. Most of them are based on directional couplers. An integrated-optics directional coupler is formed by fabricating two pararell waveguides in close proximity so that light in one waveguide can couple to that in the other waveguide via the evanescent fields. In Figs. 6.4 and 6.5, the schematic pattern of waveguided directional-coupler circuit is illustrated. Two waveguides with propagation constants β_r and β_s are brought within a distance d of one another for a length l. Over this length the waveguides are coupled so that optical energy can transfer between the two guides. If the waveguides have the same propagation constants and energy is incident in only one guide, it will transfer completely to the other guide in a distance L. The coupling length L is given as follows [49]:

$$L = \pi/2\kappa = \lambda/(\beta_{even} - \beta_{odd}),$$

where κ is a coupling constant that describes the strength of the interguide coupling and where β_{even} and β_{odd} are the propagation constants for even and odd mode induced in the coupled waveguides.

Electrically controlled optical switching can be achieved with a directional coupler because the degree of light transfer between the waveguides depends upon the difference in propagation constants, $\Delta\beta=\beta_R-\beta_S$, which can be controlled via the electrooptic effect. When we apply an electric field to one waveguide and some difference in propagation constant is induced in each waveguide, the output light intensity I_1 from the waveguide is given as follows if the incident light intensity is I:

$$I_1 = I \cdot \sin^2\{\sqrt{((\Delta\beta/2\kappa)^2 + 1)} \cdot l\pi/2L\}/(1 + (\Delta\beta/\kappa)^2).$$

When the sample length $l=(2m-1)L$ ($m=1,2,3,$—), $I_1=0$ for applied bias is zero, then, the voltage V_π that gives $I_1=I$ is given as

$$V_\pi = (\lambda d/2n_g^3\gamma_{mj}\Gamma L) \cdot \sqrt{\{4n^2/(2m-1)^2-1\}},$$

where d is the separation length of electrodes, n_g is the refractive index of the waveguide, m and n are minimum integers determined by the condition $4n^2>(2m-1)^2$, and Γ is the field correction factor that accounts for incomplete overlap of the optical and applied electrical fields. This value is 1 for the facing paralell electrodes of bulk and 0.3 ~ 0.6 for waveguiged structures. The value γ_{mj} is the relevant electrooptic coefficient which depends upon the crystal orientation and the direction of the applied field and is discussed in detail in Chapter 3.

The greatest disadvantage of the directional-coupler switches is that it is difficult to make the sample length coincide with the coupling length. The best method for controlling the coupling length is to use the stepped $\Delta\beta$ reversal switch [50] , which requires two sets of electrodes (Fig. 6.6). The electrodes divide the coupler into two equal-length sections in which phase mismatches of opposite sign can be applied via the electrooptic effect. With equal but opposite sign phase mismatches, electrical adjustment of both switching states is obtained.

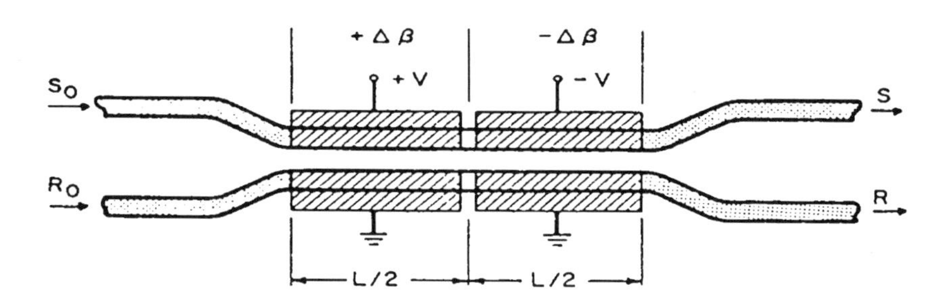

Fig. 6.6. Schematic configuration of stepped-$\Delta\beta$ reversal switch requiring two sets of electrodes.

Switches with reversal $\Delta\beta$ electrodes using GaAs waveguide structures [35, 39] and having a crosstalk level of -26 dB have been demonstrated. Recently, 4x4 [51] and 8x8 matrix switches [52] using GaAs/AlGaAs electrooptic waveguide directional couplers have been reported. GaAs/ AlGaAs material systems are attractive for use as crosspoint elements in matrix switches because of their low absorption loss in long wavelength regions, fast switching speeds, low electric power consumption, capability for wavelength-independent operations, and suitability for manufacture by advanced microfabrication technologies such as reactive ion beam etching.

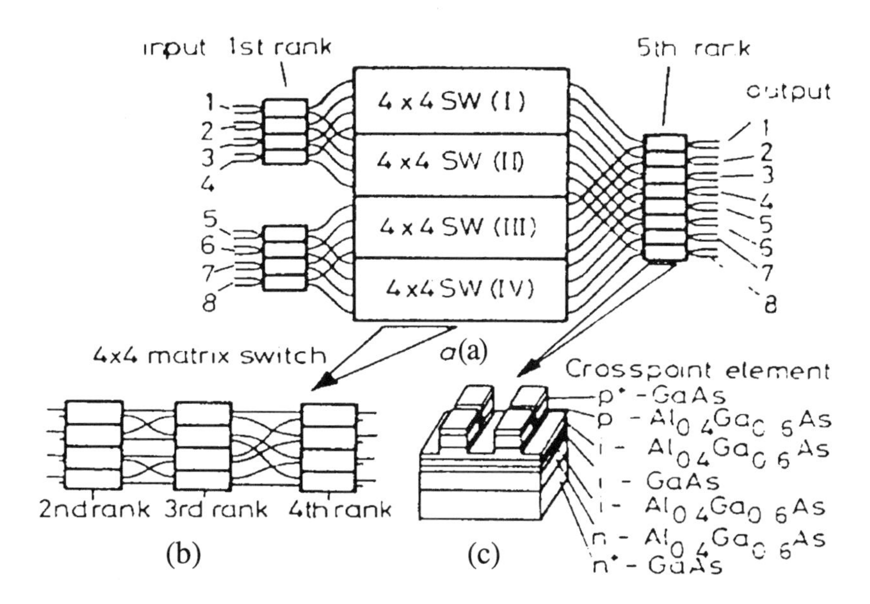

Fig. 6.7. 8x8 GaAs/AlGaAs optical matrix switch structure [51].

Figure 6.7 shows a schematic diagram of an integrated 8x8 matrix switch designed with the same tree structure, a nonblocking architecture, used for the previous 4x4 matrix switches [51]. This switch consists of four 4x4 matrix switches, each with eight switch elements at their input and output sides. An individual switch element is shown in Fig. 6.7(c). Waveguide width and waveguide spacing are both 2 μm and the coupler length is 3 mm. Under alternating $\Delta\beta$ operations with 1.3-μm incident light, the crossover state was obtained at 12.0 V and the straight-through state at 25.5 V. Typical crosstalk values are -21 dB for the former state and -23 dB for the latter state. The minimum propagation loss for the 8x8 optical matrix switch was 8.7 dB, not including coupling loss. With a 4x4 optical switch 17 mm long, a propagation loss of 1.6 dB has been achieved [53].

A 4x4 directional coupler switch with the MQW structure and only 7 mm long and 0.4 mm wide has been reported [54]. This switch consists of six 2x2 matrix switches and uses the quantum confined Stark effect of InGaAlAs/InAlAs MQWs. The switching voltage of each 2x2 switch is between 5 and 6 V, and the crosstalk is less than -17 dB. Its transmission loss is between 17.5 and 18.5 dB, and there is little variation between the losses in the various switching routes.

A fully packaged 4x4 switching module with an operation speed of 10 Gbit/s has been reported [55]. This switch (Fig. 6.8) is composed of six directional couplers forming the three-stage Benes network as shown in Fig. 6.8(b). A 0.4-μm-thick MQW layer is composed of 9-nm-thick undoped InGaAlAs wells and 5-nm-thick InAlAs barriers whose absorption edge wavelength is 1.39 μm. The directional coupler has a 1.7-μm-wide ridge with a 1.6-μm gap and has a coupling length of 1.2 mm. To reduce the coupling loss when the switch is attached to single-mode fibers, spot-size convertors are formed. The interval between the

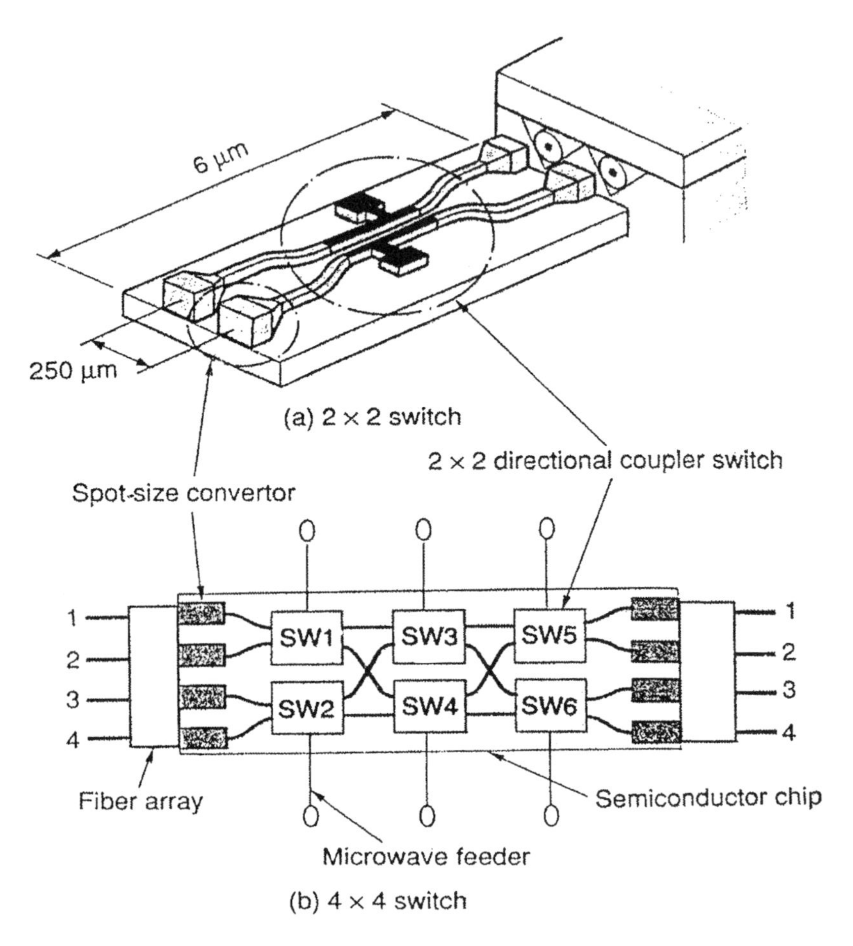

Fig. 6.8. Structure of a directional-coupler switch integrated with a spot-size converter.

convertor was chosen to match the 250-μm pitch of the array fiber. The length of the 2x2 switch chip, including bent waveguides and spot-size convertors, is 6 mm. The total length of the 4x4 switch chip is 16.5 mm. Insertion loss, switching voltage, and crosstalk of the 4x4 switch were respectively ~19 dB, -9 to -10 V, and -13 dB.

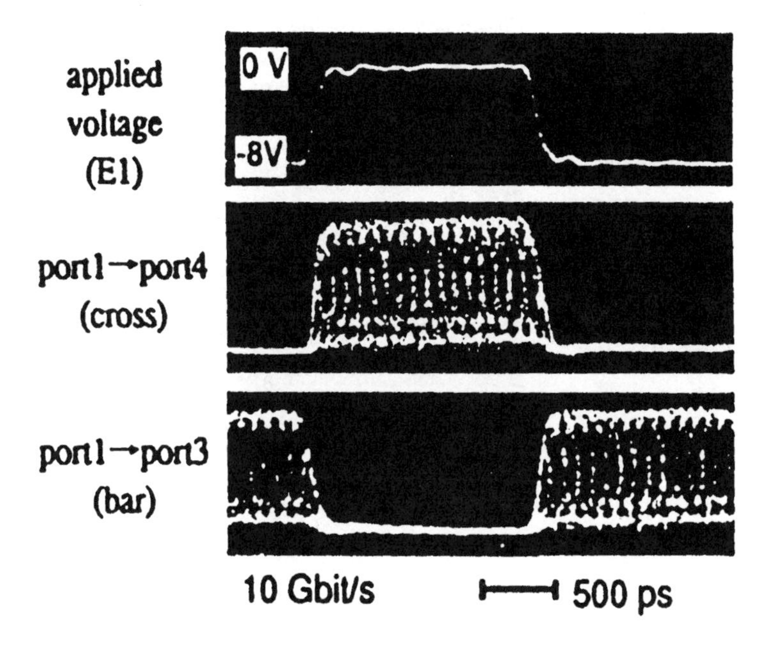

Fig. 6.9. Eye pattern of a 2x2 switch module.

A 10-Gbit/s signal was switched by changing the applied voltage from 0 V to -8 V on the 2x2 switch [56]. As shown in Fig. 6.9, it is clear that the module can be successfully switched from cross state to bar state by applying -8 V. The number of bits lost at the moment of switching was no more than one.

Figure 6.10 shows the bit-by-bit operation obtained using the 4x4 switch module [57]. The 10-Gbit/s input signal was demultipled to 5 Gbit/s at the first stage, 2.5 Gbit/s at the second stage, and 1.25 Git/s at the third stage. The 5-ps switching time, was sufficiently short, which was limited by the driver speed. This switch module has a potential for 10- Gbit/s bit-by-bit switching, and when the modules are used in a 10-Gbit/s optical asynchronous-transfer-mode (ATM) switching system, no guard time between the signal packets is required.

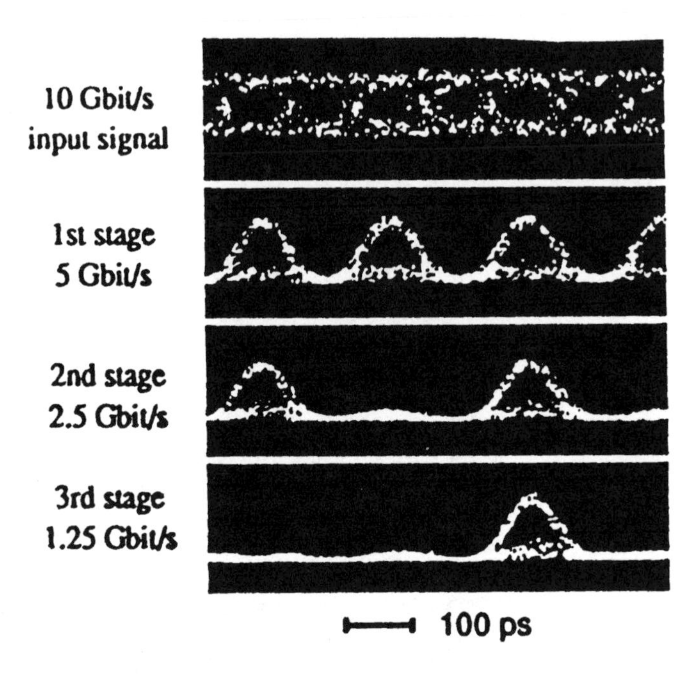

Fig. 6.10. Demultiplexing by a 4x4 switch module [57].

6.5 Intersectional Switch

The major drawback of directional couplers and the drawback for monolithic integration are their large size, typically several millimeters or more, due to large coupling lengths and to the size of the branching network that separates the access waveguides. An intersectional switch provides very compact crosspoints but requires large refractive index changes for effective switching. As described in Chapter 5, current-injection types were the first kinds of intersectional switch tried because of their large refractive index changes. Their speed, however, is not high, and power dissipation and junction heating effects still limit the bandwidth and linearity of current-injection switches. RC-time-constant-limited intersectional switches based on the depletion-edge-translation effect [58], the quantum confined Stark effect [59], and the voltage-controlled transfer of electrons into multiple quantum wells (BRAQWET) [60] are being investigated because of their high speed and relatively large refractive index changes.

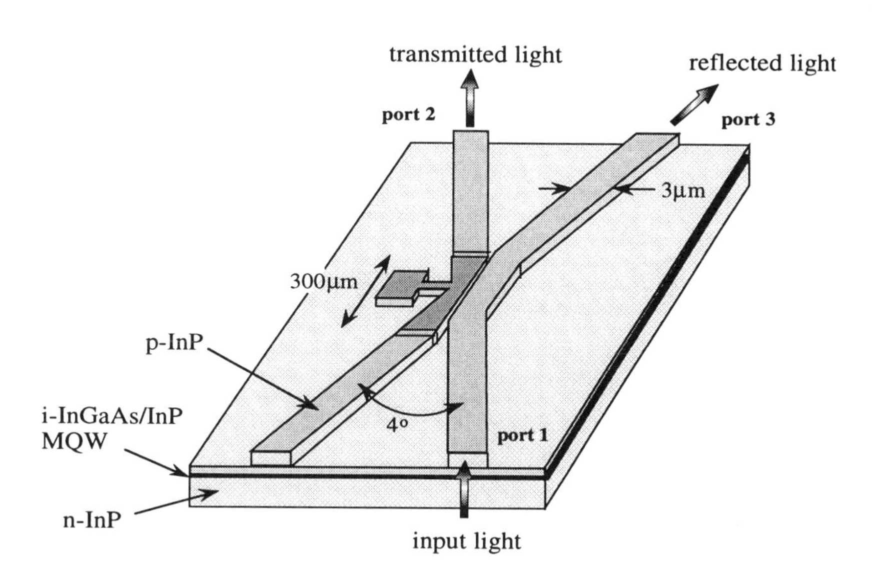

Fig. 6.11. Three-dimensional view of an intersectional switch.

Figure 6.11 shows a three-dimensional view of the intersectional switch operating at 1.3 μm [59]. The operation principle is described as follows. The light entering at port 1 gets transmitted to port 2 when no electric field is applied. The application of an electric field to only the n-region decreases the refractive index in that region of the MQW layers, causing the light entering port 1 and this get reflected to port 3. At -9 V the light intensity at port 3 increases by a factor 2.7, which increase is due to the light reflected from the straight arm at the intersectional region because of field-induced refractive index changes in the MQW layers. The intensity at port 2 decreases by a factor 10 because of reflection to port 3 as well as absorption.

InGaAs/InAlAs quantum well intersecting waveguide switches operating at 1.55 μm and based on electrorefraction in tunable-electron-density multiple quantum wells have been demonstrated [60]. The crossover (port 1 to port 3) state of the switch is achieved for negative bias and the straight-through (port 1 to port 2) state for positive bias, with an extinction ratio of 4.1 dB for the straight-through state and 3.0 dB for the crossover state at +/- 13 V . The switching performances described above are not good compared with those of switches based on carrier injection because of their preliminary stage, but these performances will be improved by optimizing the MQW structure, the carrier concentration of the MQW layers, and the cladding layers so as to concentrate the applied electric field to the MQW layers, by employing single-mode waveguides and by using fine fabrication techniques in order to prevent scattering between waveguides.

6.6 References

[1] N. Nakagawa, K. Hagimoto, S. Nishi, and K. Aoyama, Topical Meeting on Optical Amplifiers and their Applications, PD-11, 1991.
[2] Y. Ohmachi and J. Noda, Appl. Phys. Lett., 27, p. 544, 1975.
[3] F. Auracher and R. Keil, Wave Electron., 4, 129, 1980.
[4] F. J. Leonberger, Opt. Lett., 5, 312, 1980.
[5] J. P. Donnelly, N. L. DoMeo, G. A. Ferrante, K. B. Nichols, and F. J. O'Donnel, Appl. Phys. Lett., 45, pp. 360-362, 1984.
[6] J. P. Donnelly, N. L. DoMeo, G. A. Ferrante, and K. B. Nichols, IEEE J. Quantum Electron. Lett., QE-21, No.1, pp. 18-21, 1985.
[7] P. Buchmann, H. Kaufmann, H. Melchior, and G. Guekos, Appl. Phys. Lett., 46, pp. 462-464, 1985.
[8] C. Wuthrich, J. Faist, W. Baer, and F. K. Reinhart, Electron. Lett., 24, No. 16, pp. 1047-1048, 1988.
[9] R. G. Walker, Appl. Phys. Lett., 54, pp. 1613-1615, 1989.
[10] R. G. Walker, and I. Bennion, IOOC-ECOC'91, Tech. Dig., WeB7-3, 1991.
[11] H. Takeuchi, K.Kasaya, and K.Oe, IEEE Photon. Technol. Lett., 1, pp. 227-229, 1989.
[12] J. -F. Vinchant, J. A. Cavailes, M. Erman, Ph. Jarry, and M. Renaud, IEEE J. Lightwave Technol., 10, pp. 1-7, 1992.
[13] M. Bachmann, E. Gini, and H. Melchior, ECOC'92, TuB7.4, pp. 345-348, 1992.
[14] F. Kappe, C. Bornholdt, and D. Hoffmann, OFC'94, PD11, pp. 54-57, 1994.
[15] J. S. Cites and P. R. Ashley, IEEE J. Lightwave Technol., 12, pp. 1167-1173, 1992.
[16] J. E. Zucker, K. L. Jones, B. I. Miller, and U. Koren, IEEE Photon. Technol. Lett., 2, pp. 32-34, 1990.
[17] C. Rolland, R. S. Moore, F. Shepherd, and G. Hiller, Electron. Lett. , 29, pp. 471-472, 1993.
[18] H. Sano, H. Inoue, S. Tsuji, and K. Ishida, OFC'92,ThG4, San Jose,

CA, Feb. 2-7, 1992.
[19] N. Agrawal, D. Franke, C. M. Weinert, and C. Bornholdt, OFC'94, WL7, 1994.
[20] J. E. Zucker, K. L. Jones, T. Y. Chang, N. J. Sauer, B. Tell, K. Brown-Goebeler, M. Wegener, and D. S. Chemla, Electron. Lett., 26, pp. 2029-2030, 1990.
[21] N. Agrawal, D. Hoffmann, D. Franke, K. C. Li, U. Clemens, A. Witt, and M. Wegener, Appl. Phys. Lett., 61, pp. 249-250, 1992.
[22] A. H. Gnauck, S. K. Korotky, J. J. Vesselka, J. Nagel, C. T. Kemmerer, W. J. Minford, and D. T. Moser, OFC'91, San Diego, 1991, Paper PD17.
[23] H. Sano, T. Ido, S. Tanaka, and H. Inoue, OECC'96, Makuhari Messe, 1996, pp. 188-189, Paper 17D2-4.
[24] N. Putz, D. M. Adams, C. Rolland, R. Moore, and R. Mallard, IPR'96, Tup-C1, p. 152, 1996.
[25] P. J. Duthie and M. J. Wale, Electron. Lett., 27, pp. 1265-1266 , 1991.
[26] A. Ihata, H. Furuta, and H. Noda, Proc. IEEE, 60, 470, 1972.
[27] S. Somekh, E. Garmire, A. Yariv, H. L. Garvin, and R. G. Hunsperger, Appl. Phys. Lett., 22, 46, 1973.
[28] E. A. J. Marcatili, Bell Syst. Tech. J., 48, 2071, 1969.
[29] H. F. Taylor, J. Appl. Phys., 44, 3257, 1973.
[30] K. Tada and K. Hirose, Appl. Phys. Lett., 25, pp. 561-562, 1974.
[31] M. Chmielowski and D. W. Langer, J. Appl. Phys., 49, 927, 1989.
[32] J. A. Cavailles, M. Erman, and K. Woodgridga, Photon. Technol. Lett., 1, pp. 373-375, 1989.
[33] H. Kawaguchi, Electron. Lett., 14, pp. 387-388, 1978.
[34] K. Tada and H. Yanagawa, J. Appl. Phys., 49, pp. 5404-5406, 1978.
[35] J. C. Campbell, F. A. Brum, D. W. Shaw, and K. L. Lawley, Appl. Phys. Lett., 27, pp. 202-205, 1975.
[36] F. J. Leonberger and C. O. Bozler, ibid., 31, pp. 223-226, 1977.
[37] A. R. Reisinger, D. W. Bellavance, and K. L. Lawley, ibid., 31, pp. 836-838, 1977.
[38] F. J. Leonberger, J. P. Donnelly, and C. O. Bozler, ibid., 29, pp. 652-654, 1976.
[39] J. C. Shelton, F. K. Reinhart, and R. A. Logan, Appl. Opt., 17, pp. 2548-2555, 1978.
[40] A. Carenco and L. Menigaux, J. Appl. Phys., 51, pp. 1325-1327, 1980.

[41] M. Fujiwara, A. Ajisawa, Y. Sugimoto, and Y. Ohta, Electron. lett. 20, pp. 790-792, 1984.
[42] H. Takeuchi, H. Nagata, H. Kawaguchi, and K. Oe, Electron. Lett., 22, pp. 1241-1243, 1986.
[43] J. E. Zucker, K. L. Jones, M. G. Young, B. I. Miller, U. Koren, Appl. Phys. Lett., 33, pp. 2280-2282, 1989.
[44] K. Komatsu, K. Hamamoto, M. Sugimoto, A. Ajisawa, Y. Kohga, and A. Suzuki, J. Lightwave Technol., LT-9, pp. 871-878, 1991.
[45] K. Hamamoto, T. Anan, K. Komatsu, M. Sugimoto, and I. Mito, Electron. Lett., 28, pp. 441-443, 1992.
[46] M. Gustavsson, T. L. Lagerstrom, M. Jnason, L. Lundgre, A. -C. Morner, M. Rask,and B. Stoltz, Electron. Lett., 28, pp. 2223-2225, 1992.
[47] T. Kirihara, M. Ogawa, H. Inoue, S. Nishihara, and K. Ishida, Photon. Switching Tech. Dig., PMB3, pp. 25-28, 1993.
[48] J. F. Vinchant, J. A. Cavailles, M. Erman, Ph. Jarry, and M. Renaud, J. Lightwave Technol., 10, pp. 63-70, 1992.
[49] E. A. J. Marcatili, Bell. Syst. Tech. J., 48, p. 2071, 1969.
[50] R. V. Schmit and R. C. Alferness, IEEE Trans. Circuits Syst., CAS-26, 12, pp. 1099-1108, 1979.
[51] K. Komatsu, K. Hamamoto, M. Sugou, A.. Ajisawa, and A. Suzuki, J. Lightwave Technol., LT-9, pp. 871-878, 1991.
[52] K. Hamamoto, T. Anan, K. Komatsu, M. Sugimoto, and I. Mito, Electron. Lett., 28, No. 5, pp. 441-443, 1992.
[53] K. Hamamoto, S. Sugou, K. Komatsu, and M. Kitamura, Electron. Lett., 29, No. 17, pp. 1580-1582, 1993.
[54] H. Takeuchi, Y. Hasumi, S. Kondo, and Y. Noguchi, Electron. Lett., 29, No. 6, pp. 523-524, 1993.
[55] K. Kawano, S. Sekine, H. Takeuchi, M. Wada, M. Koutoku, N. Yoshinoto, T. Ito, M. Yanagibashi, S. Kondo, and Y. Noguchi, Electron, Lett., 31, No. 2, pp. 96-97, 1995.
[56] T. Ito, M. Koutoku, N. Yoshimoto, K. Kawano, S. Sekine, M. Yanagibashi, and S. Kondo, Electron. Lett., 30, No. 23, pp. 1936-1937, 1994.
[57] T. Ito, M. Koutoku, N. Yoshimoto, K. Kawano, S. Sekine, M. Yanagibashi, and S. Kondo, IOOC'95, ThD1-1, vol.3, pp. 86-87, 1995.
[58] T. C. Huang, T. Hausken, K. Lee, N. Dagli, L. A. Coldren, and D. R. Myers, IEEE Photon. Technol. Lett., 1, p. 168, 1989.

[59] K. G. Ravikumar, K. Shimomura, T. Kikugawa, A. Izumi, S. Arai, Y. Suematsu, and K. Matsubara, Electron. Lett., 24, No. 7, pp. 415-416, 1988.
[60] J. E. Zucker, K. L. Jones, G. R. Jacobovitz, B. Tell, K. Brown-Goebeler, T. Y. Chang, N. J. Sauer, M. D. Divino, M. Wegener, and D. S. Chemla, Photon. Technol. Lett., 2, No. 11, pp. 804-806, 1990.
[61] J. E. Zucker, K. L. Jones, T. H. Chin, B. Tell, and K. Brown-Goebeler, IEEE J. Lightwave Technol., 10, pp. 1926-1930,1992.

Chapter 7

Comparison with Various Modulators

7.1 Introduction

Many types of modulators, such as Franz-Keldysh intensity modulators[1-2], traveling-wave-structure Mach-Zehnder interferometric modulators[3-4], and QCSE modulators (phase[5,6] and intensity[7-11]) have been developed. However, no survey of external modulators has been reported. In this chapter we discuss the characteristics of various modulators and compare them based on the figure of merit. The figure of merit for external modulators is defined in the first section and compared with those recently developed modulators. Moreover, we provide to some discussion on the trade-off relationship among modulator characteristics. In Section two, we discuss the required RF power for 10 dB on/off ratio and/or π-phase shift versus 3-dB bandwidth relationship for recently developed modulators, and describe the transmission loss versus driving voltage in section 3. In the fourth section, we examine the frequency response degradation associated with high input optical power and its counterplots, and in the last section, we look at frequency chirping in the intensity modulators.

7.2 Figure of Merit

As discussed in the Chapter 1, external modulators have five important parameters: on/off ratio, driving power, 3-dB bandwidth, insertion loss and chirping parameter. There is some trade-off among these parameters. For example, the on/off ratio is proportional to the product of absorption coefficient change and sample length L, whereas the 3-dB bandwidth f_{3dB} is inversely proportinal to L and proportional to i-region thickness d when the device speed is limited by sample capacitance, and the operating voltage is proportional to d.

This situation is shown in Fig. 7.1 where we show the case of InGaAs/InAlAs MQW modulator. The structure has an 1-μm thick Au electrode, a p-InGaAs cap layer, p- and n-InAlAs cladding layers, an i-MQW(a 7.5-nm thick InGaAs quantum well, 5.0-nm thick InAlAs barrier), as

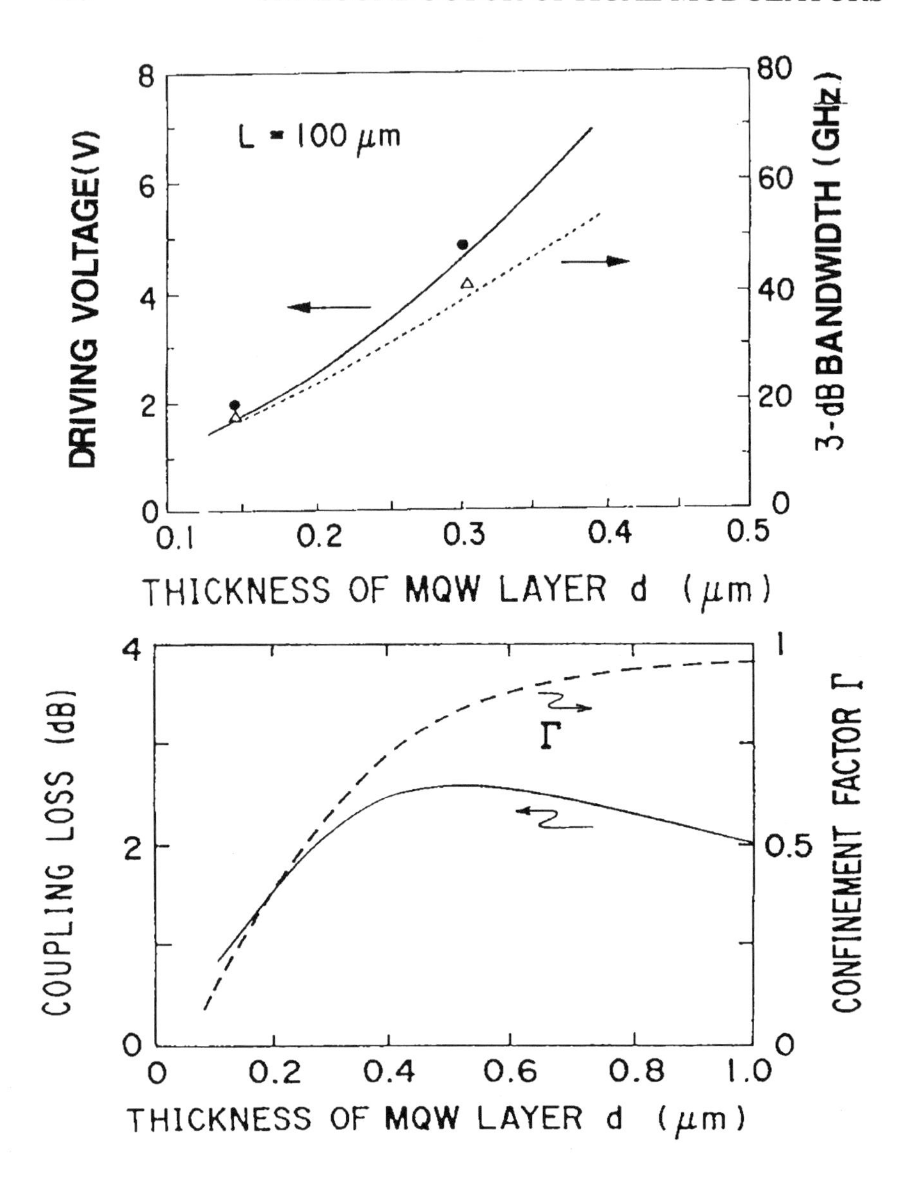

Fig. 7.1 Modulation characteristics versus absorption layer thickness for InGaAs/InAlAs MQW modulators.

(a) Required voltage for 10-dB on/off ratio and 3-dB bandwidth.

(b) Coupling loss with single mode fiber and optical confinement factor.

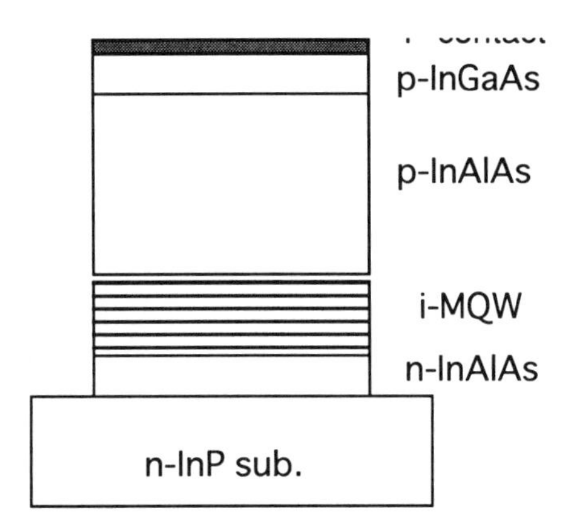

Fig. 7. 2. Schematic diagram of InGaAs/ InAlAs MQW modulators using the calculation of Fig. 7.1(a) and (b).

shown in Fig. 7.2. The coupling loss between the MQW guide and the optical fiber, which is a main insertion loss in MQW modulator, and the 3-dB bandwidth are shown as a function of an MQW guide thickness in Fig. 7.1. Therefore, we define the figure of merit as the following, based on the paper of Walker [12].

$$[\text{figure of merit}] = 2R\, f_{3dB}\, \lambda/(50+R)\, V\pi,$$

where R is the load resistance, $V\pi$ is a required voltage for π-phase shift for a phase modulator or the 20-dB on/off ratio for an intensity modulator, f_{3dB} is the 3-dB bandwidth, and λ is the wavelength. In Reference 12, a 10-dB on/off ratio is used for intensity modulators but it is too small for comparison with phase modulators, where an on/off ratio over 15 dB has been achieved. Also the system side requires an on/off ratio of over 15 dB for intensity modulators. Therefore, we have used a 20-dB on/off ratio for the above figure of merit.

Table 7.1 shows the figure of merit based on the above definition for recently developed modulators. In this table some data concerning $LiNbO_3$ modulators, which are poweful candidates for semiconductor modulators, are also listed. A significant difference in the figure of merit between the QCSE in the MQW structure and EO effect in the bulk structure is clear. However, from Table 7.1 we cannot estimate the dependence of the figure of merit on the energy difference among the used lightwave and the modulator absorption edge, and the insertion loss. In particular, for electroabsorption modulators this energy difference factor has an important effect on the transmission loss and drive voltage.

Table 7.1. Review of high-speed external modulators. Figure of merit is defined in the text. IM, intensity modulator; PM, phase modulator; TW, traveling waveguide structure.

Author(Year)	Description	λ (nm)	R(Ω)	Vπ(V)	f3dB(GHz)	Fig. of M.
Wakita(1990)	InGaAlAs/InAlAs MQW QCSE	1550	50	7	40	8.8
Wakita(1992)	InGaAs/InAlAs MQW QCSE	1530	50	1.2	15	19.6
Devaux(1992)	InGaAsP/InGaAsP MQW QCSE	1530	50	2.2	26	18.1
Aoki(1992)	MQW QCSE	1570	50	1.1	14	19.7
Mak(1990)	InGaAsP/InP Franz-Keldysh EA	1550	50	5	>20	>6.2
Kato(1991)	"	1550	50	3	16	8.2
Sato(1992)	"	1554	50	3.5	23	10.2
Suzuki(1992)	"	1540	50	1.7	9.1	8.2
Lin(1986)	GaAs TW Phase Modulator	1300	50	8	0.8	0.12
Lin(1987)	Coplanar TW Polarization	1300	50	11	8	0.95
Nees(1989)	GaAs TW Buried Coplanar	1060	none	288	110	0.4
Walker(1987)	GaAs Lumped MZ(Single-Sided)	1150	none	17.5	6.5	0.85
Walker(1989)	GaAs Lumped MZ Push-Pull	1150	none	9	6.3	1.59
Walker(1989)	GaAs MZ Loaded-Line	1150	50	4.25	22.5	6.06
"	Push-Pull TW	1150	50	4.85	25	5.94
Walker(1992)	"	1150	50	6.5	36	6.4

EA: electroabsorption, TW: traveling-wave, MZ: Mach-Zehnder

Figure 7.3 shows the figure of merit for recently reported data on external modulators versus the energy difference between the incident light wavelength and absorption band edge. The numbers in the parentheses indicate the thickness of quantum well and barrier, respectively. The energy difference results in a reduction of drive voltage but an increase in transmission loss. From this graph, InGaAs/InAlAs[7] and InGaAsP/InGaAsP MQW structures[8-11] give the best figure of merit, but there is only a small difference in the figure of merit between Franz-Keldysh bulk modulators[4] and QCSE modulators at the present stage. The difference is at most a factor of two, because the MQW structures include barrier layers that do not operate any electroabsorption effect.

Note that a thick quantum well gives a large figure of merit. This is because the optical confinement factor increases as the quantum well thickness increases and, moreover, thicker quantum wells produce large QCSE, as discussed in Chapter 2. However, the large confinement factor produces the poor optical coupling efficiency and large insertion loss[10] as discussed in Fig. 7.1. Therefore, some optimum thickness exists in the quantum well. Though the figure of merit for both InGaAsP/InGaAsP MQW and InAlAs MQW seems to be similar from this figure, the waveguide width is half for the former. That is, InGaAs/InAlAs MQW is superior to InGaAsP/InGaAsP MQW structure. We think this is due to poor exciton effect in InGaAsP/InGaAsP MQW structure, which is caused by the band offset difference in the heterointerface. The band offset has an important effect on the frequency response degradation associated with high input optical power and the details will be discussed in section 5.

7.3 Required Power vs. Bandwidth

As discussed in the section 1, modulation speed and driving voltage are not independent of each other. Figure 7.4 shows a summary of the required RF power vs. capacitance limited bandwidth for recently developed modulators and integrated light sources. A drive impedance of 50 Ω is assumed, and the operating voltage is the total voltage swing for either a π-phase shift for a phase modulator or a 10 dB on/off ratio for an intensity modulator. F-K stands for Franz-Keldysh modulators, DET[13] indicates depletion-edge-transformation modulators, and the triangles indicate $LiNbO_3$ modulators. Reduction of device capacitance results in the high-speed operation, whereas drive voltage increases. The required ratio of drive voltage to bandwidth figure of merit for an MQW integrated modulator is lower than for any other existing optical modulators.

On the other hand, the maximum 3-dB bandwidth of 52 GHz has been achieved for a traveling-wave-structure (TW) bulk GaAs/AlGaAs phase modulator[14]. As discussed in Chapter 1, lumped devices are limited by device capacitance. The next target is for lumped modulators to be free from device capacitance that limits the speed. At present, lumped devices can operate up to 50 GHz and the critical product of the bandwidth and the sample length is 4.13 GHz cm for an effective refractive index of 3.24 operating at 1.55 μm. At present, the 3-dB bandwidth is over 40 GHz and the drive voltage for a 10 dB on/off ratio is lower than 2 V. The low driving-voltage MQW modulator successfully eliminates the need for a modulator driver from an ultra high-speed transmitter. This enables us to supply a transmitter free from the bandwidth limits of the modulator driver.

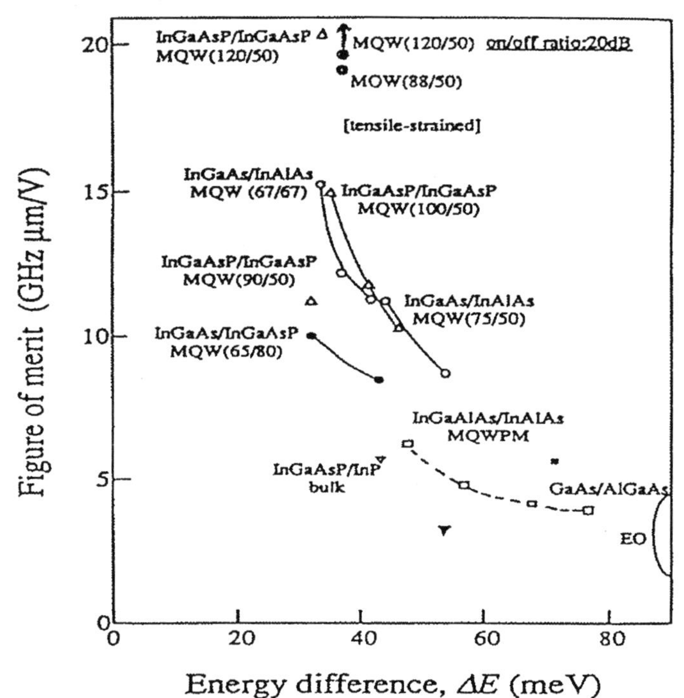

Fig. 7.3. Figure of merit versus energy difference between the incident light and absorption-band edge for recently developed external modulators. PM and EO indicate MQW phase modulator and electro-optic modulator, respectively.

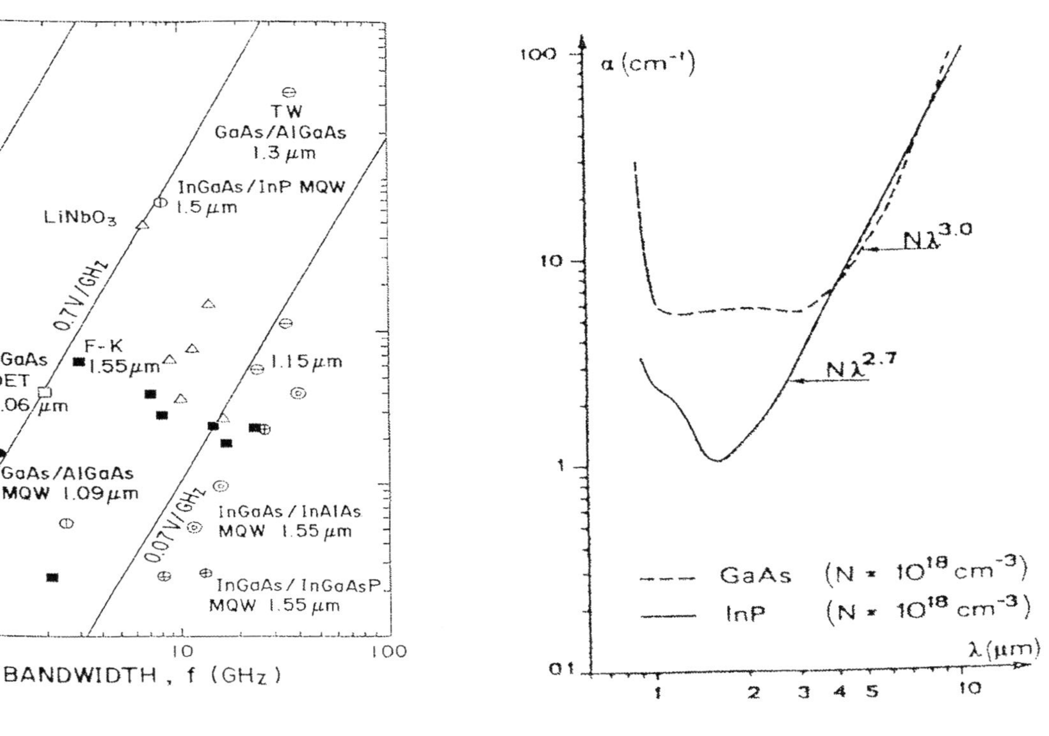

Fig.7.4. Required power versus capacitance-limited bandwidth for recently developed modulators. Solid squares, F-K, Franz-Keldysh; open triangles, $LiNbO_3$ modulator; circle with bar, TW, traveling waveguide structure; circle within circle, InGaAs(P)/InGaAsP MQW modulator; open square, DET, depletion-edge-translation modulator. Required drive power $P_{ac} = V_{ac}^2/50$, $V_{ac} = V_{pp}/2\sqrt{2}$, $V_{pp} = V\pi$ or V_{10dB}.

Fig. 7.5. Transmission loss versus optical wavelength for GaAs and InP [21].

7.4 Transmission Loss vs. Driving Voltage

Many experimental results have been reported on the propagation loss of two-dimensional waveguides. As a result of improvement in epitaxial growth and waveguide channel fabrication techniques, straight guide propagation loss has been reduced to levels comparable to that in $LiNbO_3$[15], however, the data are obtained for the non-doped heterostructure layer. In order to utilize such technology for low-loss active III-V components such as switches and modulators, the issue of electrode absorption loss must be addressed.

Modulators employing electronic control signals require a rectifying semiconductor junction based on either doping (p-n junction) or metallization (Schottky contact). Since electrode materials for active modulators must be conductive, modulators exhibit excess propagation loss due to electrode absorption above that of waveguide layers.

Although the materials are used at photon energies below the fundamental bandgap, they still exhibit some residual intrinsic absorption. For doped materials, dominant transmission loss is due to the free carrier absorption. Figure 7.5 gives experimental results for GaAs [16] and InP[17]. At wavelengths larger than 3 μm, the absorption is proportional to $N\lambda^p$ where N is the carrier concentration, λ is the wavelength and $2.5 < p < 3.5$, depending on the absorption mechanisms involved[17]. In the 1-2 μm range, the absorption is dominated by interband transitions[18] and is proportional to N only for $N > 10^{17}$ cm^{-3} [19]. At low free carrier concentrations, absorption due to the presence of deep levels can be larger than the free carrier concentration. Thus, purity of the material is essential to achieve low loss waveguides.

In an active waveguide, the dominant contribution to the losses is the absorption of the N^+ and P^+ regions. As discussed in the Chapter 2, two-dimensional optical confinement has to be realized in a waveguide in order to achieve single mode operation. Moreover, the driving voltage can be decreased by increasing the electric field in the intrinsic region, which is done by decreasing its thickness. In this case the optical confinement factor is usually not so large that the portion of the light in the the N^+ and P^+ regions will increase and so will the losses. In short, these two parameters, driving voltage and optical loss, cannot be optimized separately [20].

Figure 7.6 shows the absorption loss of the modulator α (dB/cm) vs. the normalized driving voltage (V in V cm) for several different structures reported recently, in addition to the data reported in [21]. The overall result is that homostructure modulators exhibit higher αV products than the heterostructure ones and the heterostructure modulators exhibit higher αV product than MQW structure modulators. The former can be understood because in a homostructure waveguide optical confinement is achieved by the N^+ and P^+ regions, and therefore the optical and electrical problems are closely related, whereas this is less the case in heterostructure modulators. The latter can be understood by the discussions in the previous section, where due to the QCSE the MQW structure can produce a small energy difference between the operating wavelength and the absorption egde.

It is likely that the αV products for electroabsorption modulators and the electro-optic devices are similar, although the structures and the physical effects are different even if $LiNbO_3$ modulators are included.

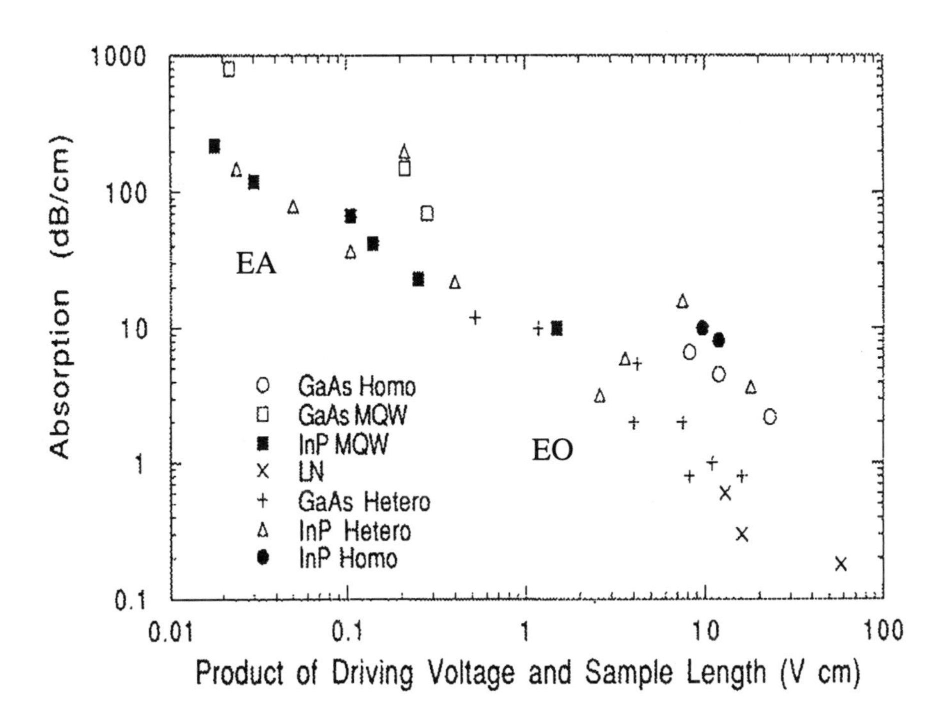

Fig. 7.6. Transmission loss versus driving voltage for reported external modulators, including MQW structures, homo- and hetero-structures consisting of InP and GaAs and $LiNbO_3$.

Because the electroabsorption devices operate near the energy band gap, they have higher absorption but lower driving voltage. The separation of the plane (α, V) into two regions: electro-optic(EO) devices and electroabsorption(EA) devices, as reported in [21], is smeared a little when MQW structure devices are included, but its general trend is similar. That is, introduction of the MQW structure allows the operating wavelength to come near the absorption edge because of their short sample length and high efficiency.

7.5 Frequency Response Degradation

The AT&T group[22] reported that the saturation of the absorption coefficient associated with the intensity of the incident light is quite

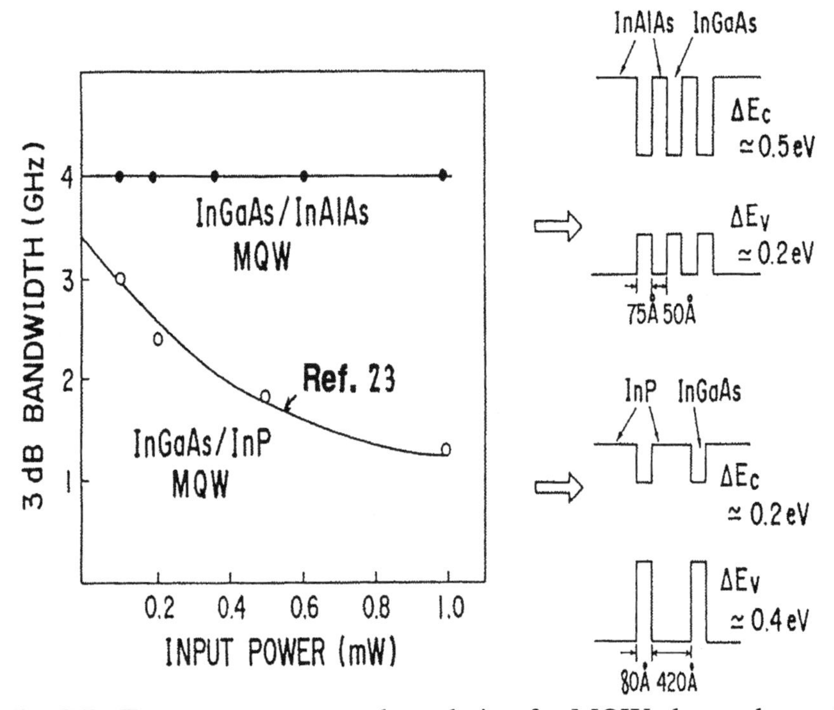

Fig. 7.7. Frequency response degradation for MQW electroabsorption modulators under high input power. The inset indicates the energy-band diagram difference between InGaAsP/InP and InGaAs/InAlAs MQW structures.

marked in MQW structures. More recently, frequency response degradation due to a high input power has been reported by the AT&T group [23] for InGaAs/InP MQWs and by the KDD group[24] for InGaAsP/InP Franz-Keldysh modulators. These results mean that, in practice, the dependence of modulation characteristics on the intensity of incident light is important.

Figure 7.7 shows the frequency response under high input optical power that was first reported by Koren et al.[23], and the energy band diagram in the MQW structures. As shown in the inset of Fig. 7.7, the InGaAs/InP MQW structures has a large band offset for the valence band, whereas the InGaAs/InAlAs MQW structure has a small offset. Therefore, accumulation of photo-generated carriers at high-input power is expected to be smaller in the InGaAs/InAlAs MQW structure. In fact, no degradation has been observed in this material systems.

Figure 7.8 shows the small-signal modulation response at different optical input power levels for InGaAs/InAlAs MQW modulators. As the power level increased up to 10 times the initial value, no decrease in the small-signal modulation bandwidth was observed. This indicates that both carrier storage in the quantum wells and reduction in electroabsorption at high carrier densities may be small for InGaAs/ InAlAs MQW modulators.

This situation is the similar to that for bulk InGaAsP/InP modulators. The KDD group reported the degradation of frequency response due to high input power and demonstrated the usefulness of introducing the buffer layer between the InGaAsP waveguide layer and the P-cladding layer. The buffer layer relaxes the heterobarriers and reduces the carrier storage at the heterointerfaces resulting in good frequency response.

In order to reduce the hole pile-up in the heterointerface associated with the absorption of incident light, which produces electroabsorption saturation and degradation of frequency response, a small band offset configuration and thin barrier have been designed for an InGaAsP/ InGaAsP MQW structure [8,11] and good results have recently been reported. However, this configuration results in a poor quantum effect. Therefore, some device redesigning is necessary from an overall point of view, in addition to introducing strained wells [25,26]. Recently, a strain-compensated structure with a compressive InGaAsP well and a tensile InGaAsP barrier, to provide small valence band discontinuities, has been proposed to reduce the hole escape time from quantum wells [27, 28] and the modulation with suppressed hole accumulation are realized [29].

On the contrary, polarization insensitive MQW modulators, which are required from system side to use optical demultiplexing, use tensilely strained quantum wells to reduce polarization anisotropy and this configuration increases the valence band discontinuities resulting in a poor saturation level [30]. An insertion of new InAsP layer produces a better result for a high input optical power [31].

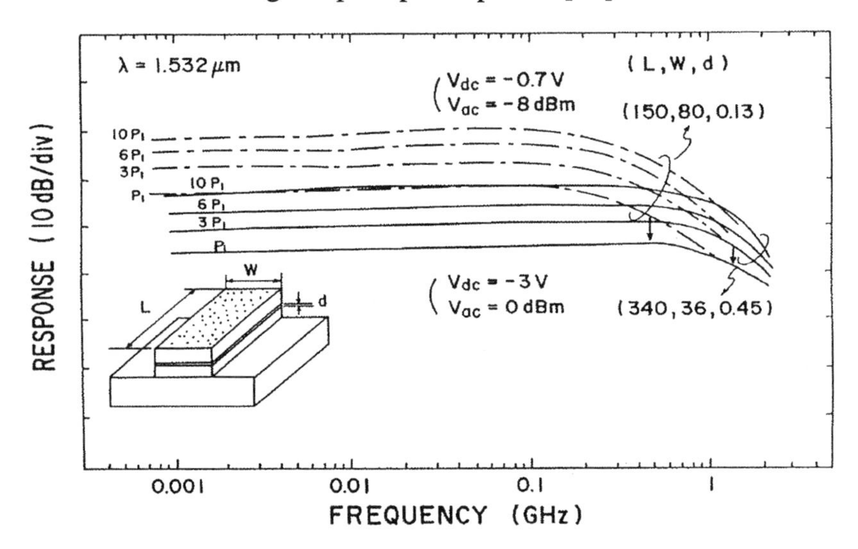

Fig. 7.8. Small signal frequency response of InGaAs/InAlAs MQW modulator under various input optical powers.

7.6 Frequency Chirping

Frequency chirping is an important parameter for intensity modulators. For a system operating at a wavelength where fiber dispertion is nonzero, frequency chirping limits the maximum transmission length or bandwidth due to intersymbol interference in the transmitted data. The magnitute of this effect is related to the linewidth enhancement factor α_C which is defined as the ratio of the change in the real part of the refractive index to the change in the imaginary part in the modulators. This value is calculated for intensity modulators, such as a loss modulator, directional coupler type modulator, Mach-Zehnder interferometer type modulator, and total internal reflection type modulator[32]. In these modulators, loss modulators using electroabsorption effect have relatively large α_C parameter, whereas other modulators use electrorefraction effect and have small (or even zero) values of α_C parameter. Table 7.2 shows the α_C parameter for various types of external modulators [32].

Measurements are also reported for both bulk type and MQW structures. (Method to obtain α_C parameter will be described in Chapter 10. 3). Although chirping parameters can vary from 0.15 to 4 depending on the material bandgap, the device structure, and the modulation index, the magnitude is significantly less than that which is obtained from direct intensity modulation of injection lasers.

The reported data are within 1.0 for intensity modulators based on the electroabsorption effect (Franz-Keldysh [33-36] and quantum confined Stark effect [37-40]).

The α_C parameters have been obtained from the calculated and measured pulse width as a function of total dispersion of the optical fibers for an InGaAsP/InGaAsP MQW modulator/DFB laser integrated light source module [40]. The pulse width broadening after transiting several dispersive fibers is related to the α_C parameter. The α_C parameter strongly depends on the applied DC voltage and the detuning energy between the absorption band edge and the incident lightwave but its value is uniquely determined at a given condition and cannot be controlled. This is a common characteristic for electroabsorption modulators and is different from that for the interferometric modulators discussed below.

A novel dispersion compensation technique that can be used to transmit a prechirped optical signal made by bit synchronized sinusoidal modulation to a laser diode prior to intensity modulation with an electroabsorption modulator has been proposed and demonstrated [36, 41]. The principle of the prechirp method is based on a pre-equalization method. The transmitter configuration for this technique is shown in Fig.7.9. At first, the DFB LD is modulated with a bit synchronized sinusoidal wave. A sinusoidall output is intensity modulated by an electroabsorption modulator with RZ format (Fig. 7.9(b)). The prechirped optical waveform (Fig. 7.9(c)) consists of lower frequency components at the trailing edge of the pulse. If this prechirped waveform is transmitted through a positive dispersion fiber, its pulse width becomes narrower at the beginning of the transmission and broadens after further transmission. As a result, the transmission length becomes longer than that for an ideal external modulator.

Table 7. 2. Review of α_C-parameter for electroabsorption modulators.

Author	firm	material	effect	α-Parameter	detuning energy(meV)	reference
Y. Noda	KDD	InGaAsP/InP	F-K	1 ~ 2	51.5	JLT,LT-4,1445, '87
T. H. Wood	AT&T	GaAs/AlGaAs	QCSE	1	12(TM)	APL,50,798, '87
K. Wakita	NTT	InGaAs/InAlAs	QCSE	0.6 ~ 0.8	22 ~ 27	JJAP,26,1169, '87
H. Soda	Fujitsu	InGaAsP/InP	F-K	1	68	EL,24,1194, '88
T. Saito	NEC	InGaAsP/InP	F-K	0*, 0.6	85.7	OEC'90,13A2-4, '90
K. Wakita	NTT	InGaAs/InAlAs	QCSE	0.7	27	PTL,3,138, '91
M. S. Whalen	AT&T	InGaAsP/InP	QCSE	0.6 ~ 0.8	24.7 ~ 28.3	PTL,3,451, '91
M. Suzuki	KDD	InGaAsP/InP	F-K	0.2 ~ 0.4**	56 ~ 71	PTL,4,586, '92
J. Langanay	Alcatel	InGaAsP/InGaAsP	QCSE	-0.2 ~ 0.2	24.7 ~ 27.4	APL,62,2067, '93
F. Devaux	CNET	InGaAsP/InGaAsP	QCSE	1.2	---	PTL,4,720, '92
F. Koyama	AT&T	InGaAs/InGaAsP	QCSE	0.4 ~ 1.0	5.4 ~ 21.2	PTL,5,1389, '93
I. Kotaka	NTT	InGaAsP/InGaAsP	QCSE	0.8	30.5	PTL,5,61, '93
F. Devaux	CNET	InGaAsP/InGaAsP	QCSE	-2.0 ~ 3.0	---	PTL,5,1288, '93
T. Kataoka	NTT	InGaAsP/InGaAsP	QCSE	0.2 ~ 1.4	30.5	EL,30,872, '94
T. H. Wood	AT&T	InGaAsP/InP	QCSE	0.5	28.3	JLT,LT-12,1152, '94
T. Ido	Hitachi	InGaAs/InAlAs	QCSE	0.6	32.2	PTL,6,1207, '94

*prechirped **underestimated, revised values 0.585~0.795 (JLT, LT-12, '94)

On the other hand, the intensity modulator based on Mach-Zehnder interferometric configuration gives various values of α_C parameter. Therefore, choosing a non-zero value for the chirp parameter can be advantageous, depending on the fiber dispersion coefficient and distance, so as to provide some amount of pulse compression, since the lowest fiber dispersion penalty is generally not obtained for a frequency chirp parameter equal to zero. This effect was first pointed out and demonstrated by Korotky for a $LiNbO_3$ intensity modulator [42].

In a Mach-Zehnder modulator we can drive two modulator electrodes in various voltage ratios and phases. For example, to obtain pure phase modulation with a modulation index of unity the two electrodes are driven in phase, and each with a voltage swing corresponding to π-radian phase-shift. The potential applications of this AM/PM modulator are manifold and include fiber dispersion compensation, optical pulse compression,

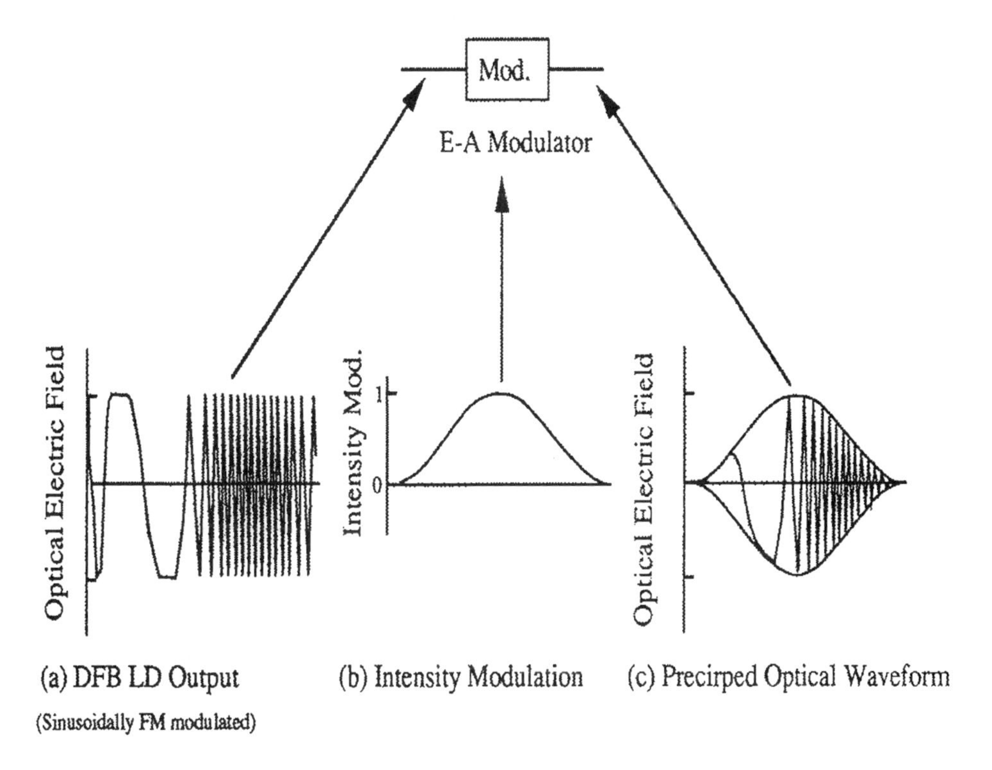

Fig. 7.9. Principles of prechirping : (a) DFB laser output (sinusoidally FM modulated. (b) Intensity modulation. (c) Prechirped optical waveform [36].

spread spectrum communication and optical frequency generation.

A highly efficient and small size Mach-Zehnder modulator using InGaAsP quantum wells was first reported [43], but its driving voltage was still too high (-10 V with only 7 dB of extinction ratio) for practical use. Recently a low driving voltage InGaAs/InAlAs MQW Mach-Zehnder modulator with high extinction ratio (>15 dB) [44], for 10-Gbit/s long haul transmission systems operating in the 1.55 μm wavelength region has been reported. For a 30-quantum well device with sample length of 500 μm, voltage required for p-phase shift is 3.0 V at 1.57 μm; a 3-dB bandwidth of 12 GHz is obtained.

An ultracompact sample length directly results in higher-speed operation. The device with only 300 μm long has a 3-dB electrical bandwidth of 18-GHz [45]. In a recent optical transmission experiment over 160 km of dispersion-shifted fiber, this modulator was used to send 10-Gbit/s data in a phase-shift-keyed digital coding scheme.

However, the intensity modulators using phase changes in quantum wells have still large operating voltages compared with those of loss modulators as discussed in the previous sections. Especially, a driving voltage lower than 2 V is required by system designers but has never achieved for such modulators with high-speed (3-dB bandwidth over 20 GHz). More improvement is necessary. Novel electrorefractive mechanisms such as transfering electrons into quantum wells, low-dimensional structures (quantum wires and quantum dots) are being pursued for device applications. The potential benefits of these mechanisms will be realized in near future.

7.7 Optimization of MQW Structure

7.7.1 Introduction

In the InGaAs multiple quantum well (MQW) modulators on InP substrates, the barrier and waveguide may be made from quaternary alloys of either InGaAsP or InGaAlAs. If a lattice-matched alloy with a particular bandgap is selected, the other properties are fixed at values which depend on the choice of InGaAsP or InGaAlAs. Although the bandgap discontinuity ΔE_g at the heterojunctions may be the same in both cases, the discontinuities in the conduction and valence bands ΔE_c and ΔE_v are different. For InGaAs/InGaAsP ΔE_c:ΔE_v is thought to be about 40:60 [46], whereas for InGaAs/InGaAlAs this ratio is about 72:28 [47]. Because the effective masses of electrons and heavy holes differ by an order of magnitude, the band discontinuities have a significant effect on the well thickness appropriate for a desired output wavelength. The well thickness must be greater for InGaAlAs barriers. In the quantum confined Stark effect (QCSE), the exciton peaks shift in proportion to the fourth power of the well thickness when the applied field is weak [48].

Moreover, the valence band offset ΔE_V has an effect on the modulator characteristics such as frequency response and absorption saturation under high input optical power because of hole pile-up at the heterointerface. Especially for InGaAs/InP MQW structures with ΔE_V of 0.38 eV, the frequency response is degraded because of hole-pile up [49]. InGaAsP/InGaAsP MQW structures with an ΔEv of 0.2 eV is used instead to reduce this degradation.

On the other hand, for the same quaternary alloys we can tailor the well thickness and well bandgap energy even though the energy band edge of the MQW is constant. Because larger absorption variations can therefore be obtained with similar applied electric fields, it is necessary to optimize the well and barrier structure for modulators.

The first proposal to improve the electroabsorption effect in MQW structures by replacing the ternary InGaAs well with a quaternary InGaAsP well was reported in 1987 [50]. The optimization of the InGaAsP/InP [51,52] and InGaAlAs/InAlAs [51] systems have also been reported.

7.7.2 Theory

Quaternary QW materials are believed to be more advantageous than the conventional ternary QW materials for enhancing excitonic electroabsorption effects. Since the bandgap of quaternary materials is larger than that of ternary materials, the exciton transition energy is also greater. The well thickness must therefore be increased for the quaternary QW's in order to keep the exciton transition energy the same as that for the ternary QW's. This is necessary for operating the device at a certain fixed wavelength, such as the 1.55 μm wavelength used for optical fiber-communications. The energy shift of quantum levels (Stark shift) induced by application of an external electric field increases with well thickness [48]. The quaternary QW is therefore thought to have field effects more pronounced than those of the conventional ternary QW's, operated at the same wavelength. The increase in well thickness, however, implies a decrease in the oscillator strength of the excitonic transition. This factor may impede the above enhancement.

A simple calculation, in which only the lowest electron-to-heavy-hole (n=1 e-hh) exciton transition was considered and the linewidth of the excitonic absorption is assumed to be Gaussian was tried [51]. An index η to describe the degree of excitonic electroabsorption is defined as follows.

$$\eta = (\alpha(F) - \alpha(0))/\alpha(0),$$

where $\alpha(F)$ is the the absorption coefficient as a function of electric field F and is calculated using

$$\alpha(F) = 4\pi^2 e^2 hf\Delta(h\omega - E_{ex})/nm_0c.$$

Here $h\omega$ is the photon energy, n is the refractive index, f is the oscillator strength given by the following equation, and the other constants represent conventional notations:

$$f = Q^2 | \int \psi_e(z)\psi_h(z)dz | /2\pi m_0 E_{ex} a_B{}^2 L_z,$$

where L_z is the QW thickness, E_{ex} is the n=1 e-hh exciton transition energy, a_B is the exciton Bohr radius, Q is the matrix element between the s-like spin-1/2 conduction-band wave function and the p-like spin-3/2 valence band wave function at the Γ point, and $\psi_e(z)(\psi_h(z))$ is the electron (hole) envelope wave function. Because the optical modulator must have a high extinction ratio and a low insertion loss, which corresponds to maximizing the above index, this index can be regarded as one figure of merit for the electroabsorption-type MQW modualtor.

Figure 7.10 shows a three-dimensional representation of the index as a function of well thickness and the bandgap (E_g) of QW material for two kinds of QW's, InGaAsP/InP and InGaAlAs/InAlAs. Parameter values used in this calculation are summarized in Table 7.3. Here, the electric field is 120 kV/cm, the incident light wavelength is 1.55 μm, and the absorption coefficient at the off-resonant wavelength is 10 cm^{-1}. The L_z plane (E_g=0.730 eV) corresponds to the usual InGaAs/InP and InGaAs/InAlAs QW's. This figure clearly shows that there is an optimum QW

Table 7.3. Parameter values used in the calculation (Fig. 7.10). Put x = 0 for conventional InGaAs/InP and InGaAs/InAlAs QW's.

	$In_yGa_{1-y}As_{1-x}P_x$ [a]	InP	$In_{0.53}Ga_{0.47-x}Al_xAs$	$In_{0.53}Al_{0.47}As$
m_e	$0.065(1-x)(1-y)+0.021(1-x)y + 0.17x(1-y)+0.077xy$	0.077	$0.041+0.072x$	0.075
m_{hh}	$0.45(1-x)(1-y)+0.32(1-x)y + 0.54x(1-y)+0.65xy$	0.65	$0.377+0.411x$	0.57
ΔE_c (eV)	$0.4(0.60-0.48x-0.12x^2)$	...	$0.71(0.70-1.49x)$	...
ΔE_v (eV)	$0.6(0.60-0.48x-0.12x^2)$	...	$0.29(0.70-1.49x)$	...
E_g (eV)	$0.73+0.48x+0.12x^2$	1.33	$0.73+1.49x$	1.43

[a] $y = (16.6+15.6x)/(31.2+x)$ for the layers lattice-matched to InP.

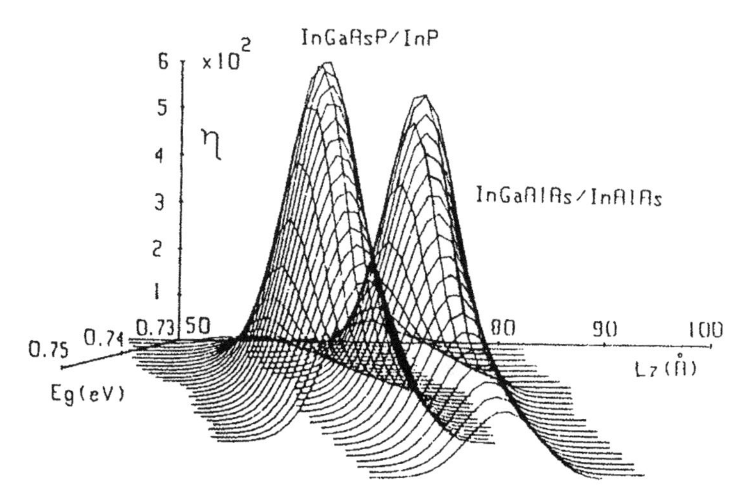

Fig. 7.10. Figure of merit $\Delta\alpha/\alpha(0)$ for excitonic elctroabsorption effects as a function of well thickness (L_Z) and bandgap (E_g) of QW materials for InGaAsP/InP and InGaAlAs/InAlAs QW's. The electric fields of 120 kV/cm are applied. The η-Lz plane (Eg=0.730 eV) corresponds to conventional InGaAs/InP and InGaAs/InAlAs QW's.

material and structure for excitonic electroabsorption. This model is too simple to be used in the quantitative discussions because it does not include transitions of the light-hole subband and 2D Sommerfeld enhanced continua, but it indicates the drastic enhancement may be obtained through the optimized balance between the Stark shift and the oscillator strength.

Many proposes and trials to improve the optimum condition and figures of merit for electroabsorption modulators have been reported [53-56].

A. Another figure of merit. A new figure of merit has been proposed and demonstrated for a Wannier-Stark localization modulator[53]. This definition includes such an important parameter as bandwidth, which was neglected in the above discussions, and introduced the parameter Γ $\Delta\alpha(F)/\Delta F$. Bigan et al. [53] also show that if any two of the three characteristics ---- extinction ratio, drive voltage, and bandwidth are equal---- the third is defined by this parameter; the extinction ratio per unit length and per unit applied electric field. Their demonstration of this is very instructive and we summarize their results as follows.

If we assume that the electroabsoption modulator speed is limited by RC time constant, the electrical 3-dB bandwidth ν_{3dB} can be expressed as

$$\nu_{3dB} = 1/\pi R C = 1/(\pi R)(d/\varepsilon W L),$$

where C is the modulator p-i-n capacitance, R is a load resistor connected in parallel with C, and ε is the intrinsic layer dielectric constant (d, W and L are respectively the waveguide thickness, width, and length). One wishes to get a large extinction ratio, a low voltage, and a high 3-dB bandwidth simultaneously. To determine a figure of merit taking into account the relationship between these characteristics, we should consider two different electroabsorption configurations determined by the following parameters:

$$(\Gamma \Delta\alpha)_{1,2},\ L_{1,2},\ d_{1,2}, \text{ and } \Delta F_{1,2}.$$

Indices 1 and 2 refer to configurations 1 and 2, and ε is assumed to display no significant change from one configuration to another. This assumption is valid for III-V materials.

If we let the extinction ratios as well as drive voltages be equal; i.e.,

$$(\Gamma\Delta\alpha)_1 L_1 = (\Gamma\Delta\alpha)_2 L_2$$

and

$$d_1 \Delta F_1 = d_2 \Delta F_2,$$

then

$$(\nu_{3dB})_1/(\nu_{3dB})_2 = (d_1/d_2)(L_2/L_1) = (\Delta F)_2/(\Delta F)_1\ (\Gamma\Delta\alpha)_1 / (\Gamma\Delta\alpha)_2$$

$$= \{(\Gamma\Delta\alpha)_1/(\Delta F)_1\}/\{(\Gamma\Delta\alpha)_2/(\Delta F)_2\}.$$

This means the higher the extinction ratio per waveguide length and per unit applied electric field, the higher the bandwidth.

Similarly, if we let the extinction ratios as well as the bandwidths be equal; i.e.,

$$(\Gamma\Delta\alpha)_1 L_1 = (\Gamma\Delta\alpha)_2 L_2$$

and

$$d_1/L_1 = d_2/L_2,$$

then

$$\Delta V_1/\Delta V_2 = (d_1/d_2)(\Delta F_1/\Delta F_2) = (L_1/L_2)(\Delta F_1/\Delta F2)$$

$$= \{(\Gamma\Delta\alpha)_2/(\Delta F)_2\}/\{(\Gamma\Delta\alpha)_1/(\Delta F)_1\}.$$

This, on the other hand, means the higher the extinction ratio per unit length and per unit applied field, the lower the drive voltage.

And if we let the drive voltages as well as the bandwidths be equal; i.e.,

$$d_1\ \Delta F_1 = d_2\ \Delta F_2$$

and

$$d_1/L_1 = d_2/L_2,$$

then

$$((\Gamma\ \Delta\alpha)_1\ L_1)/((\Gamma\Delta\alpha)_2\ L_2) = ((\Gamma\ \Delta\alpha)_1\ d_1)/((\Gamma\Delta\alpha)_2\ d_2)$$

$$= \{(\Gamma\Delta\alpha)_1/(\Delta F)_1\}/\{(\Gamma\Delta\alpha)_2/((\Delta F)_2).$$

And this means the higher the extinction ratio per length and per unit applied electric field, the higher the extinction ratio.

We have shown that if any two of three characteristics (extinction ratio, drive voltage, and bandwidth) are equal, the third is defined by the extinction ratio per unit length and per unit applied electric field expressed by $\Gamma\ \Delta\alpha(F)/\Delta F$.

This quantity, however, does not by itself suffice because it does not take into account the on-state attenuation. Both $\Delta\alpha/\alpha(0)$ and $\Gamma\Delta\alpha/\Delta F$ parameters should be considered, and these are both complimentary and depend on the device usage. For example, which parameter is preferable depends on waveguide or surface normal configuration or what characteristics are regarded as important.

B. Other figure of merit. Another proposed parameter for figure of merit is $\Delta\alpha/\Delta F^2$ [54]. The modulation efficiency is normally characterized by the drive voltage swing required to give a specified contrast ratio, say 10 dB. This voltage is analogous to V_π in the case of electrooptic phase modulators. The RF drive power is then given by $P_{ac} = V_{ac}^2/R_l$ [54], where $V_{ac} = V_{pp}/2\sqrt{2}$ and $V_{pp} = \Delta V_{10dB}$ is the peak-to-peak voltage swing required to give a 10-dB on/off ratio. The power per unit bandwidth is given by $P_{ac}/\Delta f \approx C\Delta V_{10dB}^2$. Chin and Chang [54] also propose a wide quantum well (>12 nm), but beyond certain L_z values the validity of the theoretical model becomes unreliable. That is, for very wide wells the binding energy and the quantum size effects will diminish rapidly and bulk behavior may become dominant. Furthermore, for wide wells the absorption oscillator strength may decrease faster than $1/L_z$, and the linewidth may increase with electric field rather than assumed. Chin and Chang therefore limit the L_z value to 15 nm, for which their figure of

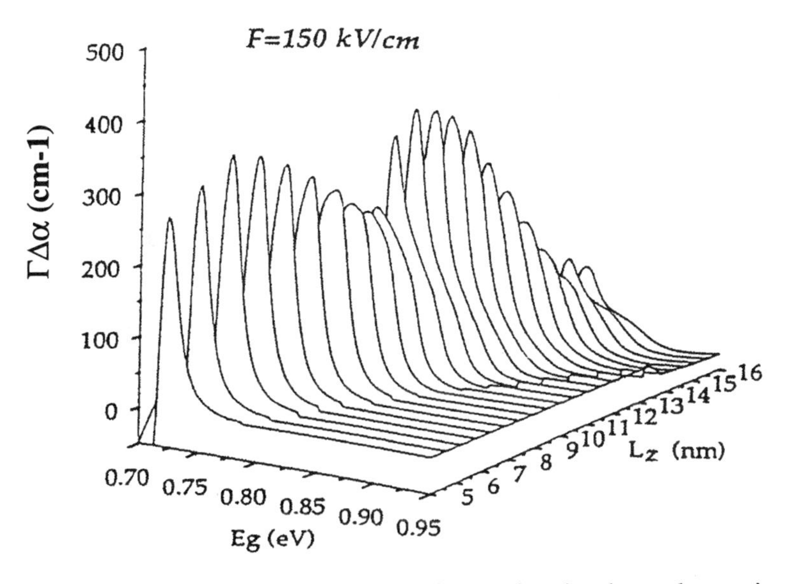

Fig. 7. 11. Figure of merit $\Gamma\Delta\alpha$ for excitonic elctroabsorption effects as a function of well thickness (L_z) and bandgap (E_g) of QW materials for InGaAsP/InGaAsP QW's. The applied electric fields are 150 kV/cm.

merit increases monotonically. In conclusion, there is no optimum condition based on their calculation and the difference between MQW structure and bulk structure is not yet clear.

We define the figure of merit as $\Gamma\Delta\alpha/\Delta F$ and calculate its value under the condition that it includes the field induced broadening and the transitions between the conduction ground subband and the topmost three valence subband (1e-1hh, 1e-1lh, 1e-2hh)[55]. The former effect, which has long been neglected, is critical for practical use in weak-confinement MQWs such as InGaAsP/InGaAsP, where under a relatively weak electric field the conduction band offset is too small to confine electrons. This field-induced spectral broadening cannot be explained quantitatively by the proposed inhomogeneous effects such as LO-phonon coupling, interface quality, well thickness fluctuation, and field-intensity inhomogeneity due to background impurities. As discussed in detail in Chapter 4, carrier tunneling associated with the applied electric fields is intrinsic [56].

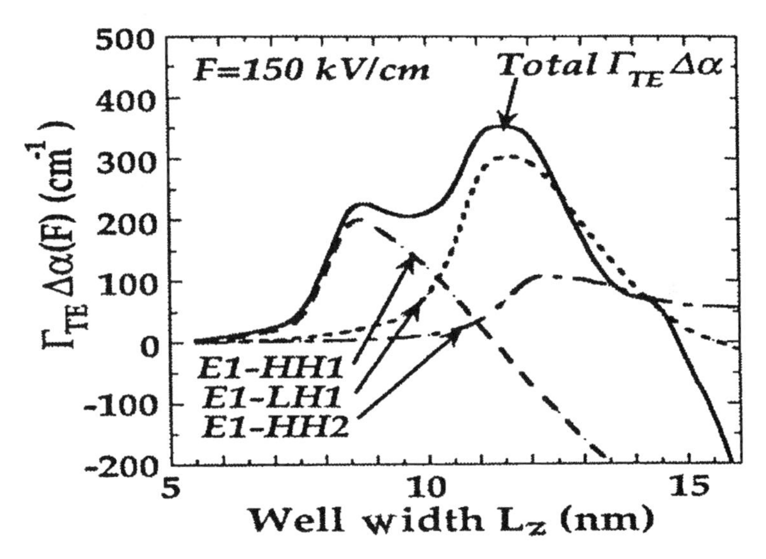

Fig. 7.12. Well-width dependence of the TE-polarized electroabsorption change at applied fields of 150 kV/cm for InGaAsP/InGaAsP MQW structures (ten wells). The bulk bandgaps of the well and barrier are kept constant at 0.8 and 1.13 eV, respectively. The barrier width of 7 nm is assumed. The solid line is total absorption and dashed lines are the contributions of the 1e-1hh, -1lh, -2hh transitions.

As a result, we obtain the optimum conditions for MQW structures, such as the well thickness and well configuration. Figure 7.11 shows the figure of merit versus well thickness and composition. The operating wavelength is 1.55 μm and the barrier thickness is 7 nm. There are two peaks: one is for a thin well and small energy gap ($E_g = 0.7$ eV), and the other is for a thick well and large energy gap ($E_g = 0.8$ eV). As shown in Fig. 7.12, the former coresponds to the first heavy hole excitonic transition and the latter to light hole exciton. The former has relatively large transmission loss because of smaller detuning energy, whereas the latter has a larger figure of merit and lower transmission loss. That is, the thicker quantum well (about 11-12 nm) is better. This is because of the larger overlap integral of 1e-1lh transition due to the light-hole character and to the enhanced joint density of state due to the valence band-mixing in the 1lh subband. Thicker wells also increase the optical confinement and lower the operating voltages.

This result gives us insight into efficient electroabsorption effect in quantum wells. The introduction of tensile strain into wells, which has been used in order to reduce the anisotropic modulation of polarization dependence, will also affect the reduction of drive voltage according to this calculated model because hh and lh transition cross under a certain tensile strain and a thicker well is used to keep the detuning energy suitable. In fact, polarization insensitive MQW modulators are operated at a lower drive voltage and high speed [57, 58]. The details will be discussed in Chapter 10.

7.8 References

[1] M. Suzuki, H. Tanaka, and Y. Matsushima, Photon. Technol. Lett., Vol. 4, Jun. 1992, pp. 586-588.
[2] H. Soda, K. Nakai, H. Ishikawa, and H. Imai, Electron. Lett., Vol. 23, 1987, pp.1232-1234.
[3] S. H. Lin, and S. -Y. Wang, Appl. Opt., Vol. 26, 1987, pp.1696-1700.
[4] R. G. Warker, I. Bennion, and C. A. Carter, Electron. Lett., Vol. 25, 1989, pp.1549-1550.
[5] K. Wakita, I. Kotaka, and H. Asai, Photon. Technol. Lett., Vol. 4, 1992, pp. 29-31.
[6] H. Sano, et al., Tech. Digest of Topical Meeting on Quantum Opto-electronics, MD1, Solt Lake City, Utah, Mar. 1991, pp. 58-61.
[7] K. Wakita, I. Kotaka, O. Mitomi, H. Asai, Y. Kawamura, and M.

Naganuma, J. Lightwave Technol., Vol. 8, 1990, pp. 1027-1032.
[8] T. Kato, T. Sasaki, N. Kida, K. Komatsu, and I. Mito, in 17th European Conference on Optical Communication ECOC'91, Paper No. WeB7-1.
[9] M. Aoki, M. Takahashi, M. Suzuki, H. Sano, K. Uomi, T. Kawano, and T. Takai, Photon. Technol. Lett., Vol. 4, 1992, pp.580-582.
[10] F. Devaux, E. Bigan, A. Ougazzaden, B. Pierre, F. Huet, M. Carre, and A. Carrenco, in Digest of Conference on Optical Fiber Communication, 1992 OSA Technical Series, Vol. 5, Optical Society of America, Washington, DC, 1992, pp. 329-332, Paper No. PD4.
[11] K. Sato, I. Kotaka, K. Wakita, Y. Kondo, and M. Yamamoto, Electron. Lett., Vol. 29, 1993, pp. 1087-1088.
[12] R. G. Walker, IEEE J. Quantum Electron., Vol. 27, 1991, pp. 654-667.
[13] J. G. Mendoza-Alvarez, L. A. Coldren, A. Alping, R. H. Yan, T. Hausken, K. Lee, and K. Pedrotti, J. Lightwave Technol., Vol. 6, 1988, pp. 793-808.
[14] GEC-Marconi catalog.
[15] R. J. Deri and E. Kapon, IEEE J. Quantum Electron., 27, 626, 1991.
[16] O. K. Kim, and W. A. Bonner, J. Electron. Mat., Vol. 12, 1983, pp. 827-836.
[17] W. Walukiewicz, J. Lagowski, l. Jastrebski, P. Rava, M. Lichten steiger, and H. C. Gatos, J. Appl. Phys., Vol. 51, 1980, pp. 2659-2668.
[18] F. Fieder and A. Schlachetzki, Solid State Electron., Vol. 30, 1987, pp. 73-83.
[19] A. A. Ballman, J. Cryt. Growth, 62, 198(1983).
[20] A. Carenco, Proc. SPIE 651, Integrated Circuit Engineering III (1986).
[21] M. Erman, Inst. Phys. Conf. Ser. No. 91: Chapter 1, Paper presented at Int. Symp. GaAs and Related Compounds, Heraklion, Greece, 1987.
[22] J. S. Weiner, D. B. Pearson, D. A. B.Miller, and D. S. Chemla, Appl. Phys. Lett., Vol. 49, 1986, pp. 531-533.
[23] U. Koren, B. I. Miller, T. L. Koch, G. Eisenstein, R. S. Tucker, I. Bar-Joseph, and D. S. Chemla, Appl. Phys. Lett., Vol. 51, 1987, pp. 1132-1134.
[24] M. Suzuki, H. Tanaka, and S. Akiba, Modulators, Electron. Lett., Vol. 25, 1989, pp. 88-89.

Chapter 8

Monolithic Integration of Intensity Modulators and Laser Diodes

8.1 Introduction and History

We have seen in Chapter 1 that semiconductor modulators have a large insertion loss, mainly in the form of coupling loss due to their small spot size and low operating voltage. Two approaches to reducing coupling loss have been investigated. One is to integrate the modulators with light sources and the other is to increase the spot size. In this chapter, we consider the monolithic integration of modulators with light sources.

Monolithic integration of optoelectronic devices has been studied for many years. Since the first concept of OEICs (opto-electronic integrated circuits) was proposed by Miller in 1969 [1], a great deal of research on developing OEICs has been done in many laboratories. Though the OEICs concept incorporates advanced technical research and development, and many devices have been fabricated and papers published, practical applications have been realized for only a few devices. This is because the practical advantages that monolithically integrated opto-electronic devices offer over hybrid devices and combinations of discrete device are not yet clear, and integration has not made the system's cost and reliability comparable. One exception, however, is monolithic integration of modulators and laser diodes. This integration many advantages; it offers low chirping, low insertion loss, high output power, and by reducing the number of parts, increased reliability. This should reduce the overall system cost.

Many previous attempts to integrate modulators and laser diodes have been reported. The first report on the integration of optical devices described the monolithic integration of waveguided light-emitting-diodes and Franz-Keldysh intensity modulators [2]. However, this was only a preliminary study and demonstration since the diodes could not lase. There are now three kinds of devices that have been integrated, as shown in Fig. 8.1.

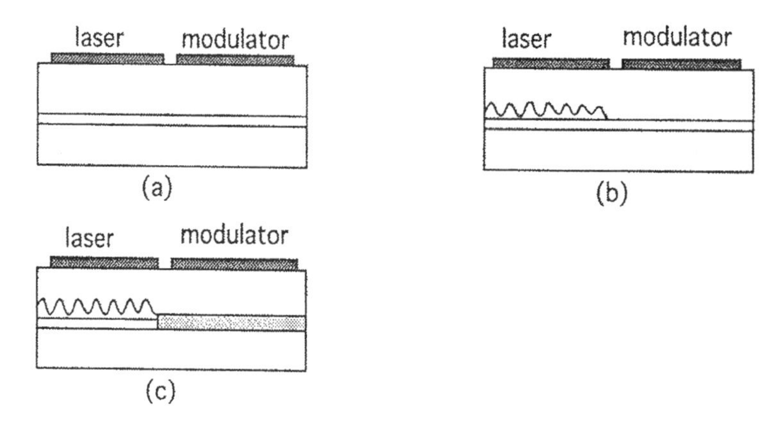

Fig. 8.1. Schematic illustration of monolithic integration of a modulator and a laser diode. (a) type (1), (b) type (2), and (c) type (3). The content is described in the text.

(1) The laser diode consists of a Fabry-Perot cavity and the modulator operates as a saturable absorber or as a Q-switching; that is, the two devices are not optically isolated though they are electrically isolated.

(2) The laser diode consists of a DFB (distributed feedback) laser or DBR (distributed Bragg reflector) diode which includes a grating in the waveguide region, and therefore can lase without mirrors. The laser and the modulator are optically isolated and the modulator operates by carrier injection or in the forward biased condition. The laser and modulator are still electrically isolated, but to a lesser degree since the laser also operates by a similar forward bias. In this case, the active layer of the laser and the modulator is the same and only the injection current intensity differs.

(3) The laser diode consists of a DFB or DBR laser and a modulator. The laser operates in the forward-biased condition while the modulator operates in the reverse-biased condition. The devices are isolated both electrically and optically. The active layer of the laser differ from that of the external modulator because the modulator operates as an absorber for the light emitted from the laser when it operates, resulting in an intensity modulator, while it is transparent for the light when it does not operate. In general, the first two types are used for special applications and the third type has been investigated mainly to provide a device with high operating speed, low driving power, and timely application. Type (1) devices have been used for optical pulse generation [3,4], and recently

have been used as wavelength switch or frequency conversion devices by exploiting the optical bistability associated with a saturable absorber [5,6].

The type (2) device preceded the type (3) device and is now seldom used for high-speed modulation. However, it is very simple and easy to fabricate and can be used for low-speed modulation. Modified structures, such as cleaved-coupled-cavity, (C^3) lasers [7], are also occasionally studied due to their specific properties such as optical bistability, optical amplification, and longitudinal-mode selectivity. Such a cavity has been used as an intracavity modulator. The first demonstration of multiple-quantum-well (MQW) structures was based on this method, where the modulator was operated in the reverse-biased condition [8]. Therefore, an overview of this devices is included in this chapter.

The type (3) device was first demonstrated by the monolithic integration of a bulk type-DFB laser and multiple-quantum-well (MQW) intensity modulator [9,10]. Since then, integrated light sources using the Franz-Keldysh effect have been reported [11-15] with the emphasis on their ease of fabrication.

This chapter is organized as follows. In Section 8.2 we briefly discuss the forward-biased modulators monolithically integrated with light sources. Semiconductor optical amplifiers monolithically integrated with modulators are also discussed. Section 8.3 deals with reverse-biased modulators monolithically integrated with DFB laser diodes. Section 8.4 describes the fabrication details of the integrated light sources. Finally, in Section 8.5 we discuss recent research on the integrated light sources.

8.2 Integration with Forward Biased Modulators

The development of DFB and DBR lasers made it possible to integrate a laser diode and an external modulator. In conventional Fabry-Perot semiconductor lasers, optical feedback is provided by facet reflections which needs an optical cavity for lasing and other devices cannot be optically isolated from the laser. On the other hand, DFB or DBR laser diodes do not need a cavity. They operate by feedback from internal gratings distributed throughout the cavity length, rather than localized at the cavity facets. This makes it possible to optically isolate other devices from the laser.

The early research used the laser diode as a modulator by dividing the laser diode contact into two areas; one to act as a laser and the other

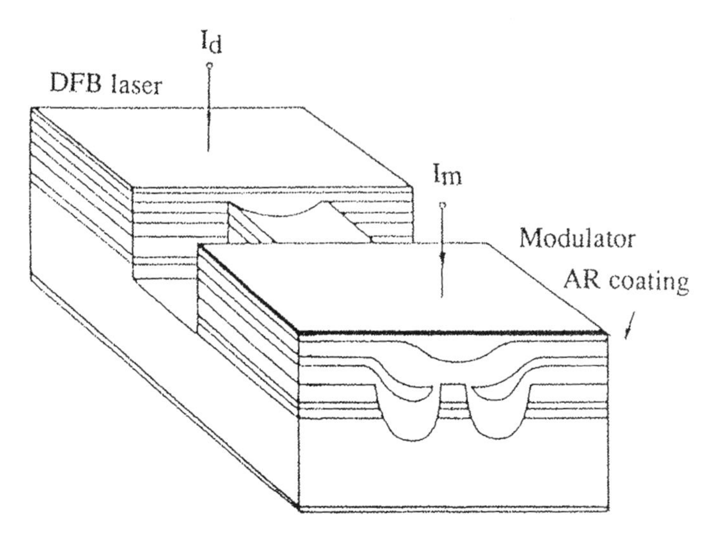

Fig. 8.2. Schematic illustration of forward-biased modulator/laser diode integration [16].

as a modulator as shown in Fig.8.2 [16]. As described in Chapter 5, the injected current or injected carriers change the refractive index whose magnitude was large in the vicinity of the energy gap, while the emitting laser wavelength is close to the energy gap. As a result, the current injection to the modulator changes the transmission loss causing intensity modulation of a laser operating at a constant current and wavelength. The response speed is limited by the carrier life time of a few ns, but the structure is so simple and easy to fabricate that it can be used as a simple intensity modulator. A loss modulator was monolithically integrated with 1.3- and 1.55-μm DFB double-channel planar buried heterostructure (DCPBH) laser diodes. Pulse modulation of 450 Mbit/s non-return-to zero (NRZ) was achieved at 1.3 μm

Figure 8.3 shows DFB lasers integrated with carrier injection optical switches [17]. The optical integrated circuit has four DFB LDs connected to each port of a 2x2 optical switch on InP substrate. The optical switch has two waveguides crossing at a 7 degree angle at the center of the device, where there is a 170-μm-long switching region. This device has a variety of applications, such as integrated modulators, switches, and optical amplifiers.

In the carrier injection switch, integration of optical amplifiers is usually used to reduce insertion loss in optical switches [18].

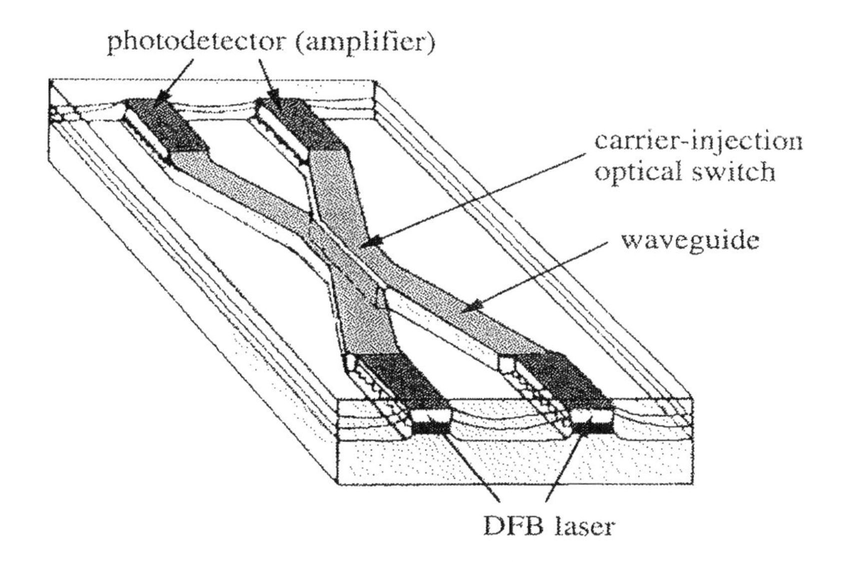

Fig. 8.3. Schematic illustration of integrated 2x2 switching semiconductor diode lasers. The device consists of an amplifier section, switching section, and a passive waveguide section [17].

8.3 Integration of Reverse Biased Modulators

8.3.1 Franz-Keldysh Modulators

The first demonstration of a light source monolithically integrated with a DFB laser and an external modulator was done in 1986 using a MQW modulator as described in Section 8. 1. Later reports, however, described the use of Franz-Keldysh intensity modulators. The Franz-Keldysh effect has been studied for a long time and the details are well known, making fabrication easier. The first demonstration of an integrated light source was reported in 1987 [9]. Since then, many laboratories in Japan have worked to develop this technology [11-15]. Recently, system experiments using integrated light sources, such as 5-Gbit/s-100-km [19], 10-Gbit/s-65 km [20], and 2.4-Gbit/s-135-km [21] optical fiber transmission experiments, have been reported. Recently laboratories in other countries have begun to investigate their use in multi-gigabit long haul transmission systems [22,23].

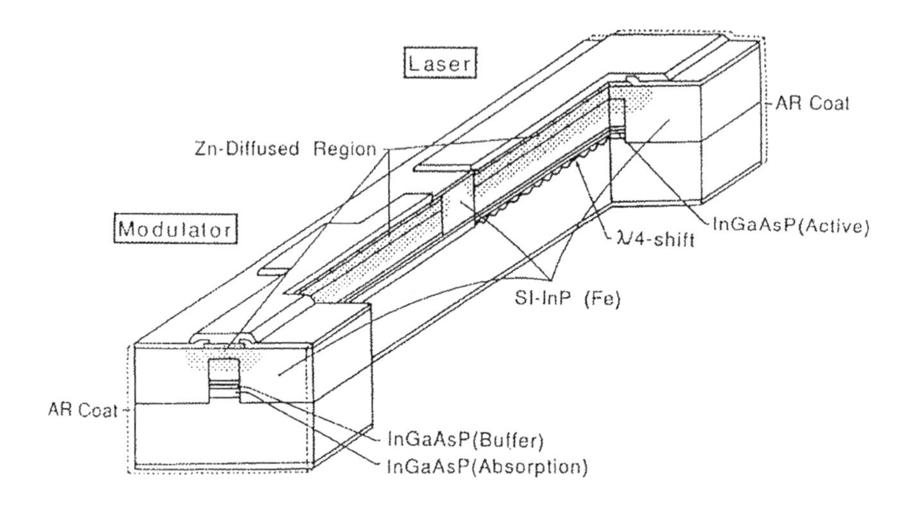

Fig. 8.4. Schematic cross-sectional view of a λ/4-shifted DFB laser/electroabsorption modulator integrated light source[21].

Figure 8.4 shows the schematic cross-sectional view of the integrated light source [21]. It consists of an asymmetric λ/4-shifted DFB laser and an InGaAsP electroabsorption (EA) modulator (which utilizes the Franz-Keldysh effect), and a semi-insulating (Fe-doped) InP buried heterostructure. A InGaAsP buffer layer (bandgap-wavelength λ_g=1.25 μm) was introduced between the InP upper cladding layer and the InGaAsP modulator waveguide layer (λ_g=1.44 μm) in order to reduce the hole pile-up effect at high optical power levels [24]. The wafer was made by three-step epitaxial growth (LPE for the laser part, and two-step hydride VPE for the modulator part and the Fe-doped InP burying layer) and selective Zn diffusion. An anti-reflection coating was deposited on both facets of the device. By optimizing the composition and thickness of the modulator waveguide, the driving voltage for a 10-dB extinction ratio was reduced to 1.4 - 3V, depending on the modulator length which ranged from 240-125 μm.

Figure 8.5 shows another integrated light source [12]. The emitted power from the modulator facet is 17 mW and 3-dB bandwidth is over 10.3 GHz. Limited wavelength broadening (chirping) within 0.01-nm under 10-Gbit/s NRZ modulation was also demonstrated. As an accelerated life test, under high-ambient temperature and high-output power operation, the integrated light source showed stable operation over

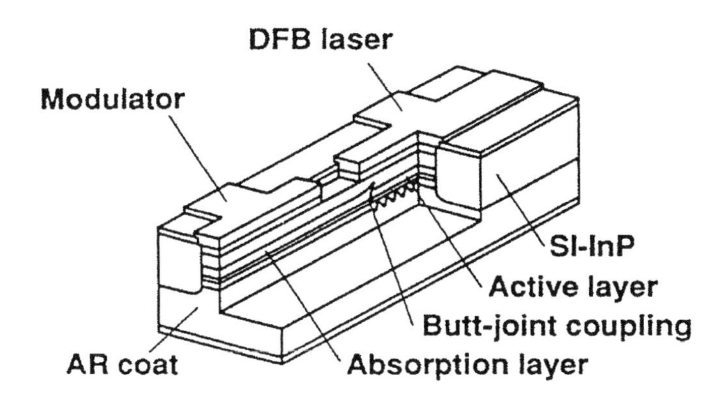

Fig. 8.5. Schematic drawing of a DFB laser/electroabsorption modulator integrated light source [12].

3000 hours without any degradation [25]. At present, this integrated light source is suitable for systems with relatively low modulation speed of about 2.4 Gbit/s.

8.3.2 Monolithic Integration with MQW Modulators

The device voltage of EA modulators utilizing the Franz-Keldysh effect in a InGaAsP/InP double-heterostructure (BH) monolithically integrated with DFB lasers was still too high for high-speed IC drivers and it was necessary to incur increased insertion loss to reduce drive voltage because of the small detuning energy.

Optical modulators made with multiple-quantum-well (MQW) structures can produce excellent performance because of their large electro-absorption effect (quantum-confined Stark effect; QCSE) as described in Chapter 4.

Moreover, recent development of DFB lasers in terms of lower threshold current densities, improved high frequency response, and narrower linewidth has been accomplished by introducing MQW structures in place of the bulk in the active layer [26]. Here we describe a low drive-voltage and high-speed monolithic light source with an MQW structure for both the DFB laser and the external modulator.

The schematic diagram of the device, fabricated by the MOVPE(metal-organic vapor-phase-epitaxy) /MBE (molecular beam epitaxy) hybrid growth technique is shown in Fig. 8.6 [26,27]. The

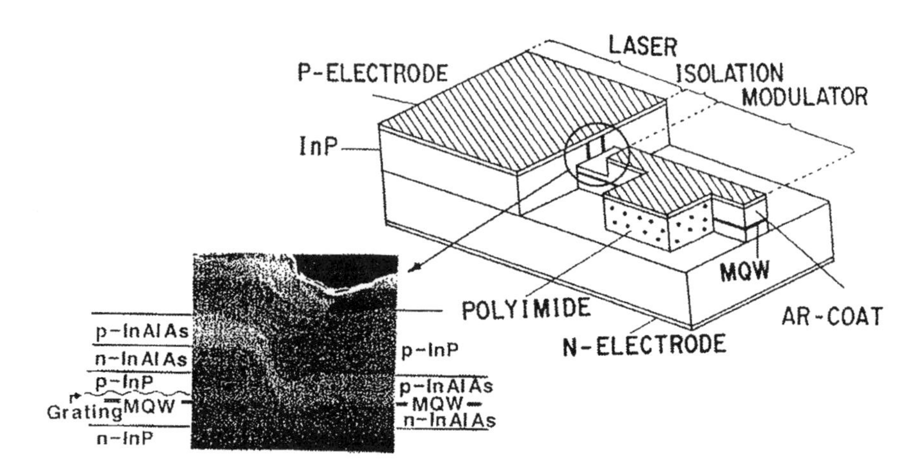

Fig. 8.6. Schematic drawing of an MQW DFB laser/MQW modulator integrated light source.

improvements due to the inclusion of the MQW structure are as follows. First, DFB lasers were fabricated with MOVPE growth and their active layers consisted of six periods of nondoped InGaAs quantum wells and InGaAsP barriers whose photoluminescense peak wavelength was 1.3 µm. This results in low threshold current and high quantum efficiency. Second, external modulators, grown by MBE, consisted of 0.13-µm-thick nondoped MQW layers whose configuraion included of 12 periods of a 7.5-nm InGaAs quantum well and a 5.0-nm InAlAs barrier. This allowed a low-drive-voltage operation for a high on/off ratio with short optical interaction length and a low insertion loss due to efficient mode matching. Third, a dry etching technique has been used to create devices with a high-mesa type geometry, which gives us precise control of the devise size and suppresses the lateral side etching associated with the wet etching[9].

A 300-µm-long laser, with a 100-µm-long modulator and a 50-µm-long isolation region was fabricated using a three-step growth procedure as shown in Fig. 8.7. During the first stage, the laser diode section was grown by MO-VPE and a grating was fabricated on the guide layer by photolithography. Then the epi-wafer was selectively etched down to the InP substrate by ECR (electron-cyclotron-resonance) dry etching and the modulator section was selectively grown by MBE using an SiO_2 mask. After the selective etching of an MBE layer on the laser section,

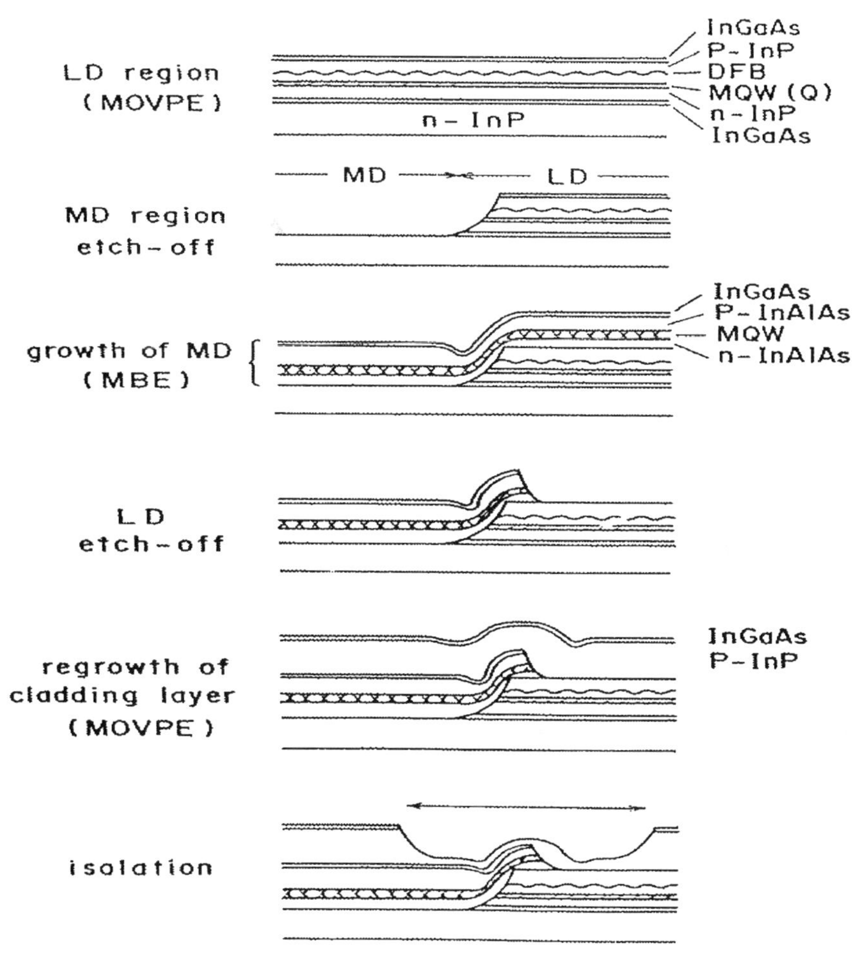

Fig. 8.7. Various steps in the processing sequence.

which was poly-crystal and easily etched off, the cladding layer and capping layer were grown by MOVPE. No space or hole were observed between the laser and the modulator and optical-coupling-efficiency of over 85 % was obtained. The epi-wafer was dry-etched to produce a high-mesa geometry with 3-4 μm wide. The degree of electrical isolation between the laser and the modulator was large enough to keep the resistance between the laser and the modulator above 500 kΩ. To minimize the stray capacitance, polyimide was spin-coated under the bonding pad over a 30-μm-square area.

The MQW modulator was mounted on a microwave stripe-line. The

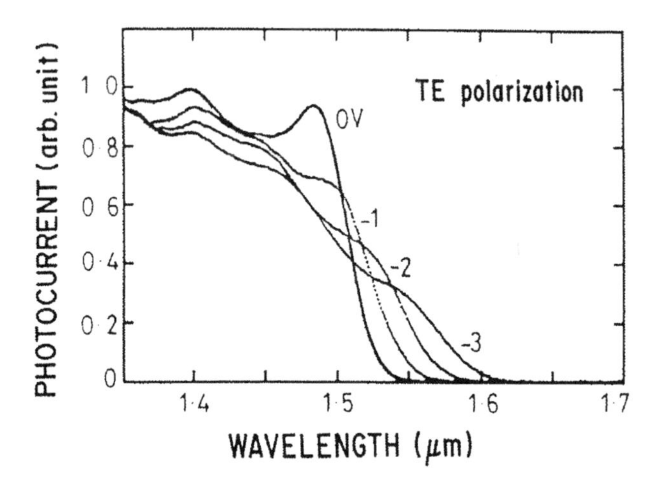

Fig. 8.8. Absorbed photo-current spectra for TE-polarized incident light in integrated MQW modulator.

bonding wire was shortened to reduce inductance, which was subsequently confirmed to be less than 0.2 nH. This value was estimated from the input-impedance of MQW modulators, which is measured by a 40-GHz-band vector-network analyzer, and from the device capacitance. The total capacitance was estimated to be 0.4 pF at a reverse bias of 2 V under a measuring frequency of 10 MHz.

The absorbed photocurrent spectra for both the MQW modulator and the MQW laser are shown in Fig. 8.8. Clear exciton resonances are observed and shift to longer wavelengths as the reverse biases increase [25]. With this monolithic device, CW operation was obtained from DFB lasers whose threshold currents were between 30-40 mA, and single longitudinal operation was observed with side mode suppression above 40 dB. The on/off ratio of the MQW modulator for the MQW-DFB laser light was investigated with a single-mode-fiber. The modulator facet was coated with a SiN film to reduce reflection loss. An on/off ratio of 20 dB was achieved with an applied voltage of -4 V. The output from the MQW modulator at zero bias voltage was 6 mW at the injection current of 120 mA for the DFB laser and decreased as the reverse bias increased to 1 mW. The coupling efficiency was estimated from the comparing the light output from the optical fiber and direct output from the modulator, and was found to be over 84 %.

The frequency characteristics were measured experimentally by applying an RF modulation signal and using an HP 8703A lightwave component analyzer. The modulator was driven by a 0 dBm RF signal at a dc bias of -1.5 V. A 50-Ω resistance was loaded parallel to the modulator. The measured frequency response is shown in Fig. 8.9. The response was fairly flat and the relaxation oscillation resonance of the laser was sufficiently suppressed due to the nearly complete optical and electrical isolation between the laser and modulator. The 3-dB bandwidth is over 16 GHz and no degradation due to high input power has observed.

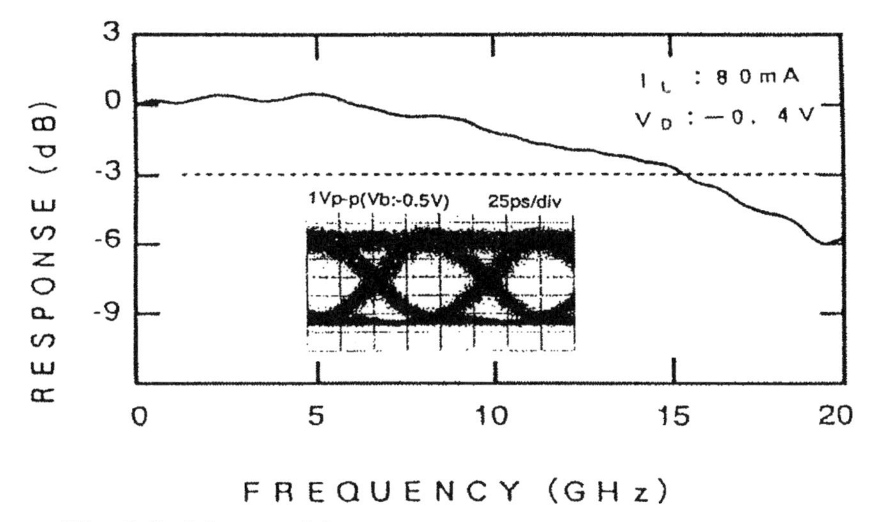

Fig. 8.9. Measured frequency response and eye-opening at an output power of 6 mW and a bias of -3 V.

The eye opening for this integrated light source is shown in the inset of Fig. 8.9. The modulator is driven by a 10 Gbit/s 2^{15}-1 pseudo-random NRZ signal with an amplitute of 1 $V_{p\text{-}p}$ and an extinction ratio of 11 dB. A clear eye opening has been observed.

8.3.3 Recent Monolithic Integration

The previous hybrid method of monolithically integrating an MQW modulator and a DFB laser is too complicated to permit easy fabrication. However, easy and simple methods using a large optical cavity (LOC) structure [28] and/or a simultaneous selective area growth method [14,15] have been reported recently. The fabrication procedures for the latter method are described in detail in the Chapter 11. In this Section, device

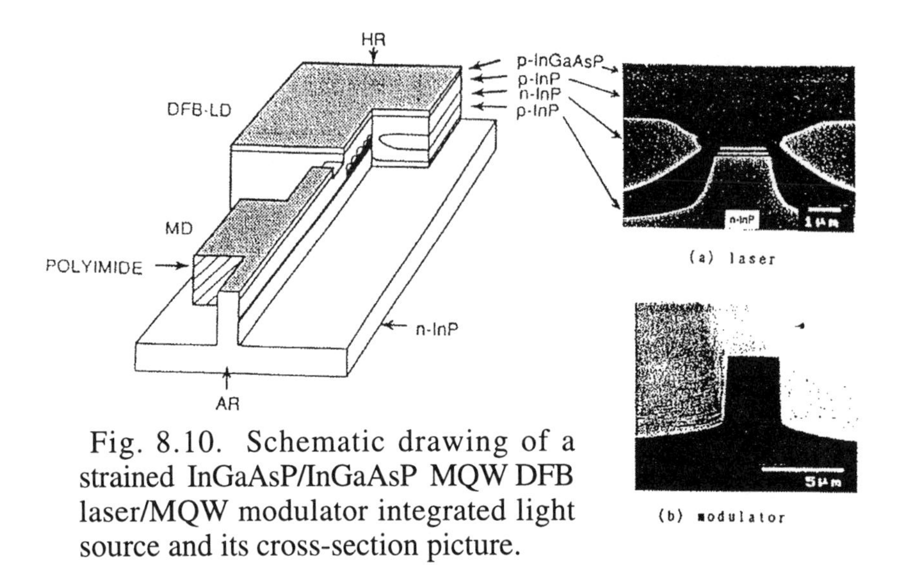

Fig. 8.10. Schematic drawing of a strained InGaAsP/InGaAsP MQW DFB laser/MQW modulator integrated light source and its cross-section picture.

characteristics are described.

Figure 8.10 shows the schematic diagram of the LOC integrated light source [28]. This structure consists of two MQW core layers: the upper is a laser active layer and lower is a modulator active layer. After the grating on the waveguide layer is fabricated, the upper core to the etching stop layer is etched off selectively by wet chemical etching and the cladding layer is regrown. The crystal quality and optical coupling efficiency are improved by the regrowth on the flat layer and two core structure, compared with those of the butt-joint-coupling method described in the previous section. The core layers consist of four and eight periods of InGaAsP quantum wells and InGaAsP barriers for a laser and a modulator, respectively. Compressive strain has been introduced in the quantum wells ($\varepsilon = 0.8$ %) in order to improve laser and modulator characteristics.

Figure 8.11 shows the current versus output power from the facet of the modulator. The threshold current is 15-20 mA and the output power is 5-10 mW at the current of 100 mA. The optical power absorbed in the front facet of the modulator is estimated to be over 15 mW from the absorbed photocurrent magnitute and no degradation of on/off ratio is observed. This is due to the small valence-band energy off-set due to the compressive strain introduced into the quantum wells which reduces hole pile-up and on/off ratio saturation when the large optical power is irradiaded.

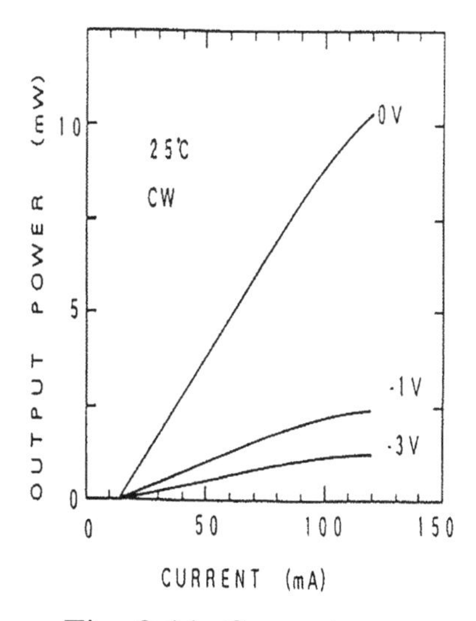

Fig. 8.11. Current versus output power from the modulator facet.

Fig. 8.12. On/off ratio as a function of applied voltage for monolithic device.

The on/off ratio, monitored with a single-mode fiber, is shown in Fig. 8.12. An on/off ratio over 30 dB has been achieved with a reverse-bias of 2.5 V and a 3-dB bandwidth of over 16 GHz. In this figure, the on/off ratio dependence on the detuning energy between lasing wavelength and modulator absorption edge is also shown.

As shown in this Fig. 8.12 and discussed in detail in Chapter 4, the driving voltage decreases as the detuning energy decreases, while the transmission loss increases. The magnitude of absorbed photocurrent is also dependent on the detuning energy and it affects the absorption saturation and the frequency response characteristics associated with the hole-pile up in the absorbed region. In this sample, the energy bandgap offset in the valence band is relatively small and such deterioration as hole-pile up has not been observed up to the emitting light power of 10 mW.

Recently, simultaneous selective area growth, which are discussed in detail in Chapter 11, has been reported [14,45,23,29-31]. With this method, both the laser section and the modulator section can be grown simultaneously by using MOVPE, based on the difference in growth speed between the dielectric mask widths. Optical coupling efficiency is almost 100 % and the fabrication process is easier than butt-joint

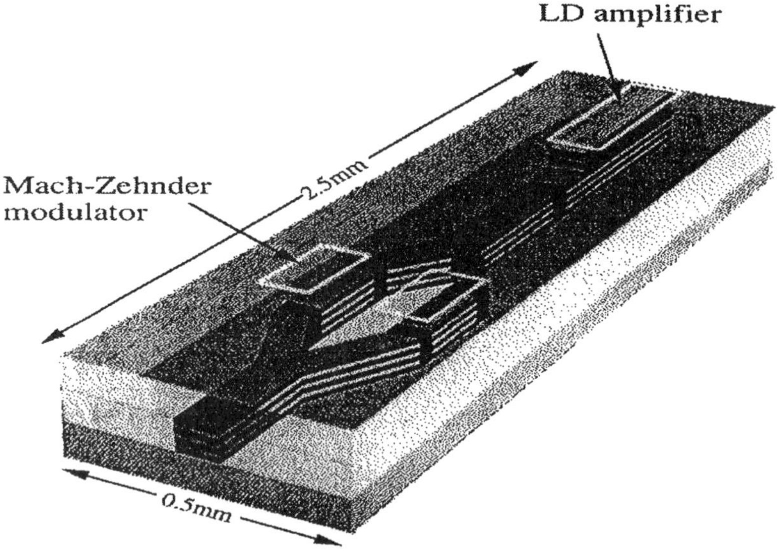

Fig. 8.13. Schematic of integrated amplifier/Mach-Zehnder modulator showing the two quantum-well gain and modulaor waveguide sections buried by semi-insulating the p-type InP regrowth [33].

processing, which has been studied for a long time. The reported device characteristics have been fair and 2.5-Gbit/s, 600-km transmission has been achieved [31].

Moreover, butt-joint coupling between an MQW laser active layer and an MQW modulator wavegude has been reported and 10- Gb/s, 120-km transmission experiment has been achieved [32]. The butt-joint coupling has previously been considered to be too difficult for applying MQW structures due to the excessive growth at the interface. That is, thick quantum wells will produce a small energy band gap causing increasimg transmission loss. However, by using MOVPE techniques, high-coupling efficiency and high-emitting power from the facet of the modulator have been obtained.

So far we have discussed only monolithic integration of electroabsorption modulators and DFB lasers. A p-i-n quantum well structure, integrated with Mach-Zehnder (MZ) modulators with a semiconductor amplifier or DBR laser has progressed much further [33-35]. Figure 8.13 shows a schematic of recently demonstrated M-Z modulator/semiconductor laser amplifier with total length under 2.5 mm [33]. The active electro-optic material in the modulator is InGaAsP/InP quantum wells with a 1.4 μm band gap which provides optical phase modulation at an operating wavelength near 1.55 μm. The amplifier section also makes use of quantum wells as the active gain medium: strained InGaAs wells with InGaAsP barriers provide a gain peak near 1.55 μm. As shown in Fig. 8.13, the heterostructure ribs in both sections are buried by InP regrowth. Here, complete process compatibility between the modulator and semiconductor laser amplifier is assumed as both the side regrowth of semi-insulating InP and the top growth of p-type InP are common to the laser and the modulator. Electrical isolation between the forward-biased amplifier and the reverse-biased modulator sections is accomplished by means of H^+ ion implantation [36]. Zero-insertion loss (overcoming all on-chip and coupling losses) is obtained at 100 mA with the 1-mm-long amplifier. A voltage-length product of 2Vmm is obtained from the modulator, which provides a 14-dB modulation depth for $V\pi$ = 3V at 1.55 μm.

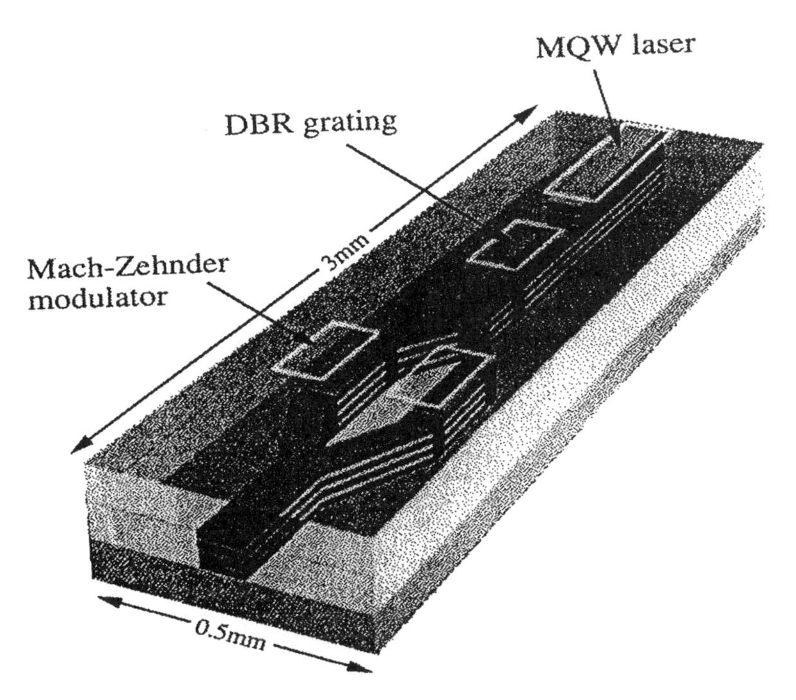

Fig. 8.14. Schematic of integrated DBR laser/Mach-Zehnder modulator [35].

At AT&T Bell Laboratories, Zucker et al. recently fabricated the first monolithically integrated laser/interferometric modulator [35] shown in Fig. 8.14. In this chip, separate sets of quantum wells are individually tailored to provide gain and optical phase modulation at 1.55 μm. The M-Z modulator has the advantages of being free from the power saturation and heating inherent in electroabsorption modulators and the chirping can be controlled. Lower driving voltage and faster operation are the next targets.

Very recently fabrication of an InP/InGaAsP based integrated DFB laser/M-Z phase modulator are also reported by other two groups [37,38] and 10 Gbit/s transmission at 1.55 μm over 100 km of non-dispersion shifted fiber has been achieved with a "push-pull" drive of 2V peak to peak per side [38].

Another recent trend of monolithic integration with modulators is to introduce a semiconductor optical amplifier for wavelength convertion. Penalty-free up- and down- conversion at 10 Gbit/s has been achieved in a wavelength range spanning the entire EDFA-window [39]. Previously, a high performance wavelength converter was realized by placing discrete semiconductor optical amplifiers in a Michelson interferometer assembled using a 3-dB fiber coupler. Complete optical wavelength convertion at 5 Gbit/s with monolithical integration of semiconductor optical amplifiers in a passive asymmetric Mach-Zehnder interferometer has also been reported [40]. Very stable, penalty-free wavelength up-conversion is obtained on InP/GaInAsP in codirectional propagation.

8.4 References

[1] S. E. Miller: Bell Syst. Tech. J., 48, 2059, 1969.
[2] J. C. Dyment, F. P. Kapron, and A. J. SpringThorpe, Inst. Phys. Conf. Ser., No. 24, 200,1975.
[3] D. Z. Tsang and J. N. Walpole, IEEE J. Quantum Electron., QE-19, 145, 1983.
[4] Y. Arakawa, A. Larrson, J. Paslaski, and A. Yariv, Appl. Phys. Lett., 48, 561, 1986.
[5] N. A. Olsson and W. T. Tsang, IEEE J. Quantum Electron., QE-20, 332, 1984.
[6] H. Kawaguchi, K. Oe, H. Yasaka, K. Magari, M. Fukuda, and Y. Itaya, Electron. Lett., 23, 1088, 1987.

[7] W. T. Tsang, Chap. 5 in *Semiconductors and Semimetals*, Vol. 22, Part B, ed. W. T. Tsang, New York, Academic Press, 1985.
[8] S. Tarucha and H. Okamoto, Appl.Phys.Lett., 48, 1, 1986.
[9] Y. Kawamura, K. Wakita, Y. Itaya, Y. Yoshikuni, and H. Asahi, Electron. Lett., 22, 242, 1986.
[10] Y. Kawamura, K. Wakita, Y. Yoshikuni, Y. Itaya, and H. Asahi, IEEE J. Quantum Electron. Lett., QE-23, 915, 1987.
[11] M. Suzuki, Y. Noda, H. Tanaka, S. Akiba, Y. Kushiro, and H. Isshiki, IEEE Lightwave Technol., LT-5, 1277, 1987.
[12] H. Soda, K. Nakai, and H. Ishikawa: ECOC'88, Technical Digest, 1988, Bringhton, pp.227-230.
[13] N. Henmi, S. Fujita, T. Saito, M. Yamaguchi, M. Shikata, and J. Namiki, OFC'90, PD8-1.
[14] T. Kato, T. Sasaki, N. Kida, K. Komatsu, and I. Mito, in 17th European Conference on Optical Communication, ECOC'91, Paper No. WeB7-1.
[15] M. Aoki, M. Takahashi, M. Suzuki, H. Sano, K. Uomi, T. Kawano, and T. Takai, Photon. Technol. Lett., 4, 580, 1992.
[16] M. Yamaguchi, K. Emura, M. Kitamura, I. Mito, and K. Kobayashi, Electron. Lett., 23, 190, 1987.
[17] S. Sakano, H. Inoue, H. Nakamura, T. Katsuyama, and H. Matsumura, Electron. Lett., 22, 594, 1986.
[18] T. Watanabe, Y. Saito, K. Sato, and H. Soda, OFC'93, WH9, pp.119-120, 1993.
[19] T. Saito, N. Hemmi, S. Fujita, 0EC'90, Tech. Dig., 13A2-4, 1990.
[20] T. Okiyama, I. Yokota, H. Nishimoto, K. Hironishi, T. Horimatsu, T. Touge, and H. Soda, ECOC'89, MoA1-3, 1989.
[21] M. Suzuki, H. Tanaka, H. Taga, S. Yamamoto, and Y. Matsushima, J. Lightwave Technol., 10, 90, 1992.
[22] P. Ojala, C. Pettersson, B. Stoltz, A. -C. Morner, and M. Janson, ECOC'93, WeC7.4.
[23] M. Gibbon, G. H. B. Thompson, D. Boyle, J. P. Stagg, B. Patel, D. J. Moule, S. Wheeler, E. J. Thrush, K. Warbrich, and P. Gurton, ibid, WeC7.5.
[24] M. Suzuki, H. Tanaka, and S. Akiba, Electron. Lett., 25, 88, 1989.
[25] Y. Arakawa, and A. Yariv, IEEE J. Quantum Electron., QE-22, 774, 1986.
[26] K. Wakita, I. Kotaka, H. Asai, M. Okamoto, and Y. Kondo, Photon. Technol. Lett., 4, 16, 1992.

[27] I. Kotaka, K. Wakita, M. Okamoto, H. Asai, and Y. Kondo, Photon. Technol. Lett., 5, 61, 1993.
[28] K. Sato, I. Kotaka, K. Wakita, Y. Kondo, and M. Yamamoto, Electron. Lett., 29, 1087, 1993.
[29] E. Zielinski, D. Bauns, H. Haisch, M. Klenk, E. Satzke, and M. Schilling, in Proc. InP and Related Compounds, 1994, ME-1, P. 68.
[30] J. E. Johnson, T. Tanbun-Ek, Y. K. Chen, D A. Fishman, P. A. Morton, S. N. G. Chu, A. Tate, A. M. Sergent, P. F. Sciortino, and K. W. Wecht, IEEE Semiconductor Conf., 1994, M. 47.
[31] K. C. Reichmann, P. D. Magill, U. Koren, B. I. Miller, M. Young, M. Newkirk, and M. D. Chien, Photon. Technol. Lett., 5, 1093, 1993.
[32] K. Sato, Y. Kotaki, K. Morito, T. Takeuchi, and H. Soda, Proc. 1994 IEICE spring Conf., C-230, P. 4-227 (in Japanese).
[33] J. E. Zucker, K. L. Jones, B. I. Miller, M. G. Young, U. Koren, J. D. Evankow, and C. A. Burrus, Appl. Phys. Lett., 69, 277, 1992.
[34] J. E. Zucker, K. L. Jones, B. I. Miller, M. G. Young, U. Koren, B. Tell, and K. B. -Goefeler, J. Lightwave Technol., 10, 924, 1992.
[35] J. E. Zucker, K. L. Jones, M. A. Newkirk, R. P. Gnall, B. I. Miller, M. G. Young, U. Koren, C. A. Burrus, and B. Tell, Electron. Lett., 28, 1888, 1992.
[36] J. E. Zucker, K. L. Jones, B. Tell, K. B. -Goefeler, C. H. Joyner, B. I. Miller, and M. G. Young: Electron. Lett., 28, 853, 1992.
[37] T. Tanbun-Ek, P.F. Sciortino, Jr., A. M. Sergent, K. W. Wecht, P. Wisk, Y. K. Chen, C. G. Bethea, and S. K. Sputz, Proc. of the 7th Intern. Conf. on InP and Related Materials, p. 713, Sapporo, Japan (1995).
[38] N. Putz, D. M. Adams, C. Rolland, R. Moore, and R. Mallard, Proc. of the Integrated Photonics Research Conf.,Tup-C1, P. 152, 1996.
[39] B. Mikkelsen, T. Durhuus, C. Joergensen, and K. E. Stubkjaer, ECOC'94, PD, P. 67
[40] N. Vodjdani, F. Ratovelomanana, A. Enard, G. Glastre, D. Rondi, R. Blondeau, T. Durthuus, C. Joergensen, B. Mikkelsen, K. E. Stubjaer, P. Pagnod, P. Baets, G. Dobbelaere, ECOC'94, PD, P. 95.

Chapter 9

Surface Normal Switch

9.1 Introduction

There is considerable interest in surface normal modulators for high-speed, wide-band, inductionless, and parallel processing optical interconnection technology. The high speed and high density in VLSI technology make clear problems such as large wiring processing, large time lag for signal transmission, the short bandwidth of transmission lines for computer systems as well as the integrated circuits. As mentioned in the Chapter 6 on waveguide type modulators, surface-normal modulators are most useful in applications where a fairly high-density of outputs are desired. Since it is easy to fabricate two-dimensional arrays, surface-normal modulators can be formed as spatial light modulators, which can be used for many interesting switching and signal processing applications [1,2]. In most surface-normal (tranverse) modulators, there is a major difficulty in obtaining a high on/off ratio, or contrast. For the same reasons, it is also difficult to obtain a high modulation efficiency, since the interaction length with an optical beam is limited to be a few micrometers.

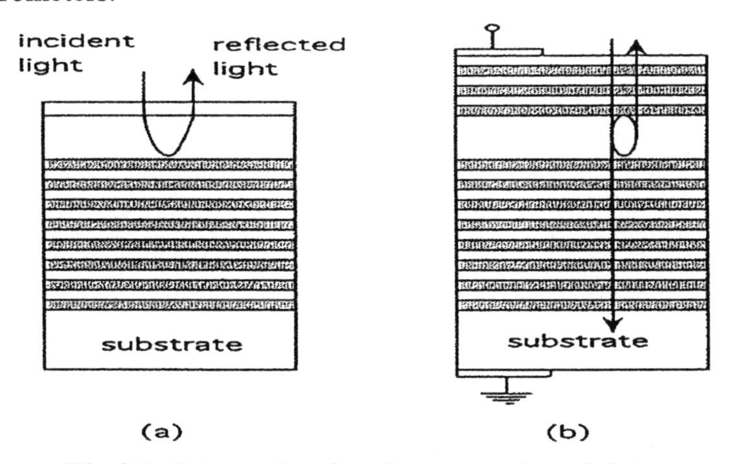

Fig.9.1. Schematic of surface-normal modulators.
(a) nonresonant type, and (b) resonant type.

There have been several attempts to address this interaction length problem by using reflectors at one or both ends of the modulator's active region[3-9](Fig. 9.1). Figure 9.1 (a) shows a single backside reflector which consists of a $\lambda/4$ thick AlGaAs/GaAs DBR (distributed Bragg reflector). The thickness of the optical absorption layer is thus effectively doubled. The contrast ratio for this single backside modulator [3] is not high and the modulation efficiency [%/V] also cannot be increased much.

In the case of two reflectors, as shown in Fig.9.1 (b), the effective interaction length is multiplied by the finesse, F, of the Fabry-Perot cavity that is formed, where $F = \pi R/(1-R)$, and R is the mean mirror reflectivity. This configuration, which consists of asymmetric multi-reflection layers, can change the reflectivity and enhance the modulation efficiency, even under a low applied voltage. The cavity resonance characteristic is the sharpest and the reflectivity is zero when both DBR reflectivities are the same and no optical absorption exists between the upper and the lower DBR. However, when an optical absorption layer is inserted between them, the reflectivity of the lower DBR must be less than that of the upper.

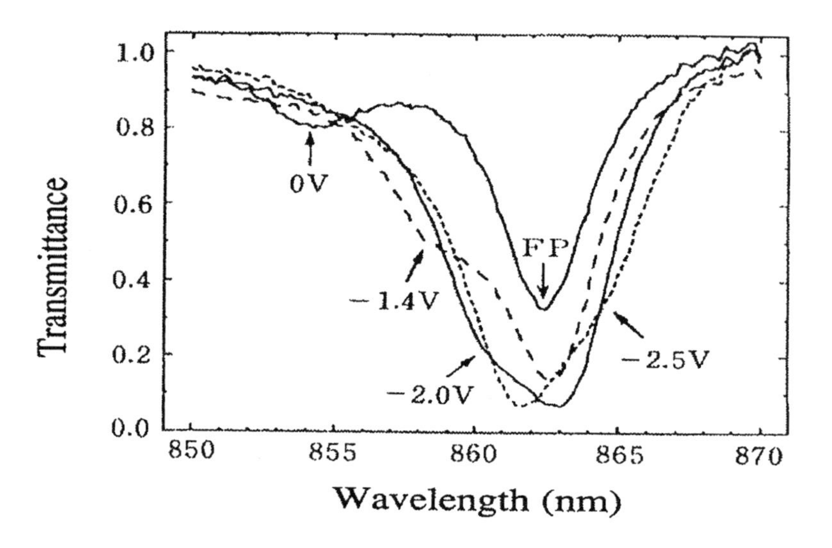

Fig. 9.2. Wavelength dependence of reflection for asymmetric Fabry-Perot cavity type modulator [9].

Figure 9.2 shows the reflectivity change resulting from such an asymmetric Fabry-Perot cavity[9]. The modulation efficiency of 20%/V and the contrast ratio of 15 have been achieved. However, the design tolerance of the applied voltage and layer thickness is small, because it uses the cavity resonance. Therefore, a key technology is high quality crystal growth with precise layer thickness control. For example, the tolerances for the incident light wavelength, applied voltage, and layer thickness are 2nm, 0.5V, 0.3%.

While high speed operation of surface-normal modulators had not been reported previously, recently asymmetric Fabry-Perot modulators with a millimeter-wave (37 GHz) frequency response have been reported [10]. Figure 9.3 shows the schematic structure of this modulator. The absorbed layer consist of undoped 8.0~10.0-nm GaAs quantum wells with 4.5-nm $Al_{0.3}Ga_{0.7}As$ barriers. The front quarter-wave stack is two periods of $AlAs/Al_{0.2}Ga_{0.8}As$ (R_f=52%), while the back mirror is 25 periods (R_b=99%).

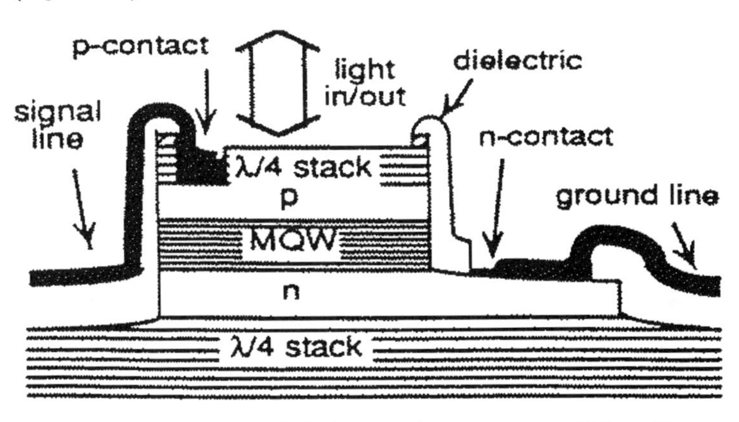

Fig. 9.3. Structure of high-speed asymmetric Fabry-Perot modulaor. Reflectivity is 52% for the front and 99% for the back, respectively [10].

The these modulators are of 16x20-μm^2 and investigated at various optical power levels (approximately 10 μW, 50 μW, 85 μW, 110 μW, and 3 mW equivalent cw power incident on the device). The modulation speed of this modulator is limited by the RC-time constant at low optical input, while at higher optical intensities it is limited by the space-charge effect in the MQW material.

Figure 9. 4 compares the reflectivity change with modulation voltage for recently developed surface normal modulators [11]. The refractive index type is better than the electroabsorption type for large-scale monolithic integration, since the latter produces little temperature rise. However, it must be operated at a longer wavelength than is used in the absorption type.

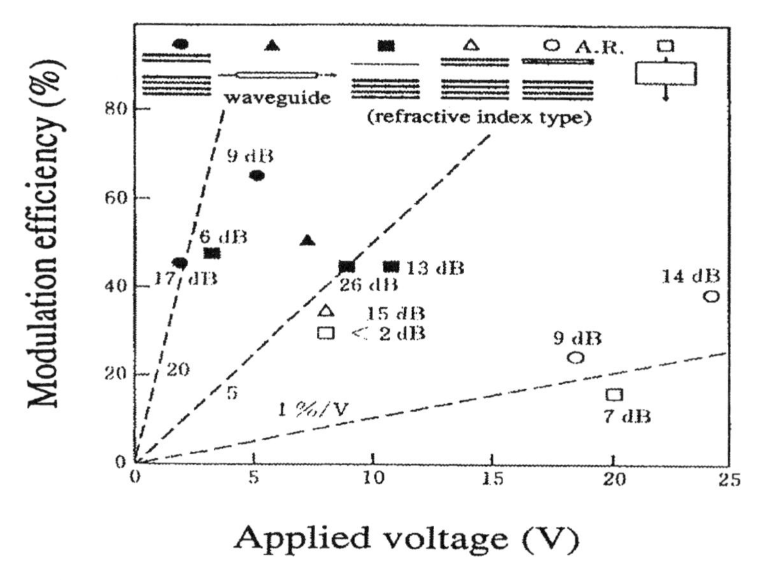

Fig. 9.4. Comparison of modulation efficiency for various surface-normal modulators [11].

9.2 SEED

(1) Operation Principles.

This device, developed by Miller at AT&T Laboratories[12], consists of a p-i-n structure and a series resistance. The i-layer consists of an MQW structure operating on a constant voltage source. SEED is an abbreviation of self-electrooptic effect device. In the simplest form of SEED, the QW diode is itself the only photodetector. The diodes used for QCSE modulators make good photodetectors[13-15], with internal quantum efficiencies near unity.

A circuit that can show optical bistability is shown in Fig. 9.5. The diode is illuminated at a wavelength near the zero field exciton peak (e.g., near 1.46 eV in Fig. 9.5) to show optical bistability. When there is no light shining on the diode, all the bias voltage is across the QWs, and the absorption is relatively low. As the light power is increased,

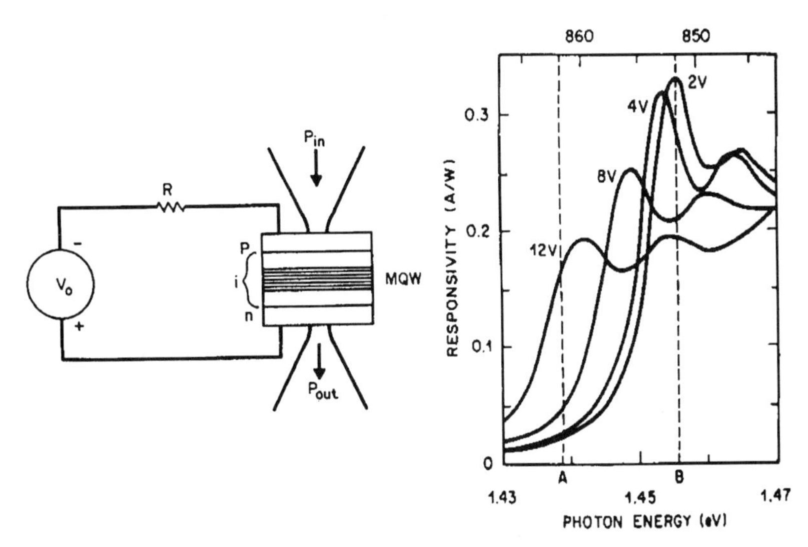

Fig.9.5. Schematic of SEED and its circuit configuration [13].

photocurrent is generated, and, consequently, the voltage across the resistor drops, reducing the voltage across the QWs and, hence, increasing the absorption. This increase in absorption results in yet more photocurrent, giving more voltage across the resistor, less voltage on the QWs, more absorption, and so on. This positive optoelectronic feedback mechanism can become strong enough to cause switching into a high absorption state.

Figure 9.6 illustrates this operation principle. The photosensitivity S for a p-i-n diode is a function of the applied bias V of the diode, $S = S(V)$. The incident light wavelength is near the exciton peak. The sensitivity S, shown by the solid line in Fig. 9.6 (a), changes by the applied voltage, based on QCSE. When the electric source voltage is V_0 and the load resistance is R, $S = (V_0-V)RP$, where P is the incident light power. The load line in Fig. 9.6 successively changes from point A to B, C, and D, as the incident light power increases from zero. Three cross points are given by the above sensitivity line between point B and C. Both side points are stable and optical bistability is induced.

The power at which switching takes place is determined by the resistor, with larger resistors giving lower power. The switching time is approximately the resistance-capacitance time constant of the load resistor and the device capacitance. The power-speed product is constant over many orders of magnitude. For example, a switching time of 30 ns

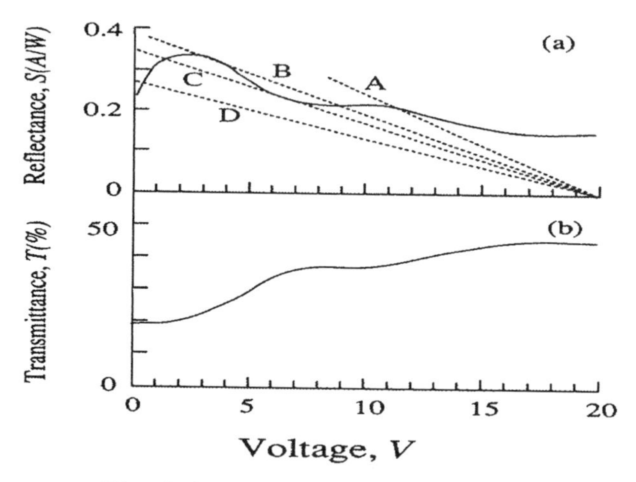

Fig. 9.6. Operation principle of SEED [13].

has been observed with an optical power of 1.6 mW in a 100-μm device with a 47-kΩ load resistor and an estimated device capacitance of 0.6 pF, whereas 20 ms with an optical power of 4.9 mW has been obtained for the load resistance of 22 MΩ [13]. The photodetection is known to be capable of operating on a timescale of a few picoseconds [16], as is the modulation; consequently, faster operation is expected in smaller devices with improved packaging.

However, this device has several disadvantages; large input power is needed to obtain high speed operation, a low contrast ratio (at most 3:1), and skillful adjustment of bias voltage is necessary. At present, advanced structures have been proposed and demonstrated as described in the following.

9.3 Various SEEDs

Various types of SEEDs are reported, in which the load resistance is replaced with other devices. The D-SEED [17] (photodiode-biased SEED) consists of a photodiode (bandgap wavelength of λ_1) monolithically integrated on the p-i-n diode (bandgap wavelength of $\lambda_2 > \lambda_1$). The optical bistability can be changed by changing the current vs. voltage characteristics associated with the control light that is illuminated on the p-i-n diode. The on/off ratio of over 10 dB has been achieved. Moreover, 6x6 spatial light modulation has been demonstrated.

The T-SEED (transistor-biased SEED) [18] and F-SEED [19] (FET-biased SEED), which respectively consist of a bipolar transistor and FET (field-effect-transistor), are both designed to reduce the optical switching energy by using the electric amplification of electron tranport device.

The S-SEED (symmetric SEED) has been proposed to improve the low contrast ratio of SEED devices [20]. This device consists of two similar p-i-n diodes, connected in series and laterally fabricated on the same substrate (Fig. 9.7). Because the photodiodes have the same structure, high-density devices are easier to fabricate than with the other SEED devices. This is considered to be a three-terminal device that can bring about a complementary operation, making possible an optical-set and reset operation.

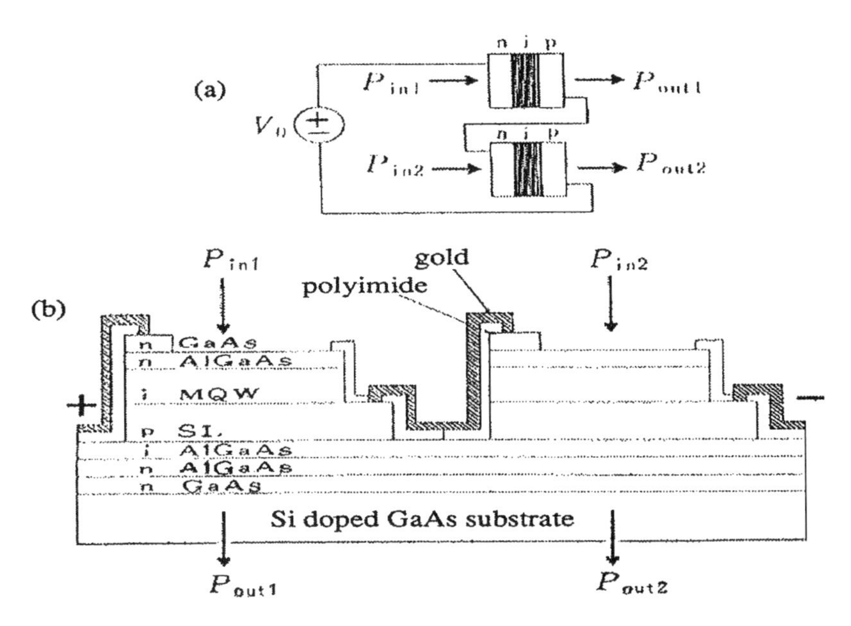

Fig.9.7. Structure of S-SEED [20].

Figure 9. 8 shows the operation principles. Assuming that the two p-i-n diodes are the same, the photosensitivity of the diode is shown in Fig. 9.8 (a). When the incident lights are the same, the stable points are the two as labeled B and D. When the diode indicated by the solid line in the Figure is under condition D and under low voltage, the transparency is low (Low state).

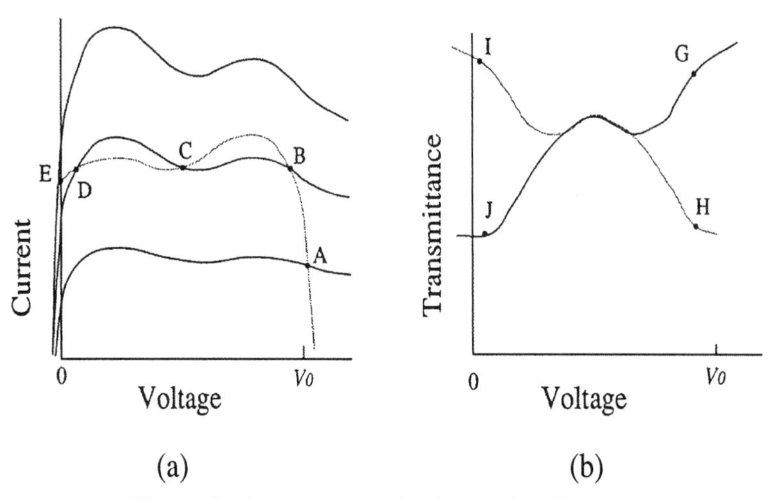

Fig. 9.8. Operation principle of S-SEED.
(a) current versus voltage performance.
(b) transmission versus voltage.

On the other hand, the diode indicated by the dotted line is under the condition in which the transparency is high. Under this condition, a light pulse is irradiated to the diode indicated by the solid line and the sensitivity of the solid line diode increases and comes back to the former position. Then, the stable point shifts to point B and the situation of 'High' and 'Low' is reversed. When the incident light intensities are the same, the stable position cannot change, even if the intensity increases. Therefore, the intensity increase of the clock light after set and reset have been achieved can produce the multiplication of the transmitted light intensity (Q and ~Q). This gives time-sequential gain. The optical switching energy E_{opt} is given approximately as

$$E_{opt} = C_{tot}V_0/S,$$

where C_{tot} is the total capacitance observed from the the electric source, V_0 is the source voltage, and S is the mean sensitivity. The electric energy E_{elect} required for the switching is then

$$E_{elect} = (1/2)\, C_{tot}V_0^2.$$

Typical values of E_{opt} and E_{elect} are 16 fJ/μm^2, 10 fJ/μm^2. The integrated structures of S-SEEDs have been fabricated,such as 8 K (128x64), 32 K (256x128) [20].

As for high-speed operation, a thinner barrier structure can produce higher speed and require less driving energy. For thick barriers (6.0-nm) the switching speed was 660 ps under the bias of 22 V, whereas it is 33 ps with a switching energy of 4+/-1 pJ for the thin barriers (3.5-nm) under 25 V [22]. This is due to the shortening of the sweep-out time of the generated carriers in the optical absorption, and the carrier sweep out depends on the barrier thickness and applied electric field intensity.

An optical module which consists of 32x32 two dimensional S-SEEDs and its 6-stage arrays of them cascaded with optical cross over connections has been reported [23] (Fig. 9.9). The optical cross over connection was achieved with mirrors and prism gratings.

An application of specific SEEDs has been reported in order to operate a more functional operation than S-SEED where the number of logic operations is limited. This device can operate accordingly as the discrete problem and reduces the number of devices [23].

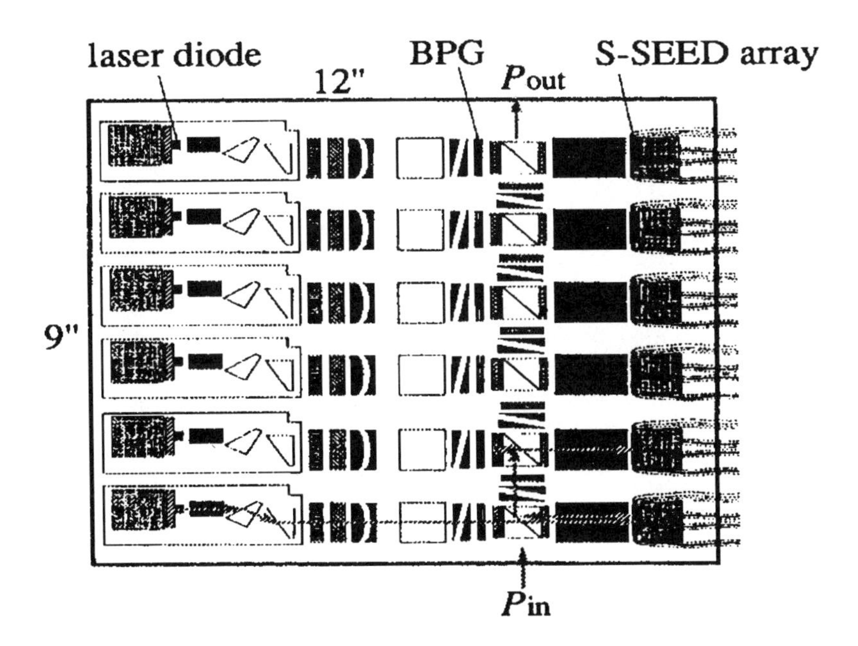

Fig. 9.9. Module of configuration consisting of 6 S-SEED arrays [23].

9.4 EARS

As described in the previous section, SEED has a low contrast ratio and no gain because it is a passive device. To solve above problems, a three-terminal optical gate, named EARS (exciton absorptive reflection switch) [24], monolithically integrated with a heterojunction-phototransistor (HPT) has been proposed and demonstrated. The HPT will produce a high contrast ratio and a high dynamic gain. Figure 9.10 is a schematic diagram of the EARS. The background impurity level in the i-MQW region is so low that the depletion layer thickness is 1 μm thick and introduction of high reflection with DBR mirrors results in doubling the interaction length. A remarkably high contrast ratio of more than 20 dB [25] and a high optical gain [26] have been reported.

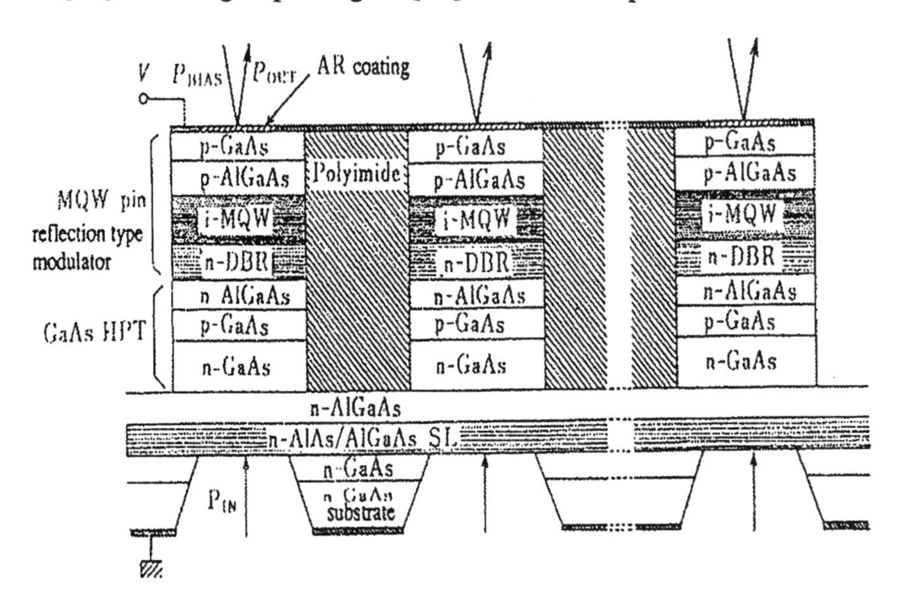

Fig. 9.10. Schematic diagram of the EARS.

The basic principle operation is that the MQW reflection modulators are electrically driven by the phototransistors, which is similar to a hybrid structure combining MQW modulators and Si-phototransistors [27] and the T-SEED [18] configuration. EARS has flexibility in the electrical connection between the MQW modulators and the HBT's, which supports a range of operation characteristics, including OR and NOR gate operations [28].

Figure 9.11 shows the contrast ratio and insertion loss as a function of wavelength [24] in a non-resonant GaAs/AlGaAs MQW reflection modulator. The incident wavelength is set at the absorption edge for the zero-biased state. This causes the modulator to suddenly change to an absorption state.

The input-output characteristics of the NOR-gate switch are shown in Fig. 9.12. The wavelength of the Ti-doped sapphire laser was set at 862 nm, the bias light power was 1 mW, and the reverse bias voltage was -30 V. When the input power was increased from 32 to 100 mW, the output power decreased abruptly from 400 to 10 mW. A high constrast ratio and optical gain were obtained. These results indicate that a photonic switch can operate as a NOR gate with a critical threshold.

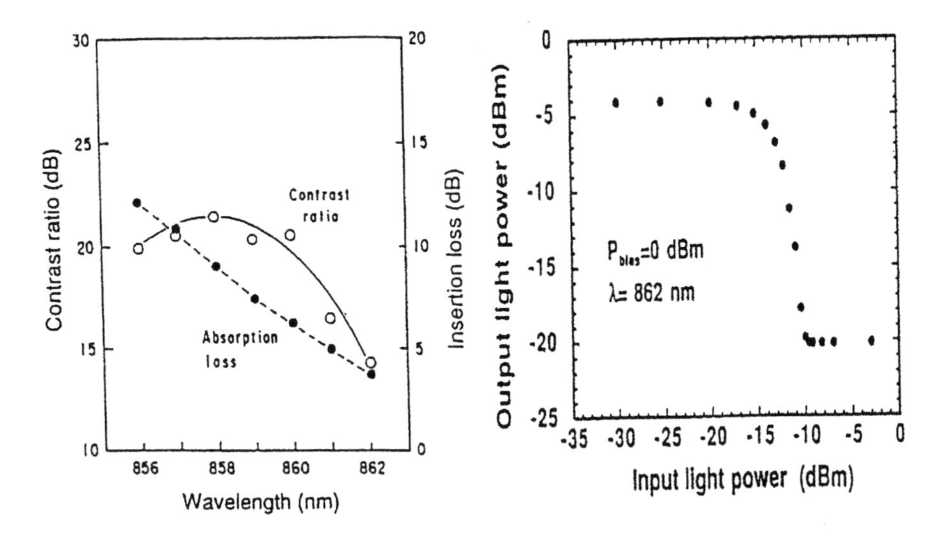

Fig. 9.11. Contrast ratio and insertion loss of EARS.

Fig. 9.12. Input versus output characteristics of EARS.

9.5 pnpn-VSTEP

Optical switching devices consisting of pnpn thyristors are called as DOES (double-heterostructure optoelectronic switch) [28, 29] or VSTEP (vertical-to-surface transmission electro-photonic devices) [30]. DOES can perform many optical functions, such as switching by using electrical drive and relatively weak optical input power without dependence on incident light wavelength, and they can emit light. This device was developed to realize the surface normal operation under small operating power, which is impossible by the optical function above [31]. A two-dimensional array (32x32) with functions such as optical switching and optical latching using pnpn-VSTEPs has been reported [32]. Lasing operation (surface normal) from LED mode with Bragg mirrors has been also achieved. This device consists of strained InGaAs MQW active layers in the pnpn structure and AlAs/GaAs multi-layer mirrors localized at the upper and lower sides of the device (Fig. 9.13). The size is 10 μm square area and the CW threshold current is 3 mA for room temperature lasing at 955 nm.

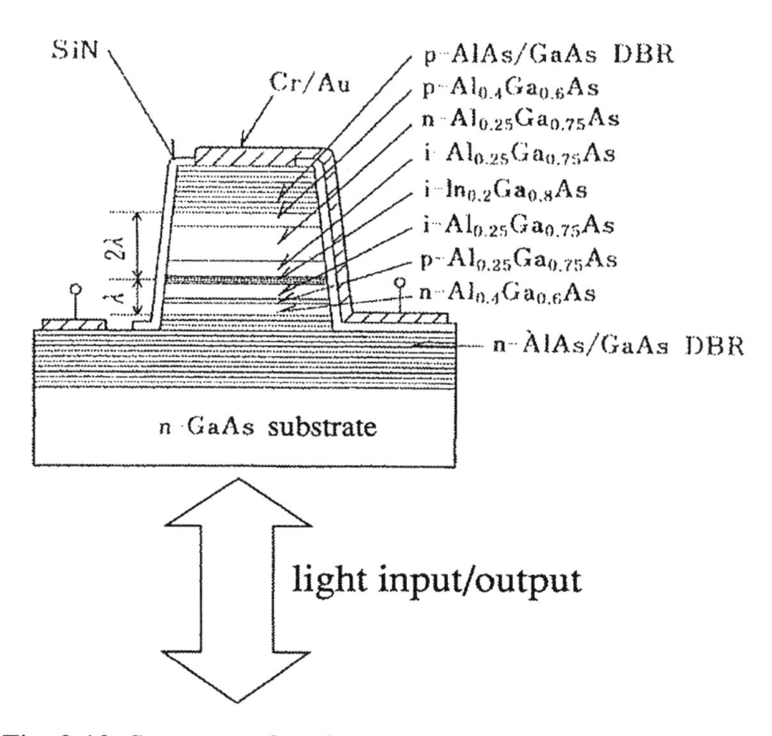

Fig. 9.13. Structure of surface normal cavity pnpn-VSTEP [32].

An optical self-routing switch in which the device itself determines its transmission direction according to a control light set in front of the signal light using VSTEPs has been reported [33].

The optical modulators described above consist of GaAs and related compounds. InGaAsP material systems are also used for surface-normal optical modulators and switches. A 32x32 two-dimensional array consisting of LED and HBT grown in the direction of epitaxial layers as a optical switch and HBT in parallel with them as a optical reset have been reported [34].

Another type of device based on MSM (metal-semiconductor-metal) photodiodes that operates as a logic gate has been reported. This device performs an exclusive EOR operation using four MSM diodes [35].

9.6 Miscellaneous

9.6.1 Electrically Tunable Polarization Rotation

As discussed in the previous sections, conventional incidence light modulators employ the QCSE to produce amplitude modulation. However, thickness constraints coupled with the maximum obtainable change in exciton absorption limit the intrinsic contrast ratio of MQW modulators to 10:1 at room temperature. The use of the interference effect associated with a Fabry-Perot cavity to increase the contrast ratio has been concentrated. Recently, Shen et al. [36] have reported a large normal incidence anisotropic excitonic absorption and birefringence in a (100) MQW structure under thermally induced in-plane anisotropic strain. Upon further study, they presented an MQW light modulator which exploits the anisotropic excitonic absorption and concomitant light polarization rotation to achieve a contrast ratio of 330:1 [37] and 5000:1 [38] in an optically adressed device.

The schematic of the device is shown in Fig. 9.14. The modulator is composed of a p-i(MQW)-n structure on (100) GaAs. The MQW thin film and the p-type annular ring were lifted off from the substrate and attached to $LiTO_3$. The $LiTaO_3$ substrate was chosen both for its transparency and for its direction-dependent thermal expansion. It was cut such that the room temperature thermal expansion coefficient matches that of the MQW while its orthogonal counterpart is different from that of the MQW. Since the sample is bonded at 150 C, the difference in thermal expansion between the $LiTaO_3$ and the MQW induces a uniaxial strain (about 0.12%) at room temperature. Uniaxial strain applied along the x (or y) direction breaks the rotation symmetry for the valence band, mixing the heavy and light hole bands in the MQW

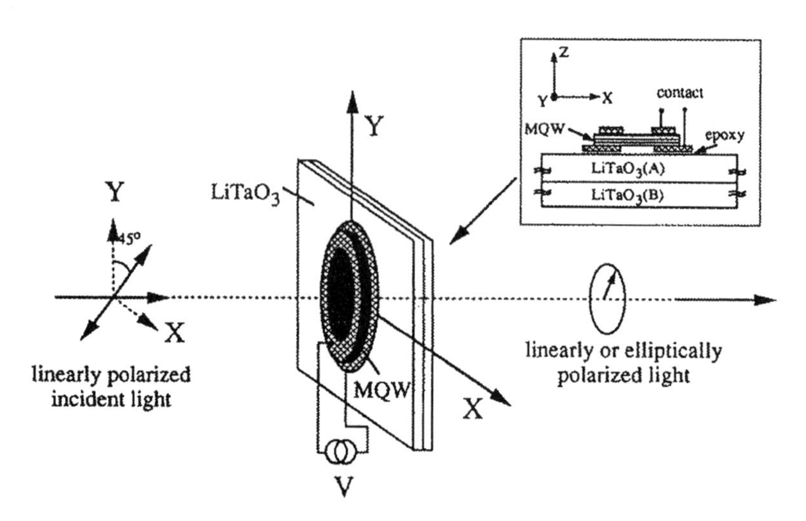

Fig. 9.14. Schematic of an MQW light modulator using anisotropic absorption and concominant polarization rotation [38].

and creating an anisotropic excitonic absorption [37]. As a result, birefringence is increased (Δn of -0.04 and 0.03 slightly below or above the heavy-hole exciton transition).

Previous optical modulators utilizing polarization rotation, such as liquid crystal and magneto-optic devices, have achieved significantly higher contrast ratios (>10^4:1), but they are hampered by poor high frequency characteristics. The speed characteristc has never been reported for this voltage-tunable phase retardation and polarization rotation modulator, but high speed operation is expected.

9.6.2 Silicon Modulator Based on Mechanically-Active Anti-Reflection Layer

A micromechanical modulator for fiber-in-the-loop applications with projected optical bandwidths from 1.3 to 1.55 μm and data rates of several Mbits/s has been reported [39]. The device is based on optical interference effects between a suspended, vertically moving membrane and the substrate and operates in surface-normal mode. Figure 9.15 shows the schematic of this device, which consists of a membrane supported by arms above an air gap. The membrane is fabricated out of silicon nitride, whose refractive index may be controlled precisely to the square root of the refractive index of silicon. When the membrane is brought into contact with the substrate by electrostatic force, an anti-reflection condition exists as a wavelength equal to four times the optical thickness of the membrane. This device is insensitive to polarization, has wide fiber alignment tolerance, and a projected insertion loss of <1 dB and contrast of >20 dB. Since this device uses standard microelectronic techniques, the cost is expected to be low.

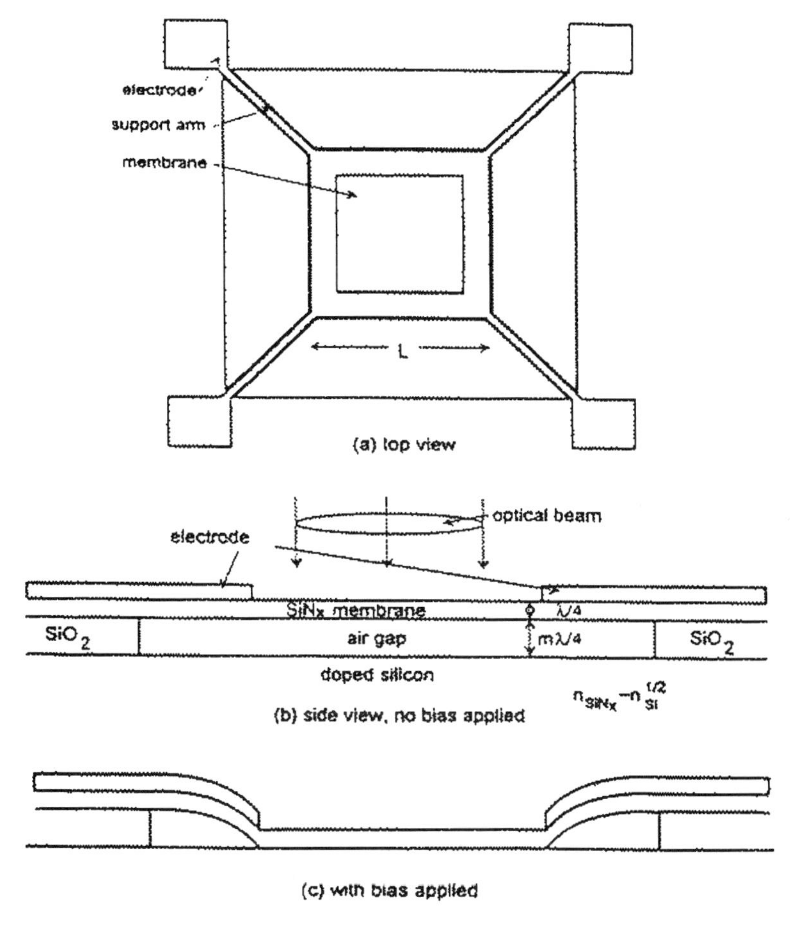

Fig. 9.15. Top (a) and undeflected (b) and side views of the silicon modulator based on mechanically-active anti-reflection layer [39].

9.7 References

[1] H. S. Hinton, IEEE J. Selected Areas in Commun., 6,1209,1988.
[2] A. L. Lenchine, D. A. B. Miller, J. E. Henry, J. E. Cummingham, L. M. F. Chirovsky, and L. A. D'Asaro, Appl. Optics 29, 2153,1990.
[3] G. D. Boyd, D. A. B. Miller, D. S. Chemla, S. L. McCall, A. C. Gossard, and J. H. English, Appl. Phy. Lett., 50, 1119,1987.
[4] R. B. Bailey, R. Sahai, C. Lastufka, and K. Vural, J. Appl. Phy., 66, 3445,1989.
[5] M. Whitehead, G. Parry, and P. Wheatley, IEE Proc.-Part J 136, 52,1989.
[6] R. H. Yan, R. J. Simers, and L. A. Coldren, IEEE J. Quantum Electron. 25, 2272, 1989.
[7] R. H. Yan, R. J. Simers, and L. A. Coldren, IEEE Phonton. Technol. Lett., 2, 118,1990.
[8] M. Whitehead, A. Rivers, G. Parry, and J. S. Roberts, Electron. Lett., 25, 984,1989.
[9] R.-H. Yan,R. J. Simes,L. A. Coldren, and A. C. Gossard, Appl. Phys. Lett., 56, 1626,1990.
[10] C. C. Barron, C. J. Mahon, B. J. Thibeault, G. Wang, J. R. Karin, L. A. Coldren, and J. E. Bowers, Device Research Conference, VIB-9, 1993.
[11] R.H.Yan et al., in Quantum Optoelectronics Topical Meeting, Salt Lake City, MB1,1991.
[12] D. A. B. Miller, D. S. Chemla, T. C. Damen, A. C. Gossard, W. Wiegmann, T. H. Wood, and C. A. Burrus, Appl. Phys. Lett., 45, 13-15,1984.
[13] D. A. B. Miller, D. S. Chemla, T. C. Damen, T. H. Wood, C. A. Burrus, A. C. Gossard, and W. Wiegmann, IEEE J. Quantum Elec tron. Lett., QE-21, 1462-1476,1985.
[14] A. Larsson, A. Yariv, R. Tell, J. Maserjian, and S. T. Eng, Appl. Phys. Lett., 47, 866-868,1985.
[15] A. Larsson, P. A. Andrekson, P. Andersson, S. T. Eng, J. Salzman, and A. Yariv, Appl. Phys. Lett., 49, 233-235,1986.
[16] K. Wakita, I. Kotaka, O. Mitomi, H. Asai, Y. Kawamura, and M. Naganuma, CLEO'90 Tech. Dig., CTuC6, 1990.
[17] D. A. B. Miller, D. S. Chemla, T. C. Damen, T. H. Wood, C. A. Burrus, A. C. Gossard, and W. Wiegmann, Appl. Phys. Lett., 49, 821-823(1986).

[18] D. A. B. Miller, Opt. Quantum Electron., 22, S61, 1990.
[19] T. K. Woodward, L. M. F. Chirovsky, and A. L. Lentine, LEOS'91, paper PD-10, 1991.
[20] A. L. Lenchine, D. A. B. Miller, J. E. Henry, J. E. Cummingham, L. M. F. Chirovsky, and L. A. D'Asaro, IEEE J. Quantum Elec tron.,25, 1928-1936,1989.
[21] L. M. F. Chirovsky et al., in Photon. Switching Topical Meeting, Salt Lake City, ThB3,1991.
[22] G. D. Boyd et al., Appl. Phys. Lett., 57, 1843-1845,1990.
[23] A. Lentine, Proceding of the Photonic Switching Topical Meeting (Salt Lake), 170-179,1991.
[24] C. Amano, S. Matsuo, and T. Kurokawa, Photon. Technol. Lett., 3, 736, 1991.
[25] C. Amano, S. Matsuo, T. Kurokawa, and H. Iwamura, Photon. Technol. Lett., 4, 31, 1992.
[26] S. Matsuo, C. Amano, and T. Kurokawa, Photon. Technol. Lett., 3, 330, 1991.
[27] P. Wheatley, P. J. Bradley, M. Whitehead, G. Parry, and J. E. Midwinter, Electron. Lett., 23, 92, 1987.
[28] C. Amano, S. Matsuo, T. Nakahara, and T. Kurokawa, IEEE J. Quantum Electron. , 29, 775, 1993.
[29] J. G. Simmons and G. W. Taylor, IEEE Trans. Electron. Devices, ED-34, 973, 1987.
[30] K. Kasahara, Y. Tashiro, N. Hamao, M. Sugimoto, and T. Yanase, Appl. Phys. Lett., 52, 679, 1988.
[31] K. Kasahara, Y. Tashiro, I. Ogura, M. Sugimoto, S. Kawai, and K. Kubota, 7th IOOC789, Kobe, 20C3-1, 1989.
[32] T. Numai, M. Sugimoto, I. Ogura, H. Kosaka, and K. Kasahara, Appl. Phys. Lett., 58, 1250, 1991.
[33] T. Numai, M. Sugimoto, I. Ogura, H. Kosaka, and K. Kasahara, Electron. Lett., 27, 605, 1991.
[34] K. Matsuda and J. Shibata, IEE Proceedings-J, 138, 67, 1991.
[35] H. Kamiyama, I. Tanaka, and T. Kamiya, in Conf. on Lasers Electro-Opt., Anaheim, CTUF2, 1990.
[36] H. Chen, M. Wraback, J. Pamulapati, P. G. Newman, M. Dutta, Y. Lu, and H. C. Kuo, Phys. Rev., 47, 13933, 1993.
[37] H. Chen, M. Wraback, J. Pamulapati, M. Dutta, P. G. Newman, A. Ballato, and Y. Lu, Appl. Phys. Lett., 62, 2908, 1993.
[38] H. Chen, J. Pamulapati, M. Wraback, M. Taysung-Lara, M.

Dutta, H. C. Kuo, and Y. Lu, IEEE Photon. Technol. Lett., 6, 700, 1994.

[39] K. W. Goossen, J. A. Walker, and S. C. Arney, IEEE Photon. Technol. Lett., 6, 1119, 1994.

Chapter 10

Evaluation of Modulator Characteristics

As discussed in Chapter 2, various modulation characteristics should be evaluated. In this chapter, we describe the chirping parameter for electroabsorption modulators, polarization insensitivities, and waveguide losses both for semiconductor modulators.

10.1 Waveguide Loss

Semiconductor waveguides are the "wires" used to route optical signals on a semiconductor chip, which provide the basis for numerous integrated optic devices suitable for applications such as lightwave telecommunications, optical signal processing, and sensors [1]. The optical waveguide is the fundamental interfacial element for the different active components of the integrated circuit. Many III-V semiconductors waveguides have shown high propagation losses compared with guides made in other materials such as $LiNbO_3$ or glass. Reduction of the propagation loss is the key issue for a practical use of semiconductor materials in integrated optics and many efforts have been made to reduce the loss. At present significant progress has occurred in the area of loss during the past several years. In general, the lowest losses have been in the range 0.1-0.2 dB/cm, and are approaching measurement limits imposed by available sample size and experimental accuracy. The realization of straight waveguides with low loss on GaAs and InP is now well established [1]. The historical progress reviews upto 1976 have been reported in detail by Garmire [2], and upto 1987 by Erman [3], and in the recent advances by Deri and Kapon [1]. In this chapter we take a bird's eye view of the semiconductor waveguide for discussing the modulators and optical switches.

Many experimental results have been reported on propagation loss of two-dimension waveguides.

10.1.1 Homostructure Waveguide

Waveguides can be realized in GaAs and InP by doping variation, since the effective mass of electrons in both materials are low enough to have a significant contribution to the free-carrier effect. The consequence on the free carriers is to lower the index of refraction as the following equation [2].

$$\Delta n_s = -N\lambda_0^2 e^2 / 8\pi^2 \varepsilon_0 n_s m^* c^2$$

Here n_s and Δn_s are the refractive index of the semiconductor and its change due to carriers at a free space wavelength λ_0, N is the carrier density, m^* is the carrier effective mass, e is the free carrier charge, ε_0 is the permitivity of vacuum, and c is the velocity of light. In n-type GaAs, for example, $\Delta n_s = 0.01$ when $N = 5\text{x}10^{18}\text{cm}^{-3}$ for $\lambda_0 = 1\ \mu\text{m}$. This refractive index difference between pure and n-type materials is large enough to permit the guiding of light if a pure layer sufficiently thick can be fabricated on an n-type GaAs substrate, and it is used in homostructure waveguides.

In a homostructure the free carrier absorption of the substrate is very strong and the only way to reduce the loss is to increase the guide thickness. In the 1-2 μm range, for photons with energy lower than the bandgap energy, GaAs and InP exhibit some residual loss that is linearly dependent on carrier concentration. The loss is larger for p-type impurities than for n-type impurities. For GaAs, the loss α at room temperature is estimated from experiments to be given by

$$\alpha(\text{cm}^{-1}) = 3\text{x}10^{-18}N + 7\text{x}10^{-18}P,$$

where N and P are the carrier concentrations (cm^{-3}) in the n- and p-type impurity.

For InP the loss is given as

$$\alpha(\text{cm}^{-1}) = 1.25\text{x}10^{-18}N + 14\text{x}10^{-18}P \quad \text{for } \lambda_0 = 1.3\ \mu\text{m}$$

$$= 1.25\text{x}10^{-18}N + 20\text{x}10^{-18}P \quad \text{for } \lambda_0 = 1.55\ \mu\text{m}.$$

Compared with the low index difference of these homostructures, the higher index variations allowed by single-heterostructures or double-heterostuctures using composition variation are generally preferred. These

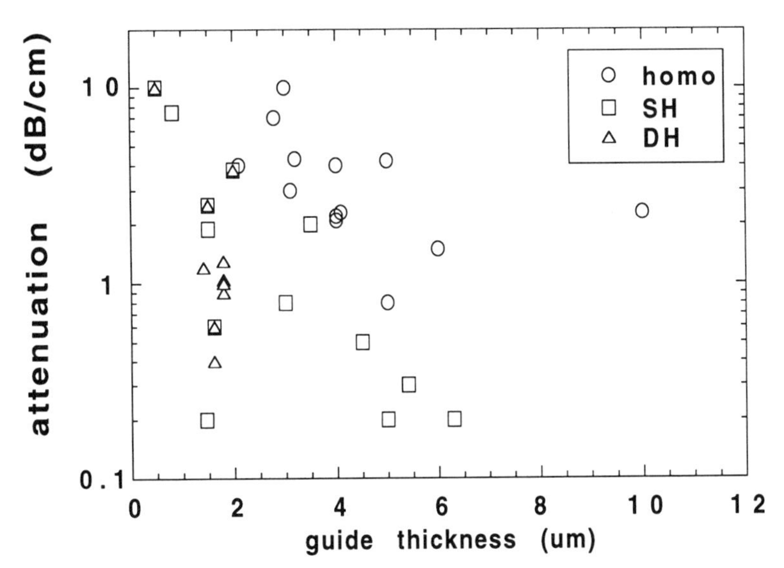

Fig. 10.1. Homostructure, single-heterostructure, and double-heterostructure waveguides based on the GaAs/AlGaAs material system.

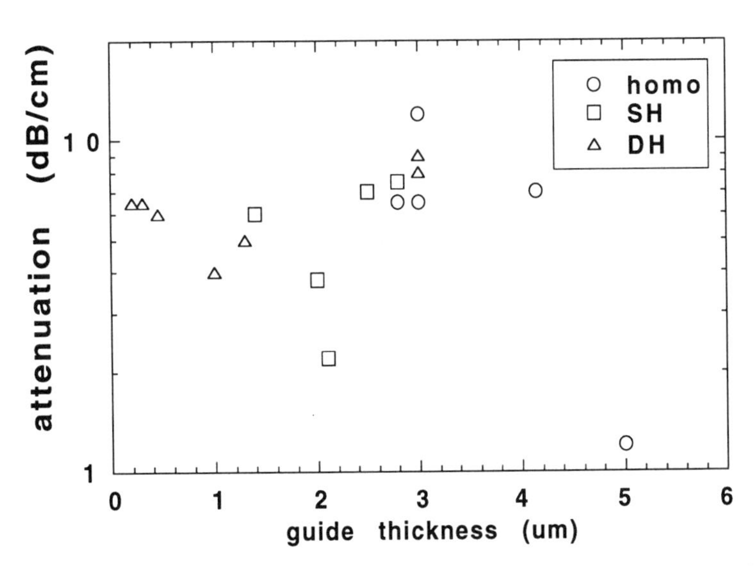

Fig. 10.2. Homostructure, single-heterostructure, and double-heterostructure waveguides based on the InGaAsP/InP material system.

heterostructures have the obvious advantage of suppressing substrate contributions to the guide loss because of their strong optical confinement, and the double-heterostructure has the additional advantage of avoiding any further absorption by electrodes. The homostructure, single-heterostructure, and double-heterostructure waveguides based on GaAs/AlGaAs and InGaAsP/InP are shown schematically in Figs. 10.1 and 10.2.

10.1.2 Heterostructure Waveguides

The experimental attenuation values reported for GaAs and InP straight waveguides [4,5], along with some new data, are shown in Figs. 10.1 and 10.2, where homostructures stand for them. It is clear that the use of single- and double-heterostructures with a very low background doping level is effective in reducing the propagation loss.

Figures 10.3 and 10.4, which were obtained by adding some new data to that in Reference 3, show the considerable reduction of waveguide loss during past few years. The continuous progress of epitaxy and processing technology has brought about definite breakthrough, such as losses below 1 dB/cm. These reductions can be explained by the improvement of epilayer surface quality, uniformity, and purity.

10.1.3 Waveguide Configuration

The lattice-matched material system can be divided into two categories:

(1) for wavelengths below 0.9 μm, GaAs/$Al_xGa_{1-x}As$ where the refractive index decreases as the parameter x is increased as shown in Fig. 10.5 [6].

(2) for wavelengths between 1.1 and 1.7 μm range, $In_{1-x}Ga_xAs_y\, P_{1-y}$/InP which is perfectly matched for x = 0.466y and $In_{1-x-y}\, Ga_xAl_yAs$ with x+y=0.468+0.017y, where the refractive index of the quaternary is higher than that of InP [4,7].

Figure 10.6 shows the refractive index of the quaternary InGaAsP crystals under the condition of lattice-matching, and Fig. 10.7 shows the refractive index of the another quaternary (InGaAlAs) lattice-matched to InP [8]. A continuous range of energy gap and refractive index can be obtained by changing x and y. In the former system, GaAs is the core of the guide and the ternary substance makes up the confining layers. In the latter system, the core is a quaternary layer and the cladding is InP or a

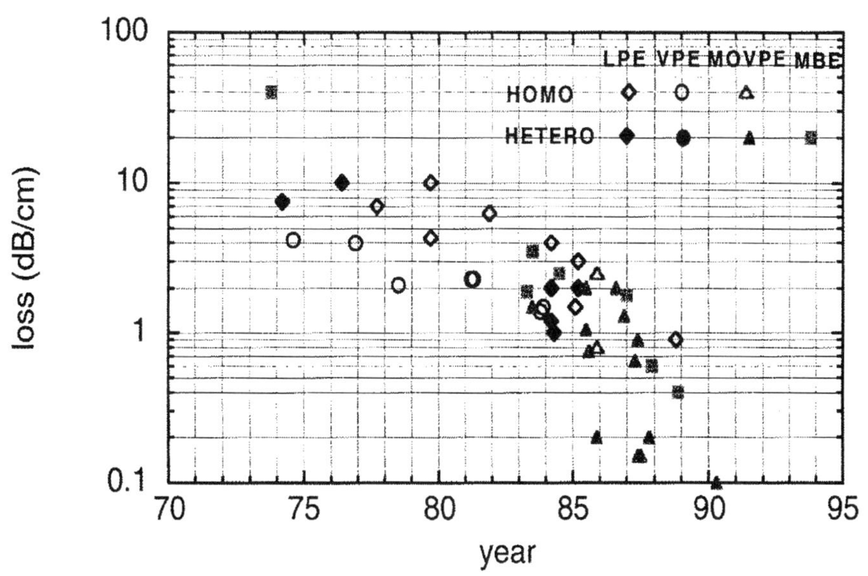

Fig. 10. 3. Recent reductions in the propagation loss of GaAs waveguides made with various epitaxial growth methods as a parameter.

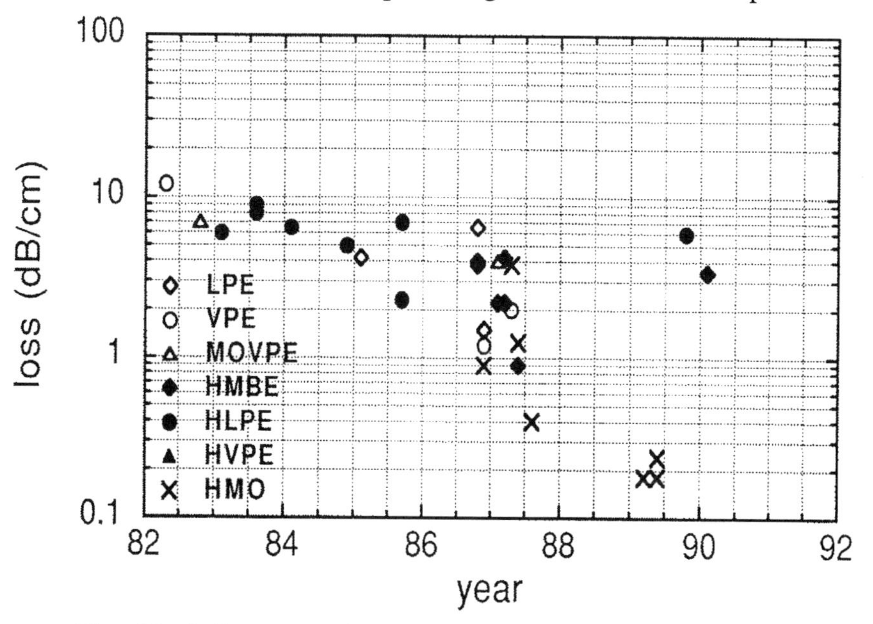

Fig. 10. 4. Recent reductions in the propagation loss of InP waveguides made with various epitaxial growth methods as a parameter.

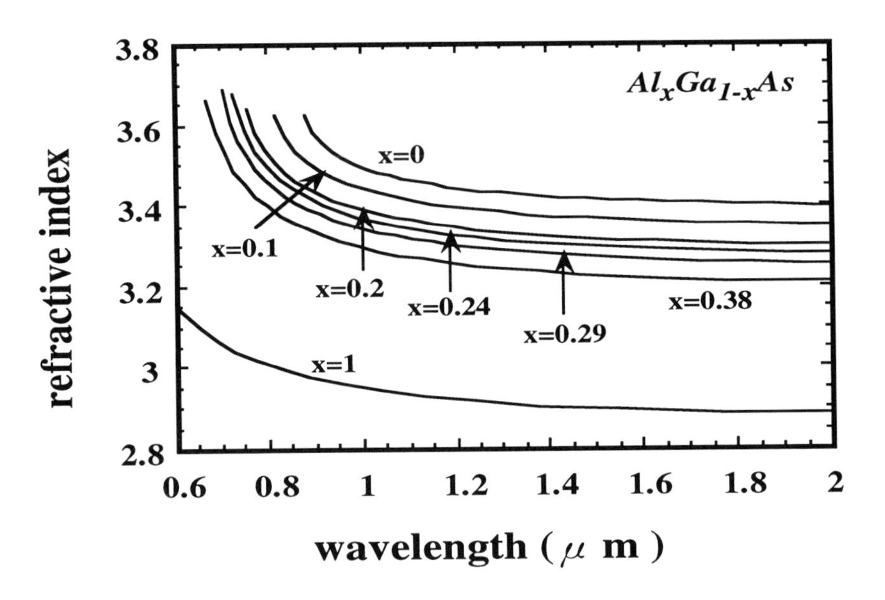

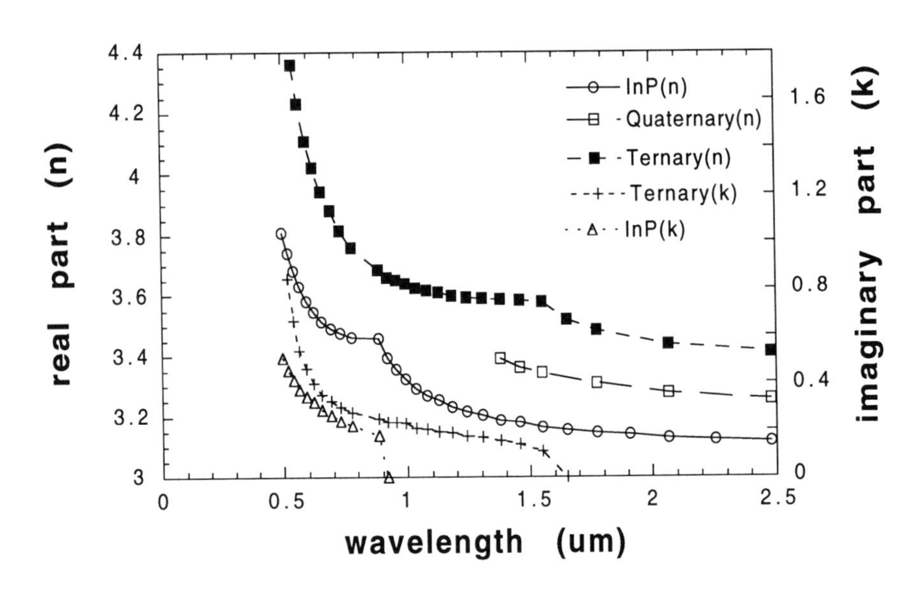

Fig. 10.6. Real and imaginary part of the refractive index of InGaAsP lattice-matched to InP substrate as a function of wavelength.

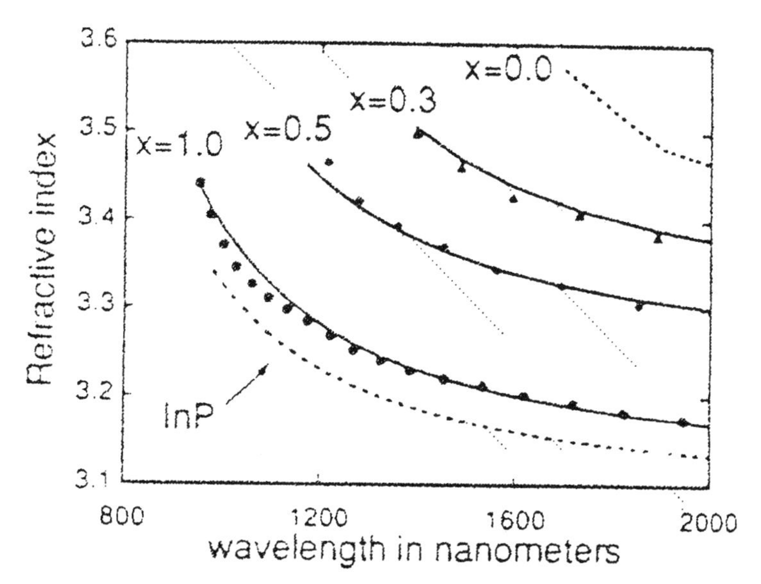

Fig. 10.7. Refractive index of InGaAlAs lattice-matched to InP substrate as a function of wavelength. x is fractional InAlAs content. That is, $(In_{0.52}Al_{0.52}As)_x(In_{0.53}Ga_{0.47}As)_{1-x}$.

quaternary with a bandgap energy higher than that of the core.

Even highly lattice-mismatched materials, where material quality generally is degraded, have yielded guides with propagation loss below 1 dB/cm when the buffer layers were thick enough to isolate the guiding layer from the mismatched interface [10,11].

The use of a superlattice or MQW structure in guiding [11,12] or cladding [13] layers has shown to be compatible with low-loss performance despite the large number of interfaces in such structures. The refractive index of the MQW structure can change over a wide range given approximately the following equation [14]:

$$n_{MQW} = \{(n_w{}^2t_w + n_b{}^2t_b)/(t_w+t_b)\}^{1/2},$$

where n_W and n_b are the refractive indices of the well and the barrier, and t_w and t_b are the thicknesses of the well and the barrier. The effective energy bandgaps of these structures also change according to the Kronig-Penny model. Moreover, strain in the MQW layers can be introduced without degradation of crystal quality more easily than it can be with bulk materials, and this offers new possibilities for engineering band structure. That is, superlattice and MQW structures give us a new degree of freedom of designing the refractive index profile for waveguide.

10.1.4 Electrode Losses in Active Devices

The discussion in the previous section focused on the problems of losses in passive straight waveguides. For active components such as switches and modulators, on the other hand, the electrode losses must be addressed. Active devices using electronic control signals require a p-n junction or Schottky contact, and p-i-n structures are used in order to obtain homogeneous distribution in the waveguide. Waveguide theory would require the i-region to be thick to reduce attenuation, whereas the thick i-region results to increase the drive voltage. In practical applications relatively thin guide layers are used to reduce drive voltage and the attenuation loss due to the free carrier loss of nonconfined optical field in cladding layers is not negligible.

This situation is similar to that with the Schottky contact. An attempt to use FET structure for optical modulators has been made, but this problem associated with the electrode is serious. Therefore some compromise between losses and drive voltage is needed.

The situation is the same for carrier-injection-type modulator.

10.1.5 Bending Loss

Directional couplers or Mach-Zehnder switches require bent waveguides as well as straight waveguides. When waveguides are bent in circular or S-bends, some guided light is lost to radiation. This distributed radiation loss is analogous to leakage [15-17] and it increases with radius R in a nearly exponential fashion [18]. This behavior is consistent with the exponential behavior predicted for bend loss in planar guides [13,15,16]:

$$\alpha = K \exp(-cR), \text{ where } c = b(2\Delta n_{eff}/n_{eff})^{3/2},$$

where K is a constant depending on the guide thickness and indices, $b = 2\pi n_{eff}/\lambda$ is the modal propagation constant, and Δn_{eff} is the difference between the modal effective index and the cladding index. This equation assumes that $\Delta n_{eff}/n_{eff}$ is small and that the bending is a small perturbation on the modal intensity distribution of the straight guide mode. There is a trade-off between small bend radius and increased scattering loss because the scattering loss is proportional to the square of Δn_{eff}.

10.1.6 Coupling Loss

Semiconductor waveguides generally have large insertion loss, most of which consists of coupling loss between a single-mode optical fiber and a semiconductor waveguide. The coupling loss is given as follows when the optical power distribution is Gaussian:

$$[\text{couplig loss}] = -10\text{x}\log\{4/((w_x/a_x)+(a_x/w_x))((w_y/a_y)+(a_y/w_y))\},$$

where a_x, a_y, w_x, and w_y are respectively the spot sizes in the direction of x and y of a single mode fiber and of the guided mode. Figure 10.8 shows the coupling efficiency, which is inverse of coupling loss, as a function of the beam spot size ratio of the optical fiber and the guide when the fiber spot size $a_x=a_y= w_f$. In a semiconductor waveguide the mode spot size is very small, especially in the y-direction (perpendicular to the layer) because of the large refractive index difference between the guide layer and cladding layers. When the guide layer is very thin, the optical confinement in the perpendicular direction is small because of the field leakage in the cladding layers, resulting in the large spot size and small couping loss. In this case the Gaussian beam approximation does not apply to the field distribution and the overlap integral between the field distributions of the optical waveguide and the single-mode fiber. Figure 10.9 shows the coupling loss as a function of core thickness [19]. As in Fig. 10.8, the thinner the guide, the less the coupling loss; and its magnitude is small compared with that calculated by Gaussian approximation.

For MQW structures, we can design the core thickness and its refractive index changing the thickness of wells and barriers. Recently, very low insertion loss of less than 5 dB from fiber-to-fiber has been reported by using strain-compensated InGaAs/InAlAs MQW structures [18] with a core thickness of 720 nm and a buried heterostructure. This structure produces quasi-spherical field pattern of the guided mode that is obtained by reducing the vertical optical confinement and increasing the transverse one as shown in Fig. 10.10, resulting in low coupling loss.

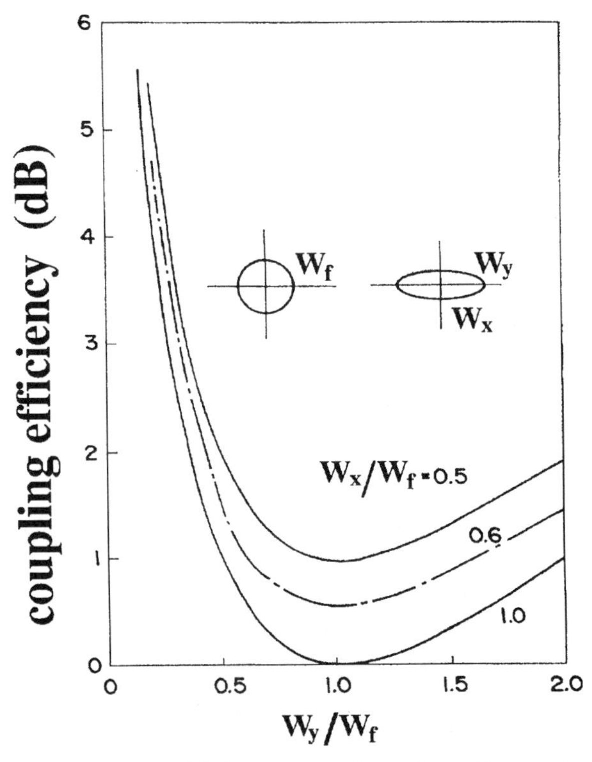

Fig. 10.8. Coupling efficiency as a function of spot size ratio of optical fiber and waveguide. w_x and w_y are spot sizes for a semiconductor guide and w_f is that for a single-mode optical fiber.

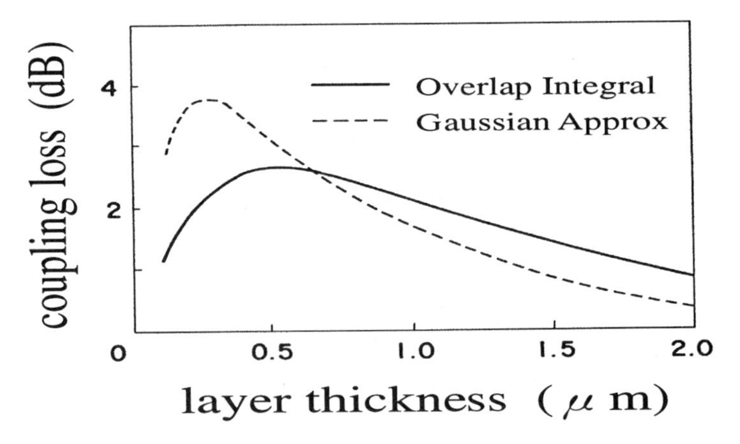

Fig. 10.9. Coupling loss as a function of core thickness.

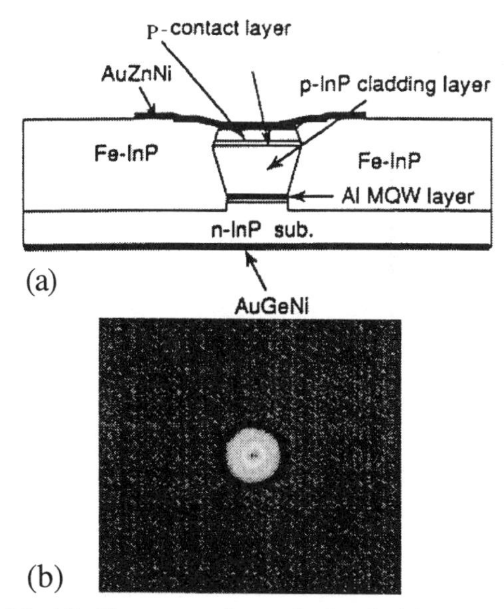

Fig. 10. 10. Cross-section of a buried heterostructure modulator (a), and its near field pattern (b).

10.2 Polarization Insensitivity

10.2.1 Introduction

External modulators operating at 1.55 μm wavelength region are expected to be used in high bit rate and long haul optical fiber transmission systems because of their low chirping characteristics. In particular, the multiple quantum well (MQW) electroabsorption modulators have advantages such as low driving voltage and high-speed operation [1,2], and monolithic integration with a DFB laser [3-4]. Strong polarization dependence of its modulation properties, however, is the next target for more practical use. To be suitable as practical optical modulators, polarization independent operation is highly desirable. In this section we discuss the polarization insensitivity. First we begin with the bulk type modulators and switches, then we take up the MQW modulators.

The source of the polarization anisotropy in MQW structures arises from the absorption anisotropy inherent in the MQWs, which will be discussed in the following.

10.2.2 Bulk Type Modulators

For bulk type electrooptic (EO) modulators, the anisotropy for the incident light polarization is generally due to the waveguide characteristics; the refractive index difference between TE and TM polarization direction. One solution for polarization insensitivity is to settle by using the isotropic configuration of crystals[23]. As described in Chapter 3, the electrooptic effect is determined by the geometrical configuration of a modulator crystal and the direction of applied electric field and incident light. That is, when the applied electric field is in the direction of (111), the modulator operates polarization-independently, though the operating voltage is large by $\sqrt{3}$ compared with that of normal configuration. This configuration needs the (111) substrate and a new fabrication technique such as dry etching with less crystal-orientation characteristics, because of cleavage difficulty.

Based on the EO effect described in Chapter 3, the field-induced refractive index change for TM polarization is zero, whereas for TE polarization the refractive index change exists when we use (100)-oriented substrate. On the other hand, refractive index changes are written by

$$\Delta n = (1/2\sqrt{3})n^3\gamma_{41}E \qquad \text{(TE polarization)}$$

$$\Delta n = -(1/\sqrt{3})n^3\gamma_{41}E \qquad \text{(TM polarization)}$$

for (111)-oriented substrates with electric field vertical to the (111) plane. Therefore, refractive index changes exist for both TE and TM polarizations. Although the amount of refractive index change differs between TE and TM polarizations, a polarization independent operation with the same switching voltage, can still be obtained using the uniform $\Delta\beta$ configuration, with the directional coupler length to coupling length ratio being 1. Figure 10.11 shows the switching diagram for a (111)-oriented GaAs/AlGaAs directional coupler with uniform $\Delta\beta$ configuration [24]. This switch operates with less than -12 dB crosstalk at a bias of 19 V.

Another solution is to use {011} cleavage planes on InP (100) substrate and to place the waveguides at 45 degrees to use only quadratic electrooptic effects [25] as shown in Fig. 10.12. This also needs low loss corner mirrors. Using this configuration, waveguided Mach-Zehnder interferometer has been demonstrated and on-off switching voltages are 30+/-1 V for both polarizations. The on/off ratios achieved are >12 dB for both polarizations. The insertion losses are below 10 dB and waveguide losses are typically 1 dB/cm.

Both devices are polarization independent but the operating voltages are relative large. In order to reduce operating voltage, introduction of MQW structures into the waveguide layer is effective as discussed on the following section.

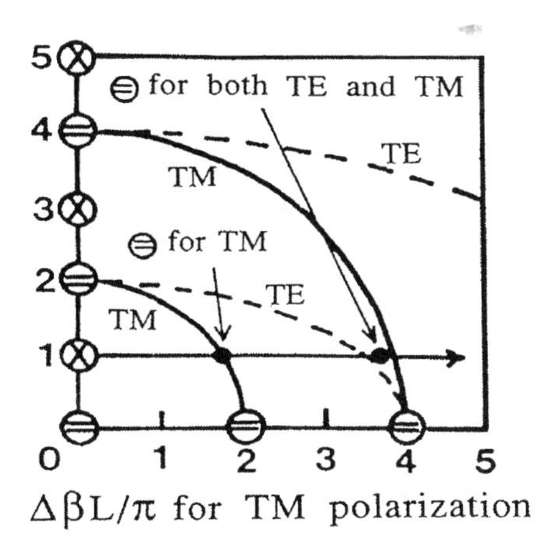

Fig. 10.11.
Switching diagram for a (111)-oriented directional coupler with uniform Δβ configuration.

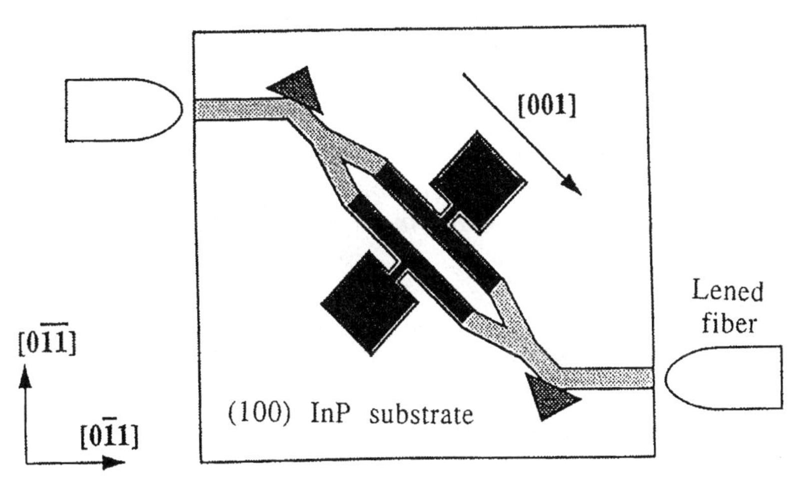

Fig. 10.12. Layout of the polarization insensitive Mach Zehnder interferometer. The length of the electrodes are 5 mm [25].

10.2.3 MQW Modulators

The source of the polarization anisotropy in MQW structures arises from the absorption anisotropy inherent in the MQWs, which is different from that in bulk types. We take up strain-compensated InGaAs/InAlAs MQW electroabsorption modulators as a typical example and mainly describe their highly efficient modulation and polarization-independent extinction ratios.

A. Theory. Since the introduction of strain into the waveguide core consisting of MQW structures has been reported for the polarization independence of optical gain in semiconductor optical amplifiers [26], many attempts to make optical switches [27] and modulators [28,29] polarization-insensitive have been done. Advances in crystal growth technologies can make it possible to fabricate thin semiconductor layers with atomically smooth interfaces. It is thus possible to introduce a precisely controlled strain and thereby tailor the energy band for devices. Recently, it is very attractive and popular to introduce strains in the quantum wells.

Figure 10.13 schematically shows the energy band change associated with the introduction of compressive or tensile strains. Compressive strains are usually used for semiconductor lasers to introduce large anisotropy in the gain for TE and TM polarization and shift the absorption wavelength to longer wavelengths. Tensile strains, on the other hand, reduce the anisotropy and the absorption band energy increases.

Figure 10.14 shows the energy band diagram of valence bands for InGaAs wells with and without strain. The introduction of a tensile strain lifts the light-hole band to a higher energy and makes it overlap with the heavy-hole band. This induces the relaxation of polarization anisotropy: under unstrained conditions, TM polarization corresponding to the light-hole excitonic absorption (e-lh) and TE polarization to heavy-hole excitonic absorption (e-hh). As a result, absorption edges merge or mix.

B. Experiments. Figure 10.15 shows photoabsorbed current spectra of a strained InGaAs/InAlAs MQW modulator for TE and TM polarization [30]. This MQW structure consisted of ten 12-nm-thick undoped InGaAs quantum wells (0.4% tensile strained) and 5-nm-thick InAlAs barriers (0.5% compressively strained). Heavy-hole and light-hole exciton peaks seemed to coincide and the absorption peak wavelength is 1.48 μm.

Polarized light (TE or TM) lasing at 1.55 μm was coupled into and out of the modulators by spherically ended single-mode fibers. Figure 10.16 shows the transmitted light intensity through the strained MQW

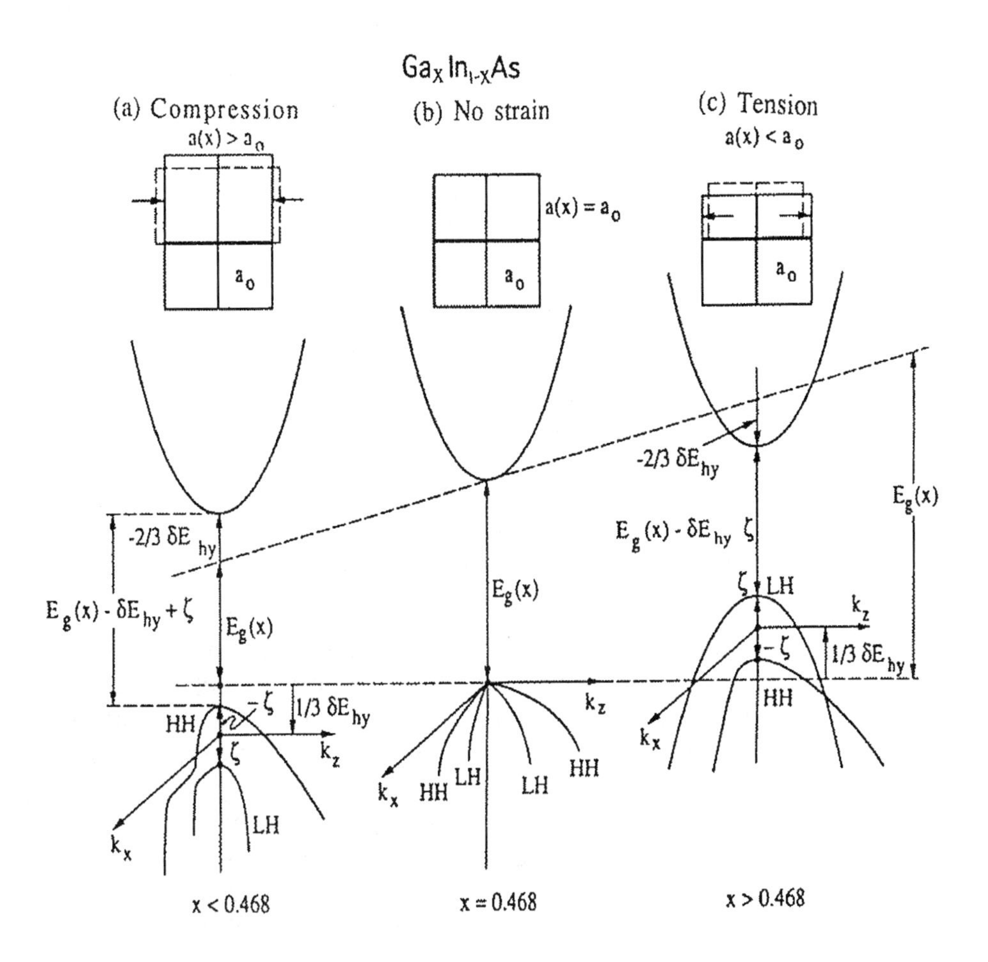

Fig. 10.13. A schematic energy band structure change of a semiconductor with and without the strain.

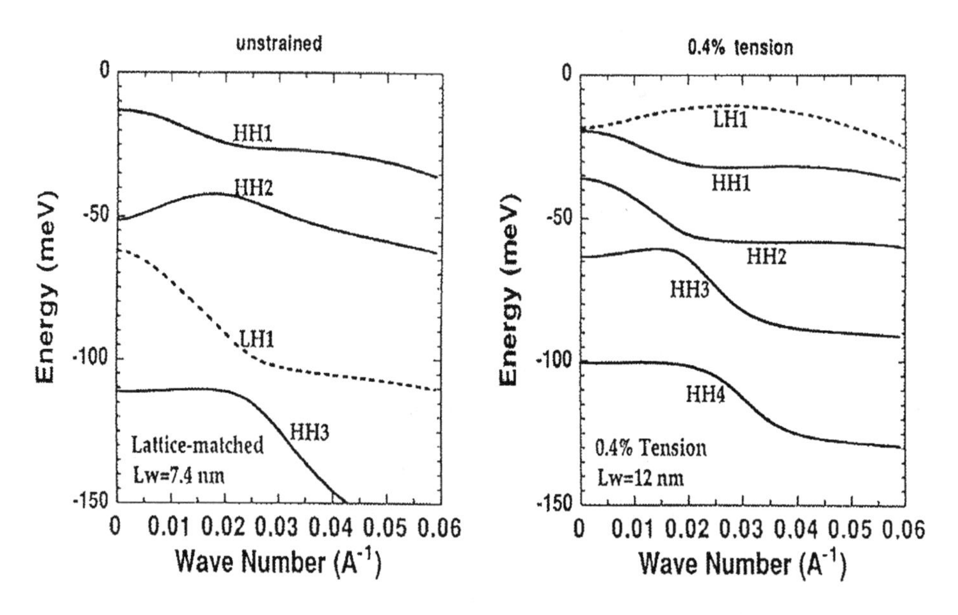

Fig. 10.14. Valence band diagram of InGaAs qantum well. (a) unstrained (b) tensile strained.

modulator as a function of reverse applied bias. The driving voltage for 20-dB extinction ratio both for TE and TM polarized light was 1.6 V, much less than that of the unstrained MQW modulator [19]. Note that the data indicate the absolute transmission optical light power, though in general the extinction ratio is shown at relative magnitude of the transmitted light intensity. The insertion loss difference between TE and TM polarized light is also less than 1 dB. That is, as far as these data are concerned, the polarization insensitivities have been carried out completely. The total insertion loss of the strained MQW modulator for TE polarized light was around 9-dB and the transmission loss was estimated to be 1-dB.

The frequency response was investigated by using an optical component analyzer (HP 8703A), and is shown in Fig. 10.17. The modulator is mounted on a microstrip transmission line in parallel with a 50 Ω terminating resistor. The modulation bandwidth is over 20 GHz, and the figure of merit, defined as the ratio of operation 3-dB bandwidth to driving voltage for 20-dB on/off ratio is larger than 13 GHz/V. This value is similar to or slightly smaller than the best value recently reported for an

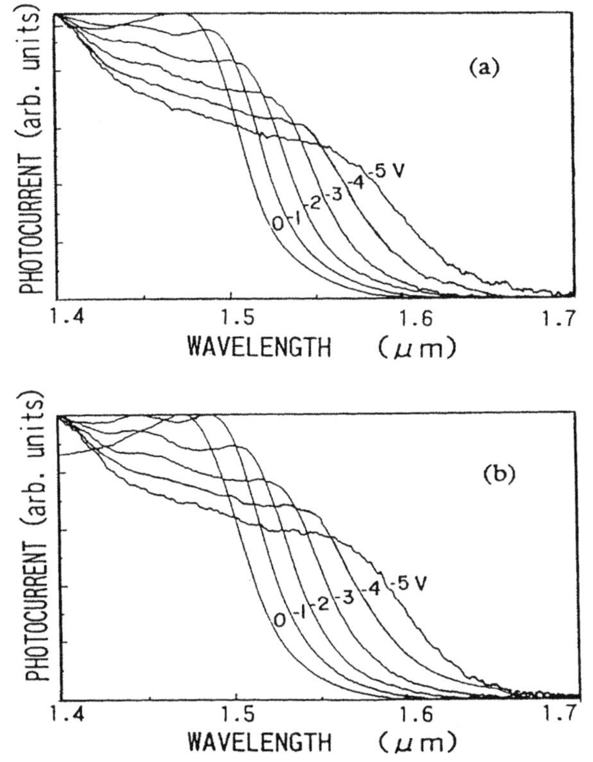

Fig. 10.15. Absorbed photocurrent spectra of strained MQW modulator. (a) TE polarization (b) TM polarization.

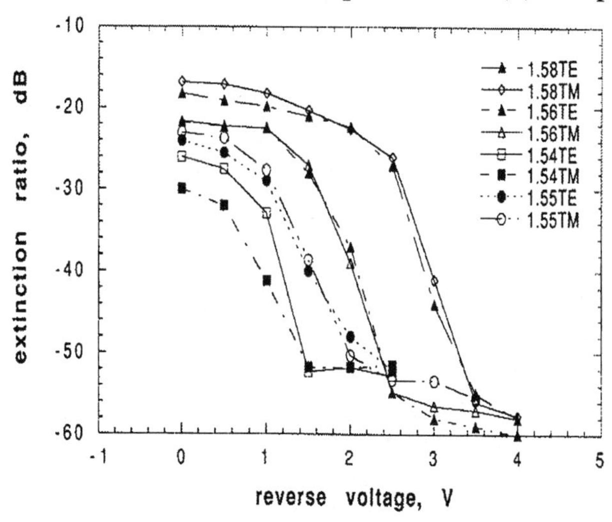

Fig. 10.16. Extinction chracteristics of the strained MQW modulauor. Incident light is TE or TM polarized with wavelengths of 1.54, 1.55, 1.56, or 1.58 μm.

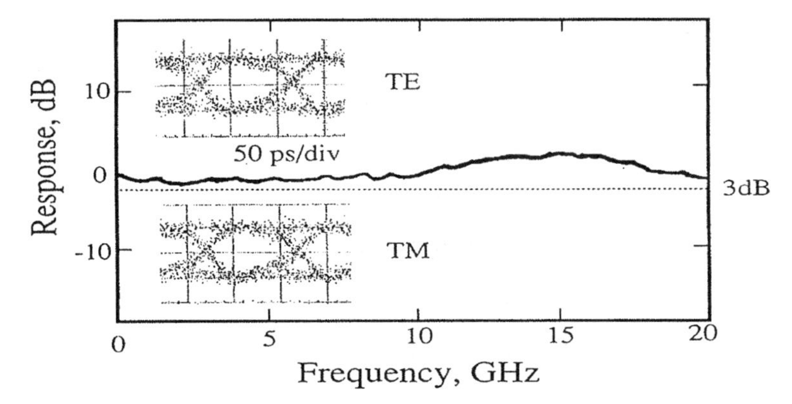

Fig. 10.17. Frequency response of the strained MQW modulator under 0.7 V DC bias and 10 Gbit/s eye diagram for TE and TM polarization of incident light.

MQW EA modulator [20],[28]. This improvement in driving voltage is considered to be due to a large band gap energy shift under applied voltage and strong optical confinement because of the wider quantum wells.

The inset of Fig. 10.17 indicates the eye diagrams for TE and TM polarization on incident light, respectively. Clear openings at 10 Gbit/s have been achieved both for TE and TM polarizations.

In conclusion, strain compensated MQW electroabsorption modulators can operate with polarization insensitivity; low driving voltage and high speed operation is demonstrated. The figure of merit of this modulator is higher than that of unstrained modulators.

10.3 Frequency Chirping

High bit-rate long-haul optical fiber transmission systems operating away from the zero-dispersion wavelength require transmitters with very low frequency chirp, such as those using lithium niobate ($LiNbO_3$:LN) waveguide electro-optic modulators. It is important to use low-chirp sources to reduce dispersion penalties, especially when Er-doped fiber amplifiers are used to overcome the limitations from fiber loss. At the same time precise evaluation of the chirp parameters of the modulators is necessary.

Previous techniques for measuring chirp in electroabsorption modulators are based on the assumption that the transmission versus voltage (T-V) property of small signals is linear. This assumption is not valid for conventional non-return to zero (NRZ) systems operated at large extinction ratios and soliton-based transmission systems, so it is very difficult to quantitatively measure the chirp of the transmitter. System engineers often had to evaluate chirp parameters by their own methods because the chirp parameters reported by device engineers, based on linear T-V characteristics, were usually so small that electroabsorption modulators were overestimated. In the following section, we summarize the various methods for measuring chirp parameters.

10.3.1 Linear method

Koyama and Iga [31] pointed out that electroabsorption modulation causes spectral broadening from the accompanying refractive index change. The index change is similar to the chirp observed in directly modulated semiconductor lasers. The first trial to determine the spectral linewidth enhancement factor α for the electroabsorption modulator was based on the sideband spectrum of the modulated output light from the Franz-Keldysh modulator [32,33]. In this method, the Fourier transform $\varepsilon(\omega)$ of the amplitude of the output light E(t) was first derived, and the power spectral ratio of the first sideband to the carrier was calculated as follows:

$$|\varepsilon(\omega_0+\omega_m)|^2/|\varepsilon(\omega_0)|^2 = |G_1/2|^2/|K_0|^2,$$

where

$$K_0 = 1 + \sum_n \frac{\gamma(\gamma-1)\text{——}(\gamma-2n+1)}{(2n)!2^{2n}} \binom{2n}{n} m^{2n}$$

$$G_p = \sum_n \frac{\gamma(\gamma-1)\text{-----}(\gamma-2n+2)}{(2n-1)!2^{2n}} \binom{2n-1}{n-p} m^{2n}$$

$$F_q = \sum_n \frac{\gamma(\gamma-1)\text{—}(\gamma-2n+1)}{(2n)!2^{2n-1}} \binom{2n}{n-q} m^{2n}$$

and

$$\varepsilon(\omega) = S_0^{1/2}\{K_0\delta(\omega-\omega_0) + \sum_p G_p\delta(\omega-\omega_0 + (2p-1)\omega_m)/2$$

$$+\sum_q F_q\delta(\omega-\omega_0 + 2q\omega_m)/2.$$

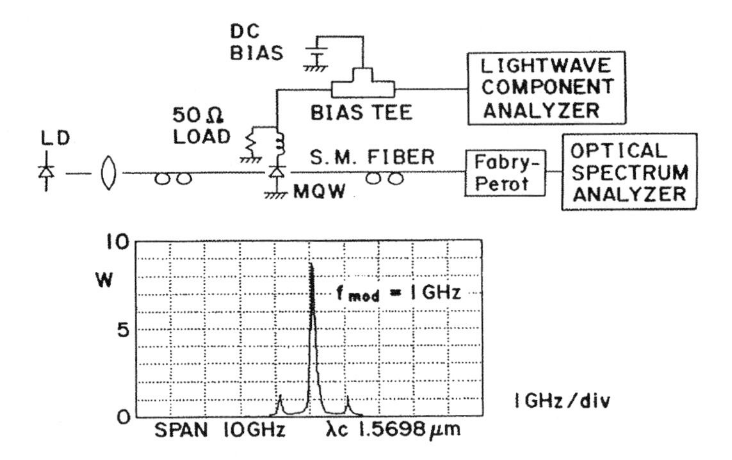

Fig. 10.18. Schematic of experimental apparatus for measuring chirp parameters.

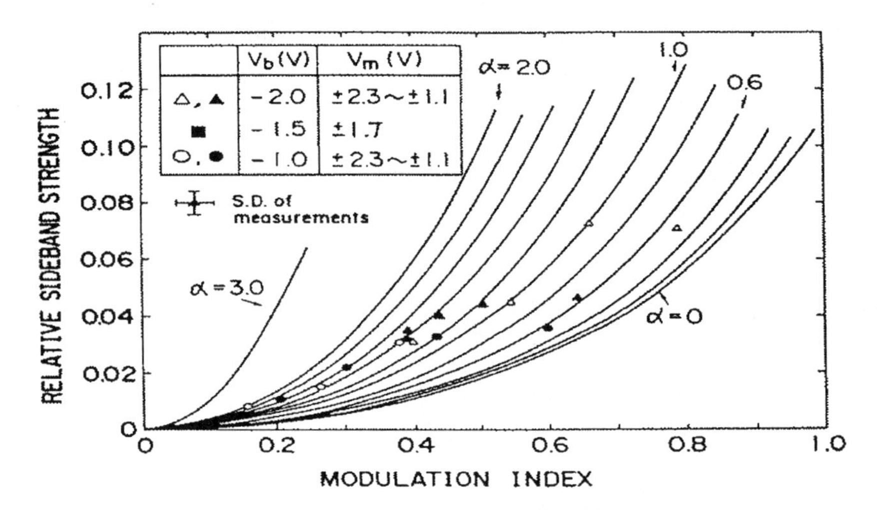

Fig. 10.19. Measured relative sideband strength and calculated curves with several α parameters.

Here ω_0 is an angular frequency of the input light to the modulator, m is an intensity modulation index, ω_m is a modulation angular frequency, and $\gamma = (1+j\alpha)/2$. By comparing the calculated results described above to the measured results, we can determine the absolute value of α.

Figure 10.18 shows a schematic diagram of experiments to measure the modulation index and signal-to-carrier ratio [34]. An rf modulation signal was applied to the modulator through an impedance-matched bias tee, and both the optical modulation depth and the relative sideband intensity in the optical power spectrum of the transmitted light were measured.

The curves for several α parameters are calculated from the relative sideband strength for MQW modulators as functions of the intensity modulation index m, and are shown in Fig. 10.19. In this experiment, the sideband strength is usually small for good modulators (small chirp) and the main peak (carrier frequency) strength depends on the extinction ratio. The signal-to-carrier ratio is at most around 0.1 when the α parameter is lower than 1.0. Usually, the power spectra of output light from modulators are measured with a scanning Fabry-Perot interferometer and indicated on an oscilloscope with a low resolution. Therefore, the accuracy of the measurement is not high. Moreover, as described in the introduction, the transmitted light intensity versus applied voltage characteristics for electroabsorption modulators are generally nonlinear, especially when the applied voltages are large. At that condition small chirp is produced. The above calculation assumes that the output intensity waveform is sinusoidal when the modulator is driven with a sinusoidal electrical signal. The α parameter results we obtained are underestimated because of reasons described in the following sections.

10.3.2 Nolinear Theory

The simple theory described above does not give exact results for electroabsorption modulators. The importance of nonlinearity in the T-V characteristics for evaluating α parameters has been reported [35-37]. The difference between parameters obtained from linear theory and those from transmission experiments was clarified when the MQW modulators were first tried in the transmission experiments. The α parameter was estimated at 1.5 from the transmitted waveform change as a function of the optical fiber length [38]. The α parameter for an MQW modulator was estimated at 0.8 [3] based on linear theory. At that time the difference was thought to be caused by the drive condition. Transmission experiment system-engineers used modulators with small dc biases to obtain high signal levels, while device engineers evaluated modulators with large biases to obtain small α parameters because they strongly depend on dc bias. Devaux et al. pointed out clearly [39] that the small α obtained by large dc biases is meaningless from the system point of view, and α should be evaluated above a 3 dB level below the maximum for transmitted light intensity because of the nonlinear relation between α and the bias.

An analytical theory that includes nonlinearities was reported [35] ,but it is limited to the simple cases in which transmitted light intensity and applied voltage have the following relation:

$$P_{out} = P_0 \exp\{-(V/V_0)^a\},$$

where P_0 is the effective intensity of the input light, V is the reverse

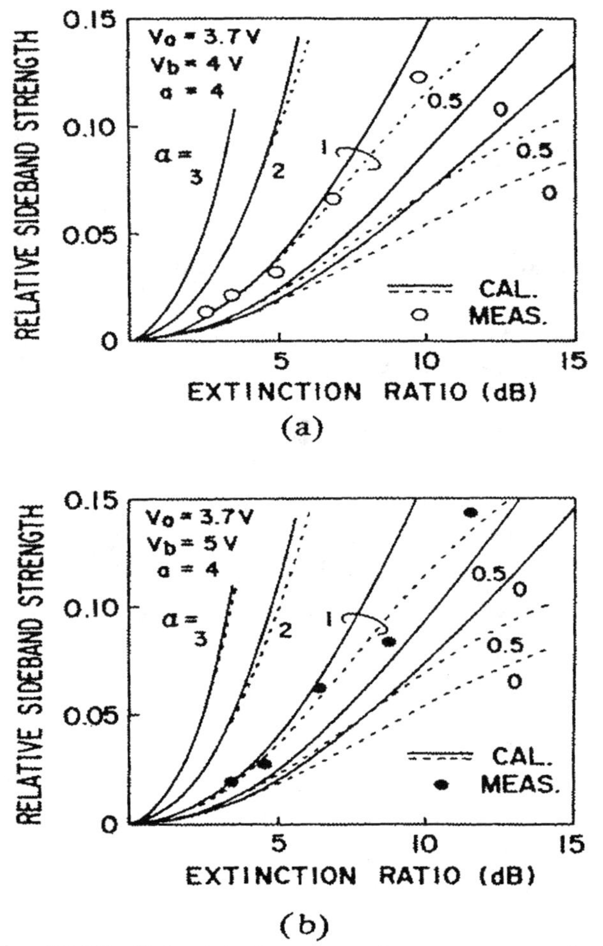

Fig. 10.20. Calculated curves of α parameters based on the theory with (solid lines) and without (dotted lines) nonlinear effect.

applied voltage, and V_0 and **a** are constant. The parameter **a** becomes approximately 2 for Franz-Keldysh modulators and 3-4 for MQW modulators. The calculation assumes the parameter **a** is an integer. This case is too simplified for real modulators, but the importance of the nonlinear effect and its difference from the linear theory are made clear. As a result, α parameter results are slightly larger than the results from the linear theory, and the difference increases as the modulation index increases. This is shown in Fig. 10.20.

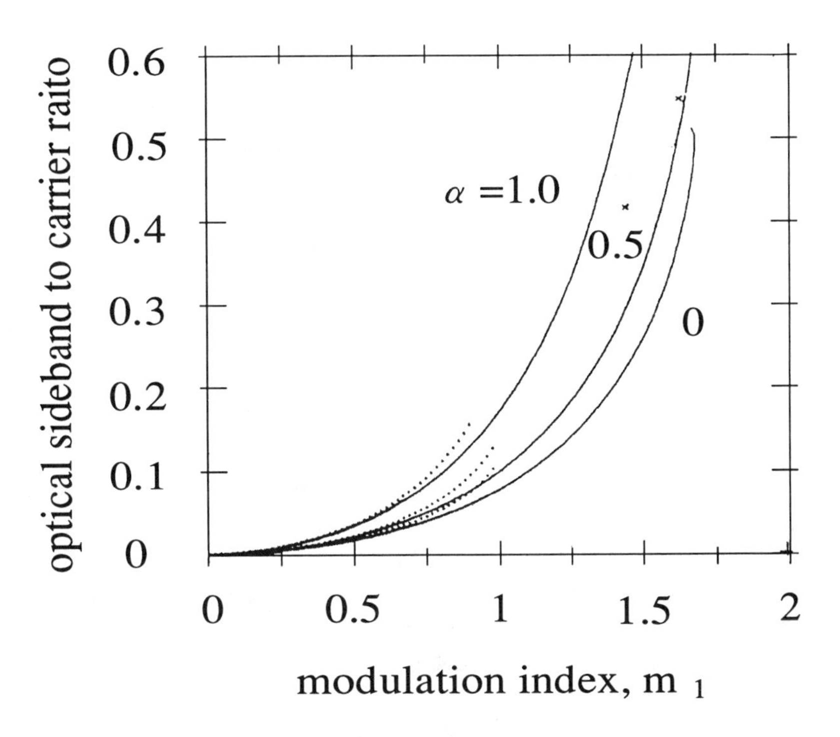

Fig. 10.21. Sideband-to-carrier ratio calculated as a function of intensity modulation index m_1. The dotted curves show the results of sinusoidal theory. The x's indicate experimental data points which are fit by α = 0.585 and 0.795 [36].

Recently Wood et al. demonstrated a numerical technique for measuring chirp that uses the actual T-V function of electroabsorption modulators [36]. The intensity of the optical output can be represented as

$$\mathbf{I}(t) = \mathbf{I}_0\{1 + m_1\cos(2\pi f_e t) + m_2\cos(4\pi f_e t) + \text{-------}],$$

where f_e is a sinusoidal signal frequency. It is possible to have m>1 without driving the optical intensity negative because the higher order terms in the above equation add components that keep $\mathbf{I}(t)>0$. In the proposed measurement technique, the α value of the peak-to-peak rf drive voltage to the modulator, V_{pp}, is estimated. The technique uses V_{pp}, the measured T-V, the dc bias value generated numerically, and the optical intensity waveform. The value of m_1 is measured directly using an HP 71400C lightwave signal analyzer, and V_{pp} is adjusted in the calculation to reproduce this value. The α parameter is defined as the ratio $\Delta n_{real}/\Delta n_{imag}$, where Δn_{real} and Δn_{imag} are the real and imaginary parts of the change in the refractive index caused by the applied electric field. A higher-frequency optical waveform that has the desired intensity and spurious phase modulation generated by Δn_{real} is generated numerically. A numerical Fourier transform of this optical waveform generates theoretical sideband-to-carrier ratios, which can be made equal to the observed values by adjusting α .

These procedures result in a remarkable difference from conventional calculations as shown in Fig. 10.21 [36]. The three full curves are calculated from the sideband-to-carrier ratio as functions of the modulation index m_1. The dotted curves that end at $m_1=1$ show results of sinusoidal theory. The two x's show experimental data points, and α is estimated at 0.795. It was estimated at 0.2-0.3 based on sinusoidal theory [32]. These results, clearly different for linear and nonlinear theory, were reported by Mitomi et al. [35] and the results obtained by Wood et al. [36] are similar to the analytical results reported by Mitomi et al. [37].

10.3.3 Other Methods for Evaluating Chirp

A. Differential absorption spectra. The simplest method to measure chirp parameters is to measure the absorption coefficient change and calculate the refractive index change through Kramers-Kronig relations. This is an indirect method and usually the Kramers-Kronig transformation needs wide-range wavelength information of the absorption spectra.

If a light source with a suitably wide wavelength range and a relatively large emitting power exists, then the transmission loss and refractive index change can both be measured by using the Fabry-Perot interferometric fringe change. This method is based on transmission loss measurements and gives us static chirp characteristics where propagation losses due to absorption are relatively small. In the large absorption loss range, where electroabsorption modulators are usually used, precise measurement is difficult. Monolithically integrated modulators with semiconductor optical amplifiers enable us to estimate α parameters in the relatively large propagation loss range [43].

B. Streak cameras. The methods described in A are static. For practical use, external modulators are driven by high-speed and large-signal modulation. Streak cameras enable us to evaluate the dynamic chirp of external modulators; the intensity change as well as the frequency change of light transmitted through modulators is measured at the same time.

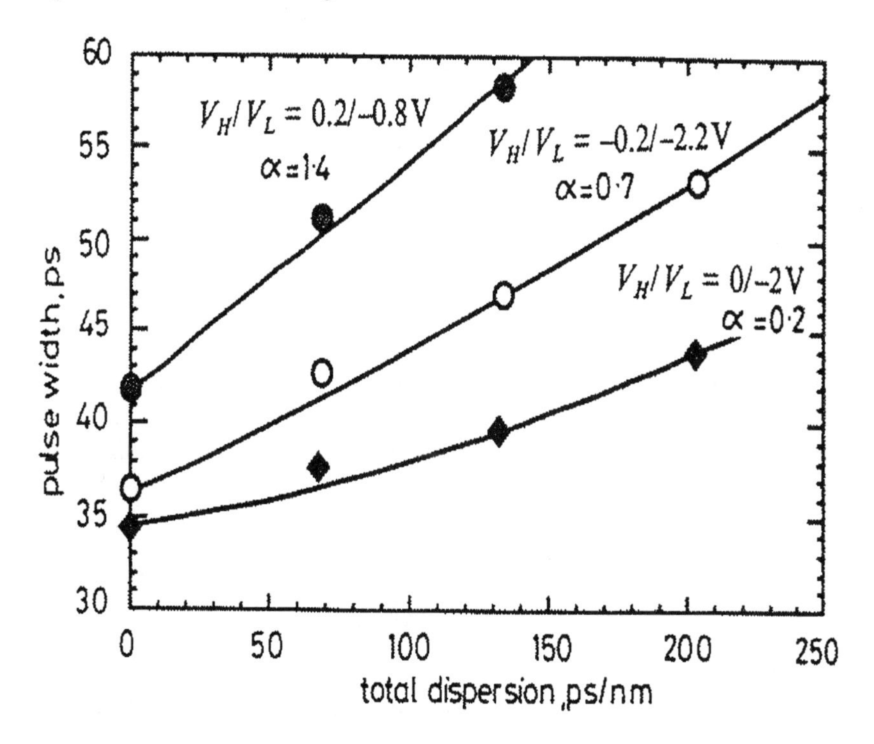

Fig. 10.22. Optical pulse width as a function of fiber length [43].

The time resolution limit is usually 10 ps, but for recent improved versions it is 2 ps. This method was first reported by Soda et al. [42] for a Franz-Keldysh modulator and its integrated light source. The streak camera is convenient, but high-priced.

C. Optical fiber. The optical pulse width broadens after transiting several dispersive fibers. Since the chirping of the modulators can be assumed to be linear, the chirp parameter can be estimated by comparing the measured and calculated pulse widths. This method is necessary for preparing many different length fibers. The results obtained show positive proof, thus, system engineers usually use this method. It was first used for evaluating α parameters of MQW electroabsorption modulators [38], and then for monolithic integrated light sources[43].

Based on the above method, an integrated InGaAsP/InGaAsP MQW electroabsorption modulator/DFB laser module was investigated. The large dependency of α on the bias voltage was measured; +1.4, +0.7, and +0.2 at the driving voltages of 0.2/-1.8V, 0/-2.0V, and -0.2/-2.2V [44]. Figure 10.22 shows the results.

Another method that uses a given length of standard fiber at 1.55 μm was reported [44]. This method is based on the small-signal bandwidth measurement of a fiber whose response exhibits sharp peaks arising from interferences of modulation sidebands (see Fig. 10.23). From small-signal computation, it can be shown that these resonance frequencies follow a very simple law [39].

$$\nu_u^2 L = (c/2D\lambda^2)\{1+2u-2(\tan^{-1}(\alpha))/\pi\},$$

where ν_u is the u-th resonance frequency with u = 0,1,2,3,...., L and D the length and dispersion of the fiber, c the speed of light, α the chirp parameter and λ the wavelength. From the linear regression of the resonance frequencies the fiber dispersion D and $\tan^{-1}(\alpha)$ are obtained with an uncertainty of less than +/-1%.

Mach-Zehnder modulators using $LiNbO_3$ were also evaluated based on this method [46]. Usually external modulators of the Mach-Zehnder type are considered to be chirp-free [31]. However, in multigigabit systems on standard (nondispersion shifted) fibers the residual chirp from Mach-Zehnder modulators cannot be neglected [46], and the dispersion penalty calculated for a chirp-free transmitter can be reduced considerably when using Mach-Zehnder modulators having small chirp [47].

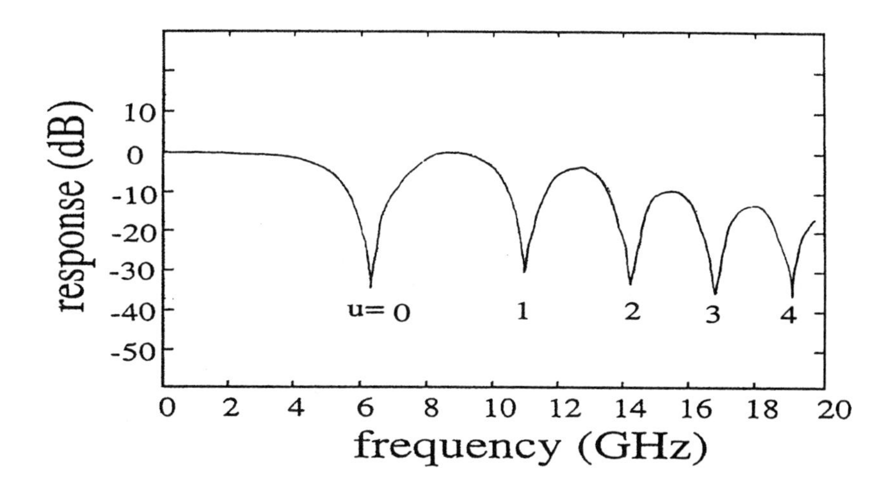

Fig. 10.23. Typical small-signal frequency response of standard fiber when an external modulator is used as a source of intensity modulated light.

D. Opening-eye evaluation. Recently, open-eye measurements and simulations of transmitted light through MQW modulators driven by high-speed modulation were compared with long-distance transmission, and their α parameters were estimated [48]. This method requires long-optical fibers but the results are certain. The frequency response and extinction ratio of modulators are also needed. By using these parameters we can calculate open-eye characteristics with α as a parameter. Electroabsorption modulators using strained InGaAs/InAlAs MQW structures were investigated and their α parameters are estimated at 0.6.

10.4 References

[1] R. J. Deri and E. Kapon, IEEE Quantum Electron., 27, pp. 626-639, 1991.

[2] E. Garmire, in "*Integrated Optics*", T. Tamir, Ed., New York: Springer-Verlag, 1985, pp. 243-304.

[3] M. Erman, GaAs and Related Compounds, Heraklion, Greece, 1987, pp. 33-40.

[4] B. Broberg and S. Lingren, J. Appl. Phys., 55, 3380, 1984.

[5] A. Carenco, Proc. ECIO87, Glasgo, 1987.
[6] H. C. Casey and M. B. Panish, *Heterostructure Lasers,* Part A, Academic Press 1978.
[7] D. Olego, Appl. Phys. Lett., 41, 476, 1982.
[8] D. I. Babic, M. J. Mondry, L. A. Coldren, and J. E. Bowers, LEOS'91, OE9.6, 1991.
[9] Y. S. Kim, S. S. Lee, R. V. Ramaswamy, S. Sakai, Y. C. Kao, and H. Shichijo, Appl. Phys. Lett., 56, pp. 802-804, 1989.
[10] R. J. Deri, R. Bhat, J. P. Harbison, M. Seto, A. Yi-Yan, L. T. Florez, M. Koza, and Y. H. Lo, IEEE Photon. Technol. Lett., 2, pp. 116-117, 1990.
[11] R. J. Deri, N. Yasuoka, M. Makiuchi, A. Kuramata, and O. Wada, Appl. Phys. Lett., 55, pp. 1495-1497, 1989.
[12] R. J. Deri, E. Kapon, R. Bhat, and M. Seto, Appl. Phys. Lett., 54, pp. 1737-1739, 1989.
[13] E. Kapon and R. Bhat, Appl. Phys. Lett., 50, pp. 1628-1630, 1987.
[14] S. Ohke, T. Vineda, and Y. Cho, Opt. Commun., 56, pp. 235-239, 1985.
[15] E. A. H. Marcatili, Bell Syst. Technol. J., 48, 2103, 1969.
[16] M. Heiblum and J. H. Harris, IEEE J. Quantum Electron., QE-11, 75, 1975.
[17] K. Kawano, K. Wakita, O. Mitomi, I. Kotaka, and M. Naganuma, IEEE J. Quantum Electron., 28, 224, 1992.
[18] K. Wakita, K. Yoshino, S. Matsumoto, I. Kotaka, N. Yoshimoto, S. Kondo, and Y. Noguchi, OFC'97.
[19] I. Kotaka, K. Wakita, K. Kawano, H. Asai, and M. Naganuma, Electron. Lett., 1991, 27, pp. 2162-2163.
[20] F. Devaux, F. Dorgeuille, A. Ougazzaden, F. Huet, M. Carre, A. Carenco, M. Henry, Y. Sorel, J. -F. Kerdiles, and E. Jeanney, Photon. Technol. Lett., 1993, 5, pp. 1288-1290.
[21] M. Aoki, M. Takahashi, M. Suzuki, H. Sano, K. Uomi, T. Kawano, and A. Takai, ibid., 1992, 4, pp. 580-582.
[22] K. Sato, I. Kotaka, K. Wakita, Y. Kondo, and M. Yamamoto Electron. Lett., 1993, 29, pp. 1087-1089.
[23] K. Tada, and H. Noguchi, IOOC'89, paper 18D2-2.
[24] K. Komatsu, K. Hamamoto, M. Sugimoto, Y. Kohga, and A. Suzu ki, Top. Meet. on Photon. Switch., P. 24, 1991.
[25] M. Bachmann, E. Gini, and H. Melchior, ECOC'92, TuB7.4, 345-348.

[26] K. Magari, M. Okamoto, H. Yasaka, K. Sato, Y. Noguchi, and O. Mikami, Photon. Technol. Lett., 1990, 2, pp. 556-558.
[27] J. E. Zucker, K. L. Jones, T. H. Chiu, B. Tell, and K. Brown-Goebeler, Integrated Photon. Research, PD7, 1992.
[28] F. Devaux, F. Dorgeuille, A. Ougazzaden, F. Huet, M. Carre, A. Carenco, M. Henry, Y. Sorel, J. -F. Kerdiles, and E. Jeanney, Electron. Lett., 1993, 29, pp. 1202-1203.
[29] T. Ido, H. Sano, D. J. Moss, S. Tanaka, and A. Takai, Fifth Optoelectronics Conf. (OEC'94) Technical Digest, July 1994, 15B3-1, pp. 342-343. Photon. Technol. Lett., 6, 1994.
[30] S. Kondo, K. Wakita, Y. Noguchi, N. Yoshomoto, M. Nakao, and K. Nakashoma, J. Electron. Materials,1996, 25,pp. 385-388.
[31] F. Koyama and K. Iga, Electron. Lett., 21, 1065, 1985. and J. Lightwave Technol., 6, 87, 1988.
[32] M. Suzuki, Y. Noda, Y. Kushiro, and S. Akiba, Electron. Lett., 22, 312, 1986. In this letter, the coefficient of sideband strength G1:(γ-2n-2) should be correctd to (γ-2n+2).
[33] T. H. Wood, R. W. Tkach, and A. R. Chraplyvy, Appl. Phys. Lett., 50, 789, 1987.
[34] K. Wakita, Y. Yoshikuni, M. Nakao, Y. Kawamura, and H. Asahi, Jpn. J. Appl. Phys., 26, L1629, 1987.
[35] O. Mitomi, S. Nojima, I. Kotaka, K. Wakita, K. Kawano, and M. Naganuma, J. Lightwave Technol., 10, 71, 1992.
[36] T. H. Wood,L. M. Ostar, and M. Suzuki, J. Lightwave Technol., 10, 1926, 1994. The same authors reported these results in CLEO'93, CWA3, where they misunderstood the results reported in Reference 5 and described an opposite conclusion. In this paper they reached the same conclusion as those described in Reference 5.
[37] O. Mitomi, K. Wakita, and I. Kotaka, IEEE Photon. Technol. Lett., 6, 205, 1994.
[38] T. Kataoka, Y. Miyamoto, K. Hagimoto, K. Wakita, and I. Kotaka, Electron. Lett., 28, 897, 1992.
[39] F. Devaux, Y. Sorel, and J. F. Kerdiles, J. Lightwave Tech- nol., LT-11, 1937, 1993.
[40] M. Suzuki, H. Tanaka, and Y. Matsushima, IEEE Photon. Tech nol. Lett., 4, 586, 1992.
[41] J. Weiner, D. A. B. Miller, and D. S. Chemla, Appl. Phys. Lett., 50, 842, 1987.
[42] H. Soda, K. Kakai, and H. Ishikawa, ECOC'88, P.227, 1988.

[43] T. Kataoka, Y. Miyamoto, K. Hagimoto, K. Sato, I. Kotaka, and K. Wakita, Electron. Lett., 30, 872, 1994.
[44] F. Devaux, Y. Sorel, and J. F. Kerdiles, Electron. Lett., 29, 815, 1993.
[45] M. Schiess and H. Carlden, Electron. Lett., 30, 1524, 1994.
[46] A. Djupsjobacka, IEEE Photon. Technol., 3, 41, 1992.
[47] A. H. Gnauck, S. K. Korotoky, J. J. Veselka, J. Nagel, C. T. Kemmerer, W. J. Minford, and D. T. Moser, IEEE Photon. Technol. Lett., 3, 916, 1991.
[48] T. Ido, H. Sano, D. J. Moss, S. Tanaka, and A. Takai, IEEE Photon. Technol. Lett., 1994.

Chapter 11

Crystal Growth and Device Fabrication

In this chapter we introduce some crystal growth methods, crystal evaluation methods and processing techniques used to fabricate optical modulators. These are similar to the methods and techniques used to fabricate semiconductor laser diodes (for details see Reference [1-3]).

11.1 Crystal Growth

Optical devices are generally used within a limited wavelength region, and the crystals used in these devices consist of heterostructures and multiple layers. It is important that the methods used to grow crystals for optical devices enable thickness and composition of the crystals to be controlled, that they produce uniform crystals of high quality, and that they allow selective growth and regrowth for buried heterostructures. The operating wavelengths of the main semiconductors and their related mixed crystals are given in Fig. 11.1. The epitaxial layers are grown on the substrate to be lattice-matched, except when strained layers are needed. And to fix the operating wavelength, it is important that the composition of the mixture in epitaxial layers can be controlled.

For conventional optical devices, thickness to within 0.1-0.05 μm has been necessary to control but for recent quantum well or strained supperlattice devices thickness must be a few monolayers (about 1 nm).

Some devices, such as optical switching devices, and monolithically integrated optical devices, are very large and the epitaxial layers used for these devices are required to be uniform on thickness and composition in the plane. Because device characteristics and reliability are sensitive to defects within or at the interfaces of epitaxial layers, lattice-matching conditions and defect-free layers are necessary.

Table 11.1 shows the main crystal growth methods used to make optical devices. Accordinng to the phase of growth source, these methods can be classified into three kinds: vapor phase epitaxy (VPE), liquid phase epitaxy

Table 11.1. Main epitaxial growth methods for compound semiconductors.

vapor phase epitaxy (VPE)	halogenated	chloride VPE hydride VPE
	metal organic VPE	atomospheric-pressure Low-pressure
liquid phase epitaxy (LPE)		
molecular beam epitaxy (MBE)	solid source MBE	
	gas source MBE	MOMBE GS-MBE chemical beam epitaxy CBE

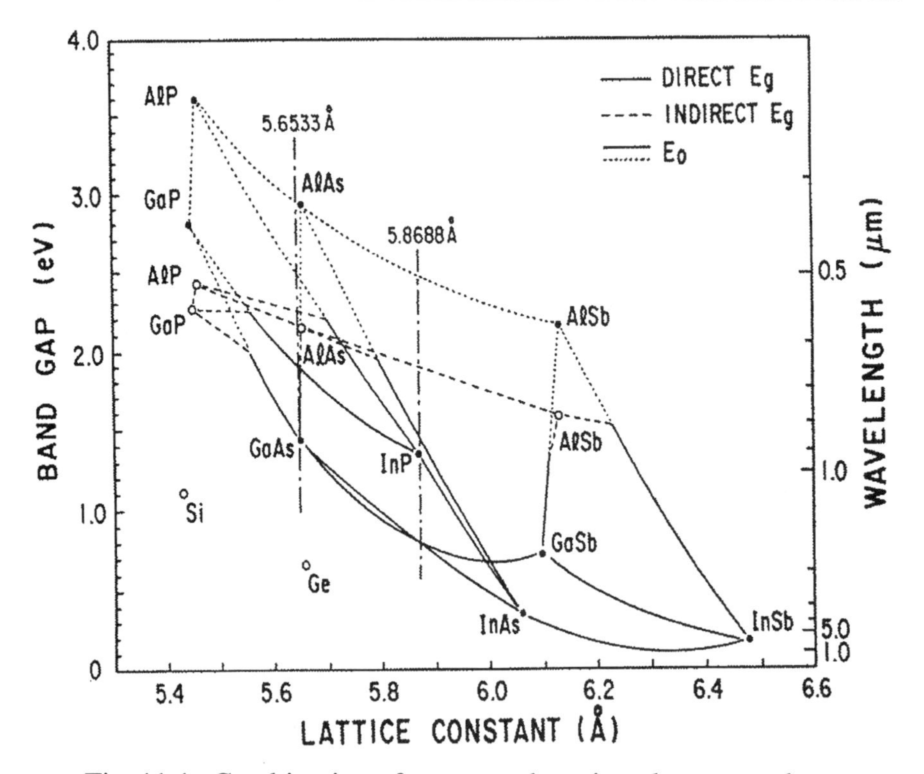

Fig. 11.1. Combination of compound semiconductors and operating wavelength as a function of lattice constant.

(LPE) and solid phase epitaxy (SPE). These kinds can be in turn be subclassified according to the vacuum degree in the growth chamber.

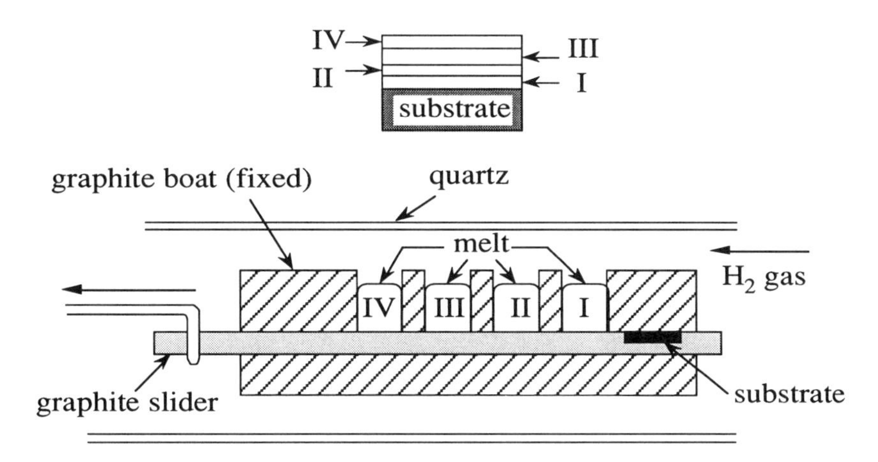

Fig. 11.2. Cross section of an apparatus used for liquid phase epitaxy.

11.1.1 LPE

LPE was the method most frequently used for making optical devices and nowadays it has been used for commercial base optical devices. The LPE technologies were developed in 1970s and the first half of 1980s in concert with semiconductor laser technologies. Some advantages of this method are that LPE equipment is simple, easy to use, stable, and less expensive than MBE and MOVPE equipment. And because epitaxial layers are grown rapidly, this method is suitable for making thick layers. It can be used to produce selective growth and can also be used to melt-back the surface damage induced during the initial stage of growth.

Some disadvantages of this method are that it is difficult to use when fabricating large-scale epitaxial wafers because the thickness in the wafer plane and the content uniformity in the growth direction are not uniform, and it is lacking in reproducibility and mass production. Moreover, it is difficult to use when applying for the quantum size effect and strained superlattice structure because precious thickness control within 10 nm and steep interface are required. That is, this method is powerful and is still the main technology used to make practical devices for which it is suitable, but it is difficult to use to make the newest devices. Figure 11.2 shows the cross section of an instrument used for LPE.

11.1.2 MOVPE

MOVPE is metal-organic vapor phase epitaxial growth, and there are two different methods: atmospheric-pressure MOVPE and low-pressure MOVPE. The other kind of VPE is also subdivided into two methods: chloride VPE and hydride VPE. These differ according to whether chloride or hydride is used as the growth source. Chloride VPE is a powerful method for making electron devices but cannot be used to make optical devices including Al, such as those made of GaAlAs, because of the reaction between the quartz tube and transport materials. MOVPE is the method most frequently used to make practical optical devices.

MOVPE was invented in 1968 and activated since the room temperature continuous operation was reported for GaAs/AlGaAs double-heterostructure lasers in 1978 by Dupuis and Dapkus. Metal-organic materials such as $(CH_3)_3In$ and $(C_2H_5)_3Ga$ are used as growth sources. These materials are liquids or solids at room temperature but have relatively high vapor pressure and can be supplied stably as gas sources when crystals are grown. As a result, both III and V-compound element can be treated as gas even at room temperature, and vacuum is not necessary for crystal growth. This results in many advantages, such as thickness and content homogeneity, steep hetero-interfaces, and good controllability. This method is currently used for the industrial production of AlGaAs/GaAs for HEMT (high electron mobility transister) and semiconductor lasers operating at wavelength of 0.7-0.98 μm.

For long-wavelength optical devices using InGaAsP/InP materials, on the other hand, LPE is the predominant method used industrially, and MOVPE is only beginning to be used in manufacturing. The growth details have been already reported [4-5], and in this section we describe the MOVPE used to make the InGaAs/InAlAs compound materials used in laser diodes and in modulators operating at long wavelengths.

InGaAs/InAlAs multiple quantum well (MQW) structures have a larger conduction band offset (ΔE_c=~0.5 eV) and a smaller valence band offset than do other candidate material systems based on InGaAs(P)/InP. This means they are very promising materials for high-speed, highly efficient opto-electronic devices such as lasers, detectors, modulators, and switches.

Low driving voltage is an important requirement for electroabsorption modulators because the output voltage of high-speed electric drivers is limited. Low-drive-voltage operation is attained by using thick well layers that enhance the quantum-confined Stark effect (QCSE), where the shift in the exciton resonance peak is approximately proportional to the fourth power of the well thickness. There are two ways to create such a modulator without changing the guiding wavelength. One is to use quaternary InGaAlAs materials and the other is to introduce strained layers. InGaAlAs makes it possible to optimize the bandgap energy and well thickness independently, and these techniques have been used in making optical modulators [6,7] and optical switches [8]. Introducing strain into InGaAs well layers also enables independent control of the wavelength and well thickness [9]. Furthermore, introducing tensile strain can reduce the polarization dependence of the modulator because exciton resonances are associated with the absorption of TE and TM polarized incident light, respectively, and tenside strain induces the degeneracy heavy hole and light hole band [10].

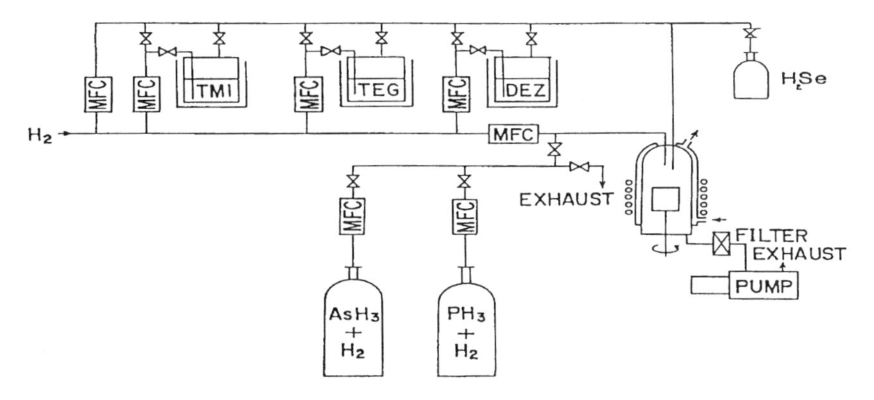

Fig. 11.3. Metal-organic vapor phase epitaxy system and growth chamber.

Figure 11.3 shows the typical MOVPE equipment. Generally, the organometallic sources are easily dissociated at a several hundred degrees centigrade and to dissociate them only at the crystal substrate the reactive quartz tube are cooled down with water. RF heating and infra-red heating are often used to heat up the substrate. The gas flow is controlled, with a mass-flow controller (MFC), to an accuracy within 1%.

InGaAs/InAlAs MQWs are grown at 70 Torr with a flow rate of 5 LSM in a low-pressure MOVPE apparatus with a vertical reactor and leak-tight system. TMI, TEG, TMA, and DEZn are suppied as the In, Ga, and Zn sources. Arsine and phoshine are also used as group-V sources. The growth temperature is 665 °C and the V/III ratio for the MQW layer growth is about 100. The growth rate for the MQW layer is typically about 1.2 μm/h.

11.1.3 MBE

MBE (molecular beam epitaxy) is a kind of VPE because it uses a vapor-phase source, but the growth is done under an ultra high vaccum ($<10^{-3}$ Pa) and the growth mechanism differs from the standard VPE growth mechanism. The mean free path of the molecule is so long (>1 m) that the source molecule is free from interacting with each other in the vacuum chamber and the growth is determined only by the substrate surface condition. That means the growth is done without processing chemically and relaxing the energy of molecule interaction. Initially, the sources was a solid- state source, but now vapor-phase sources are also used. These processes are called gas source MBE or chemical beam epitaxal growth (CBE).

So little impurities remain, because of the ultrahigh vacuum used, that high-quality epitaxial layers are grown even at very low growth rates (lower than 0.1 nm/sec). The interaction of molecules is so small that the growth can be controlled by a shutter. In the initial stages of MBE development, excessive amounts of remaining gases prevented the formation of epitaxial layers of device grade. A three-section apparatus consisting of growth, retreatment, and exchange sections has therefore been developed and designed to be free of water vapor and carbon, which are extrinsic contaminants. Since the middle of the 1970s laser diodes with low threshold current density comparable to that of LPE-grown laser diodes have been demonstrated, and steep growth in the hetero-interfaces has been obtained also. Superlattice and modulation-doped structures have been grown by MBE and new devices using those structures are now being investigated.

Figure 11.4 shows the schematics of an MBE growth system. MBE has many advantages: it enables many combination of semiconductor materials and hydrides and metal organic sources to be used as not only as source materials but also as dopants. The growth mode is also simple and many materials can be grown more easily by MBE than by MOVPE.

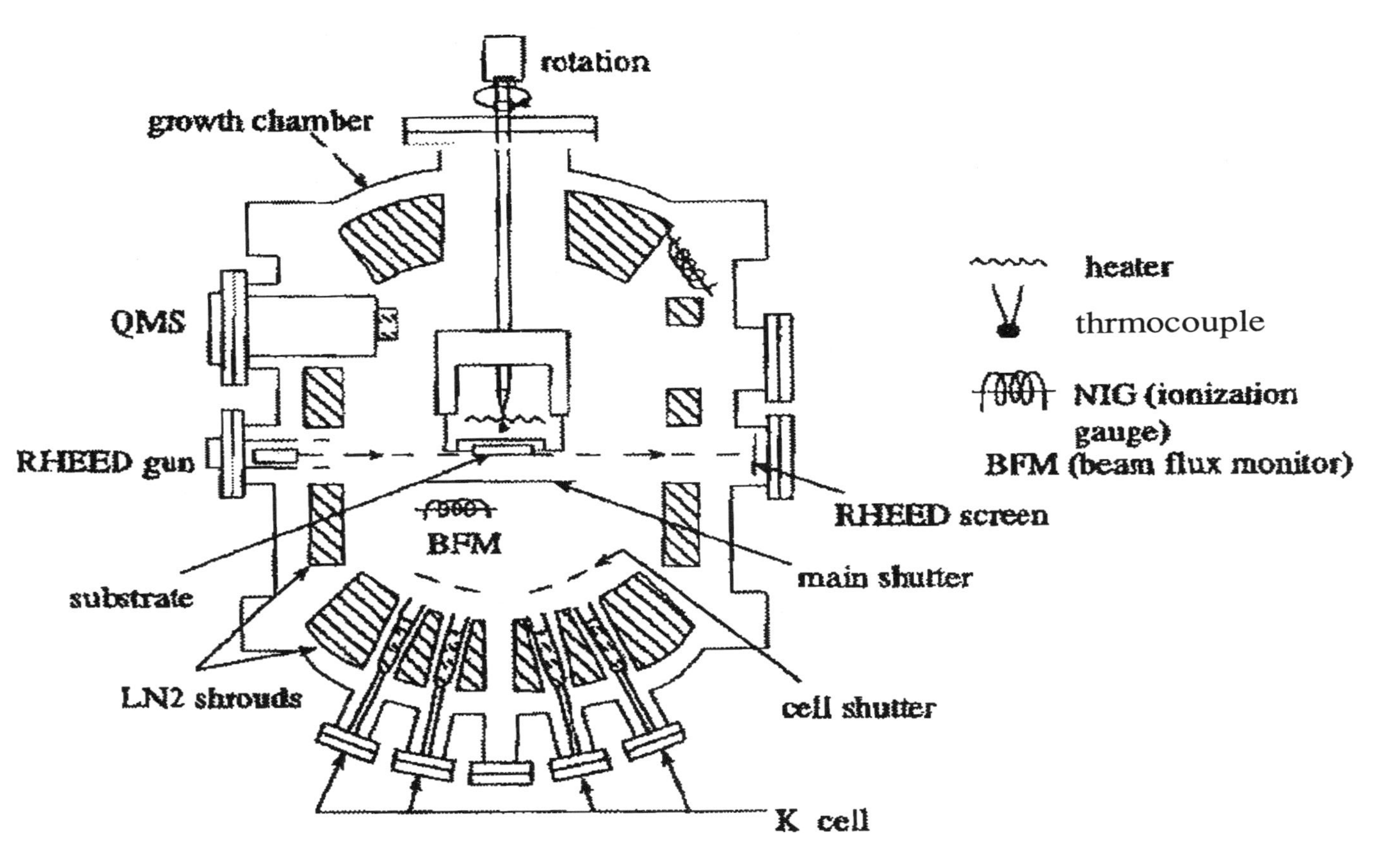

Fig. 11.4. Molecular beam epitaxy system and growth chamber.

11.1.4 Selective Growth

In the epitaxial growth condition, atoms supplied from sources grow on the substrate step by step, resulting in layers. When part of the surface of the substrate is covered by oxides or nitrides, crystal growth occurs only on the uncovered regions. A SiO_2 or SiN_x film deposited by sputtering or CVD (chemical vapor deposition) is usually used as a mask. In principle, all crystal growth techniques could be used for selective growth, but LPE and the VPEs are most selective. The next most selective is MOVPE, and the most difficult to use is MBE.

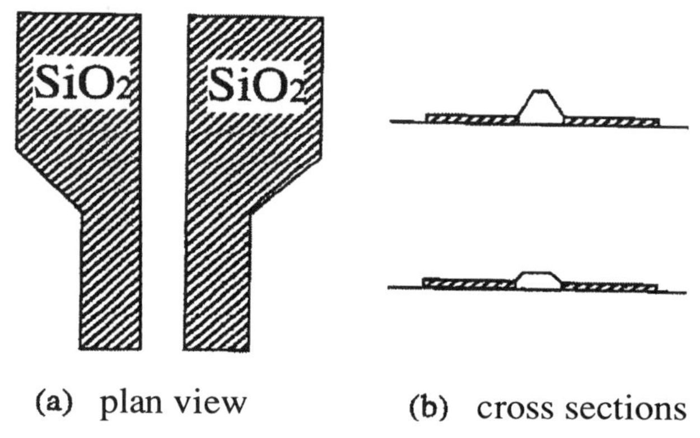

Fig. 11.5. Principle of selective-area growth.

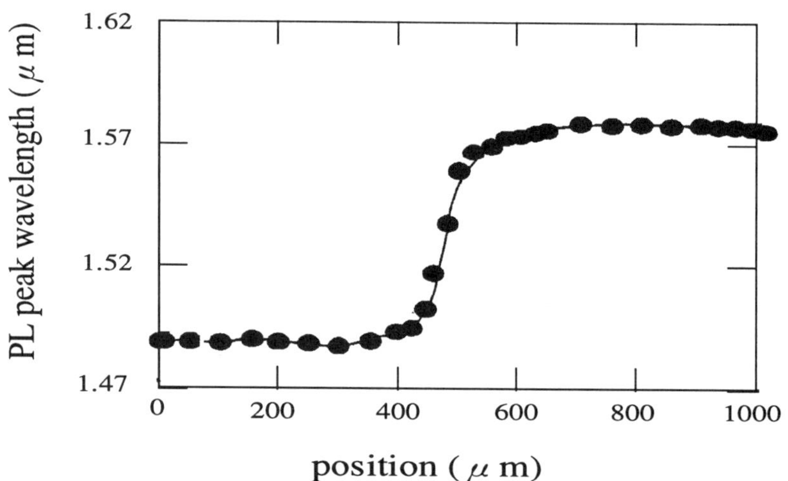

Fig. 11.6. Photoluminescence peak wavelength profile for MQWs grown by selective area growth method [14].

Recently, with regard to the monolithic integration of different devices, selective growth techniques have been very effective in controlling the bandgap energy and in burying heterostructures [11-13]. They have therefore been investigated by many groups. Figure 11.5 shows plan and cross-sectional views of a selective growth mask. The sources supplied on the mask diffuse to the bare region because of the concentration gradient, and they contribute to the crystal growth. The growth in the region adjacent to the selective growth mask is thus faster than that away from the mask (because of the increase in source supply). This is a powerful technique, when a MQW or supperlattice structure is grown because the thickness difference in quantum well layers results in an increase in the energy bandgap difference. This selective area growth by MOVPE controls the in-plane local energy of MQWs simultaneously [11-13]. This technique offers the inherently high optical coupling efficiency and reproducible fabrication needed when making integrated optical devices.

Figure 11.6 shows the photoluminecsence (PL) peak wavelengths of MQWs measured along the optical optical axis [14]. The PL wavelengths for the two sections are precisely controlled (by the selective-area-growth mask dimensions) to 1.49 and 1.57 μm. The length of the transition region between devices, which is peculiar to this method and was found to be less than 70 μm. This region operates as losses for guided light, but it is not a severe problem. This technique has been used for monolithic integration of DFB laser diodes and electroabsorption modulators, for making arrayed optical devices for wavelength-division multiplexing with multiwavelength DFB lasers, and for making spot-size converted lasers. Details of the application of this technique is discussed in 8.3.3.

11.2 Crystal Evaluations

11.2.1 X-ray Double Crystal

The double-crystal x-ray diffraction (sometimes called the X-ray rocking curve) technique is used for studying epitaxial crystals including superlattices. The full-width at half maximum (FWHM) indicates the quality of crystal. This method is mainly used for evaluating the condition for a lattice-matching between epitaxial layers and the substrate. The technique also provides imformation on supperlattice composition, layer thickness, and strain. A kinematical or a dynamical diffraction theory is

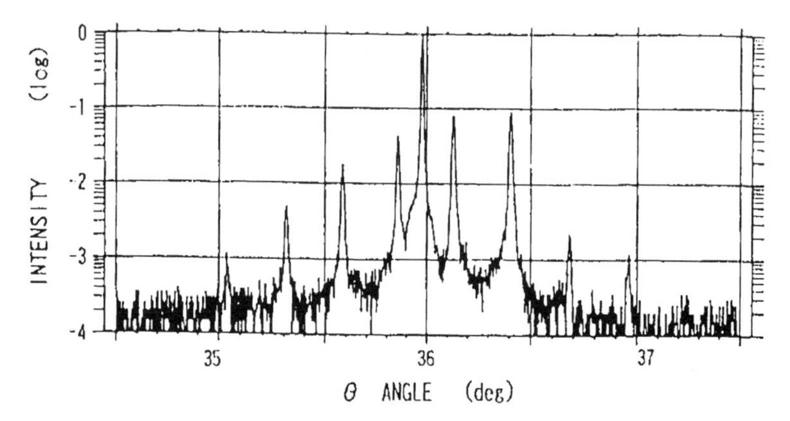

Fig. 11.7. X-ray rocking curve from a strained InGaAs/InAlAs MQW structure.

is used to analyze the x-ray rocking curve. For example, the measurement of the difference $\Delta\theta$ of the Bragg angle of a strained structure by means of high-resolution diffraction technique is a well established method for determining the parallel **e** and the perpendicular **e** strain components of the layer with respect to the interface.

For supperlattice structures, there are many satellite peaks in the diffraction pattern and these peaks are used to estimate the number of periods, each lattice-constant, and each layer thickness. Figure 11.7 shows the typical x-ray diffraction pattern for a strained superlattice structure [7].

11.2.2 Photoluminescence

Usual III-V compound semiconductors are direct transition groups and enable us to make heterostructures, which results in highly efficient photoluminescence (PL). The PL observation is easy even at room temperature and the peak energy of the PL intensity is associated with the bandgap energy. The intensity, full width at half maximum (FWHM), and the shape provide important imformation about the epitaxial layers.

The PL topograph and mapping of peak wavelength, intensity, and FWHM are easily obtained by using a personal-computer controlled instrument. This instrument was developed for evaluating long-wavelength DH wafers and at a scanning speed of 1.3 μm/sec can map the PL topograph for a whole 2-inch wafers in 20 minutes. Figure 11.8

shows a PL experimental block diagram and 2-inch-wafer PL topographs from MQW layers grown on an InP substrate [15].

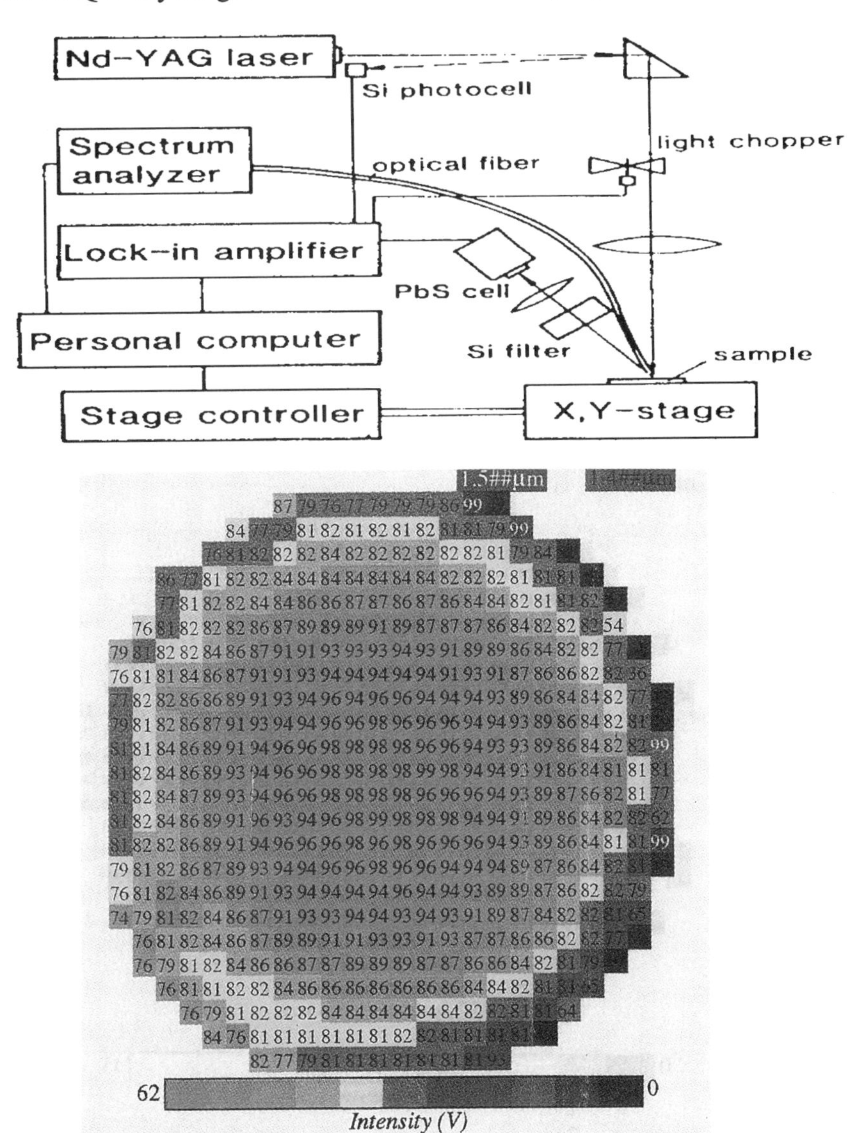

Fig. 11.8. Photoluminescence apparatus and topographs of intensity and wavelength for a 2-inch InGaAsP/InGaAsP MQW epitaxial wafer.

11.2.3 Transmission

The exact absorption band edge is evaluated in an optical transmission experiment. However, the optical interaction length is not so large because the transmission experiment uses the configuration of surface normal mode. It needs the different wafers with thicker layers than those of devices to obtain high contrast ratio. Usually optical devices are used for waveguided structures with thin layers and photocurrent experiment is used in place of transmission experiment, although device processes such as ohmic contacts, chipping, bonding, etc. are necessary. The absorption band edge and electroabsorption effects, such as peak or tail energy shift, and the change associated with applied bias are estimated by the photocurrent spectra. The schematic diagram of a photocurrent spectrum experiment is shown in Fig. 11. 9.

The polarization effect is also investigated by changing the incident light polarization with a Glan-Thomson prism. The longer the absorption region gives the higher the sensitivity in the photocurrent spectra, while the difference between the photocurrent and the absorption coefficient increases. A short optical waveguide (a few micrometers long) is used to obtain a good representation of electroabsorption spectra because the propagation length is very short.

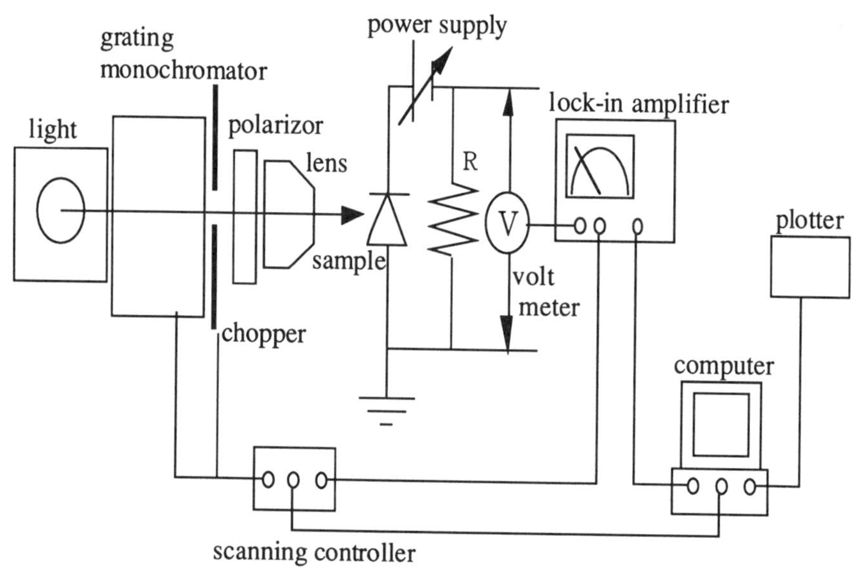

Fig. 11.9. Photocurrent experimental block diagram.

11.2.4 Loss Evaluation

In general, semiconductor modulators as well as other semiconductor devices are subject to insertion losses ranging from 6 to 18 dB. Most of the loss is coupling loss (to optical fibers). The material absorption loss is not negligible, since the QCSE or Franz-Keldysh effect is effective only near the exciton resonance or the energy band edge.

The Fabry-Perot etalon method is popular and often used to evaluate the loss of a transparent waveguide because of its simplicity and high accuracy [16]. This method is based on the interference between an incident light and a reflected light from the cavity facet. When the incident light wavelength is changed by changing the temperature of the laser light source, the transmitted power changes as the etalon resonance changes. The ratio of the maximum and the minimum of transmitted light intensity gives the propagation loss. That is

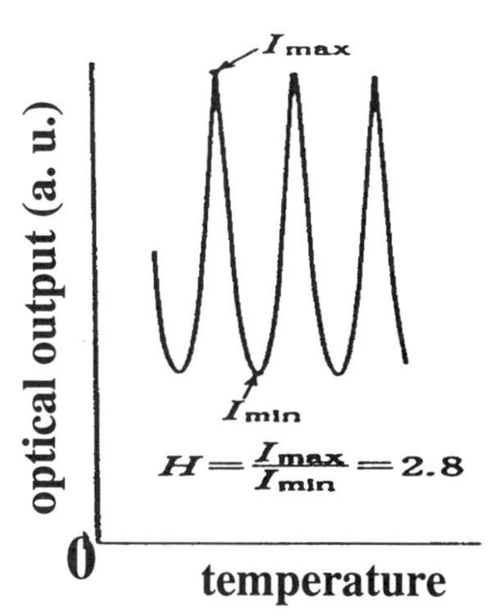

Fig. 11.10. Typical result of a Fabry-Perot resonance method.

$$H = (1+Re^{-\alpha L})^2/(1-Re^{-\alpha L})^2,$$

where R is power reflectivity, α is a propagation loss and L is a sample length. From this equation, we obtain propagation α. Figure 11.10 shows the principle of a Fabry-Perot interferometric method. The reflectivity R is given as follows by using the refractive index n of the medium when the medium is infinite.

$$R = (n-1)^2/(n+1)^2$$

For the medium with a finite and waveguided structure, R is modified depending on whether the incident light polarization is TE or TM, and the depth of the core thickness and the optical confinement.

The propagation loss for a lossy waveguide can be measured adequately by the conventional cutback method, but this method is tedius and not very reproducible. Wood [17] proposed a simpler alternative based on the photocurrent response as well as the optical transmission function and Chin [18] modified this method more accurately without a guess of the absorption coefficient.

11.3 Fabrication Process

The following sections briefly review the processing technology for fabricating semiconductor modulators and switches. Many kinds of processing are necessary: lithography, dielectric and metal deposition, etching, bonding, and packaging for microwave operation. It describes the lithography, deposition of highly reflective and antireflective films, disordering techniques, and dry etching. Disordering is described in greater detail because it is an especially powerful technique for creating a lateral energy bandgap and refractive index profile without using a regrowth process.

11.3.1 Lithography

The lithography technologies used to make optical devices are mainly optical, but electron-beam lithography is used for extreamly precise fine processing. The characteristics of various lithography techniques are listed in Table 11.2.

In photolithography, three methods are mainly used. The proximity method puts the sample in contact with the mask and is the simplest. The instrument it uses is inexpensive and can operate to any sample size. This method is therefore suitable for device fabrication in laboratories. The mirror projection method is for large samples and is suitable for production. The sample size can also be changed by changing the substrate installer. The reduction projection method has very good reproducibility, and is suitable for mass production and for making microstructures.

11.3.2 Dielectric Film Deposition

The techniques usually used for depositing a dielectric film for photolithograpy as well as for surface passivation and applying antireflection (AR) or highly reflective (HR) coatings are sputtering and plasma chemical vapor deposition (CVD). It is necessary to take into account the adhesion between the dielectric film and semiconductor surface, especially when we deposit film for AR or HR coating. Such a

Table 11.2. Lithography techniques

	photo-lithography			electron beam
	contact	mirror	reduction projection	
resolution	1-3 μm	1-5 μm	0.3 μm	10 nm
accuracy	1 μm	0.5 μm	0.1 μm	40 nm
surface irregularity	not enough	not enough	not enough	good
substrate structure	good	not enough	poor	very good
reproducibility	not enough	very good	very good	very good
productivity	good	very good	very good	good
comments	ease common	large area massproduction	high-resolution	multi-use high-resolution

film is often deposited on a pn junction, and the leak current due to deposition damage must be supppressed [19]. Electron-cyclotron-resonance (ECR) plasma has been used in low-temperature processing because of producing active and dense ions [20]. Microwave power (2.45 GHz) is introduced into a plasma chamber through a rectangular waveguide and a window made of fused quartz (Fig. 11.11). Magnet coils are arranged around the periphery of the chamber for ECR plasma excitation, and the ECR condition enables the plasma to absorb the microwave energy effectively. Highly activated plasma is therefore easily obtained at low gas pressures: 10^{-5} to 10^{-3} Torr. High-quality thin-film deposition is obtained at room temperature [21] .

The reflection coefficient is often one of the most important properties of a material used in optical devices. Semiconductor waveguides generally have relatively large insertion loss, and an AR coating is indispensable in reducing this loss. In general when a film with refractive index n_l and thickness d is deposited on a material whose refractive index is n_s , the reflectivity R at wavelength λ for the thickness $d = \lambda(2m+1)/4$ (m = 0, 1, 2, ----) is given by

$$R = (n_s - n_l^2)^2 / (n_s + n_l^2)^2.$$

Refractive indexes for III-V compound semiconductor materials are 3.2-3.4 (effective index, not bulk index) and n_l should be about 1.8. The changes in effective refractive index with changes in waveguide configuration are described in Chapter 3, and the refractive index of a film is adjusted to obtain refractive index matching. This is usually done

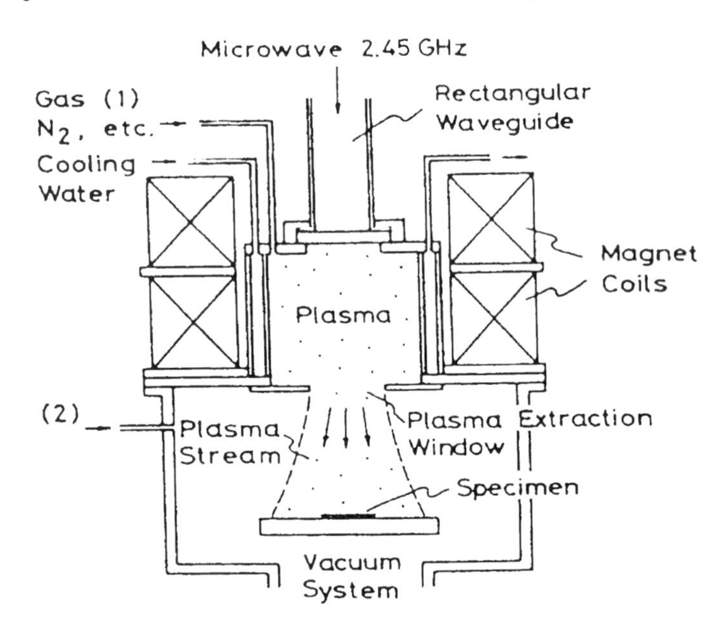

Fig. 11.11. ECR plasma deposition apparatus [20].

by using a SiO_x or SiN_x film and controlling the composition ratio. The AR condition tolerance of single film deposition for film thickness and refractive index is not large and AR coating using two films is used to obtain very low AR condition below 0.1%.

11.3.3 Disordering

Forming different devices on the same substrate, a technique often used in fabricating photonic integrated circuits, usually involves the etching and regrowing of semiconductors. Conventionally formed heterostructure waveguides have at least one interface between etched and regrown semiconductors, and this can lead to variability in processing and reduced device yield. Any sidewall roughness from the etched interface causes scattering propagation losses. As discussed in the Section 1.4, selective area growth is one of the most promising techniques for making many kinds of photonic integrated circuits, but the energy difference between two devices is limited and the transition region has an intermediate energy bandgap that results in relatively large propagation loss.

In quantum well structures, disordering techniques that create a lateral energy bandgap and refractive index profile by selective intermixing [22-24] have been investigated for formation of buried heterostructure waveguides without etching in GaAs/AlGaAs QWs [25, 26] and InGaAsP based QWs [27,28]. Disordering was first exploited to form buried heterostructure lasers by using impurity diffusion. The impurity diffused region is not used for passive waveguides because of high free carrier absorption. The thermal annealing method is free from impurity and the resultant propagation loss is relatively low [29, 30]. This method has been used to make such devices as a TE/TM mode splitter [31, 32], integrated external cavity InGaAsP/InP lasers with low loss waveguides [33], and an integrated MQW DFB laser with an external modulator [34].

Figure 11.12 shows the operation principle schematically. After a dielectric film such as SiO_2 or Si_3N_4 is deposited, the film is selectively etched before photolithograph techniques are used. Then rapid thermal annealing forms the waveguides, where the disordered thickness is not deep and depends on the annealing temperature and time. The over

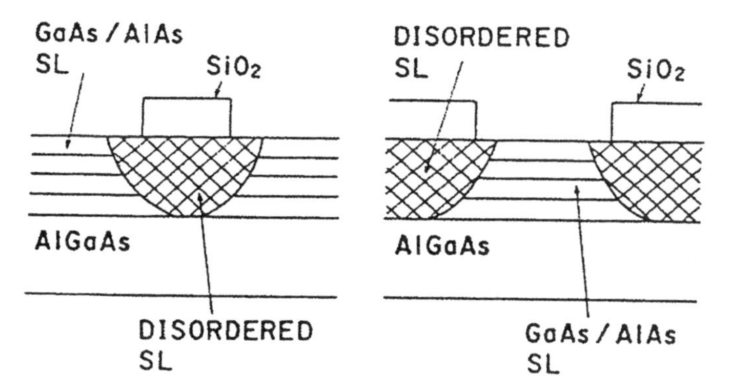

Fig. 11.12. Operation principle of disordering without impurity diffusion. The shading region under dielectric film indicates the disordred supplerlattice.

cladding layer is regrown on the disordering layer. It is possible to induce the disordering through the upper cladding layer but the degree of disordering is weak because the layer is thick.

Figure 11.13 shows the change of the photoluminescence spectrum before and after the rapid thermal annealing, where the photoluminescence change with and without the cap of dielectric film is also indicated. Some absorption coefficient change is seen even under the film, but it is small compared with that in the uncapped region.

11.3.4 Ion Implantation

Among the most promising single growth step techniques for making planar guided devices are ion implantation and the compositional disordering of supperlattices described above. One of the advantages of ion implantation for electrical isolation in waveguide devices (instead of etched isolation grooves, for example) is that planarity is maintained during implantation. This results in higher yield in the following processing steps and, as long as implant dose is low, no increase in propagation loss. Since a photoresist is used as the implantation mask, implantation is reproducible and highly controllable. Zucker et al. used implantation techniques to monolithically integrate interferometric modulators with semiconductor optical amplifiers [35, 36] and DBR Iasers [37].

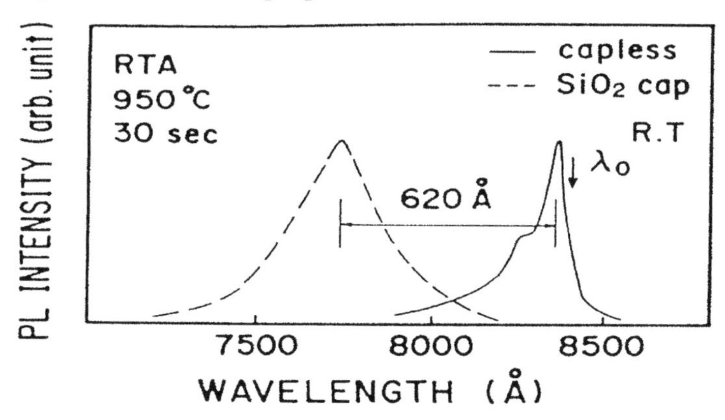

Fig. 11.13. Photoluminescence spectra of SLs and disordered SLs by SiO_2 cap annealing. λ_0 is the PL peak position of the as-grown SL structure.The solid line is for the capless region after annealing and the broken line is for the cap region after annealing.

Buried InGaAsP/InP quantum well waveguides formed by means of phosphorus ion implantation were used to make an electrooptic phase modulator operating at 1.55 μm [36]. Normal incidence absorption spectra before and after the implantation were measured and showed that at 1.55 μm the refractive index step was of the order of 1%.

This technique has been used to make InGaAs/InGaAlAs barrier reservoir and quantum well electron transfer structures (BRAQWETS) and resulted in the absorption edge being blue-shifted by 100-nm after implanting and in a complete recovery of the electroabsorption after an appropriate rapid thermal annealing cycle [38]. The implanting and annealing result in a degree of rib waveguide propagation loss, at 1.523 μm, of from 79.4 dB/mm to 6.2 dB/mm.

11.3.5 Photoelastic Effect for Waveguiding

The use of the photoelastic effect is also suitable for lateral optical confinement [39]. This technique is based on the photoelastically induced index change produced by using oxide and metal stressor layers and has been demonstrated in GaAs/AlGaAs [40-44] and InGaAsP/InP [45] material systems. The large photoelastic coefficients for III-V semiconductors make this approach attractive for producing planar guided-wave devices. A 1.3-μm Franz Keldysh modulator with an extinction ratio greater than 10 dB at 2 V and a 1.55-μm device with an extinction ratio greater than 10 dB at 7 V have been reported [45] .

11.3.6 Etching

III-V semiconductors such as GaAs, InP, and their related compounds are ideal for the development of novel two dimensional (2-D) integrated optics working in the 0.8-μm and the 1.3- or 1.55-μm wavelength regions. Chemical etching is usually used to etch off semiconductor layers. Depending on the crystallographic orientation of the form as well as on the chemical etching solution, the sides of the etched groove will be limited by different crystallographic planes. The other primary requirement for 2-D optics is to fabricate mirror elements etched into the waveguiding layer forming a semiconductor air interface. Examples of such a device are reflection type optical switches, demultiplexers, beam splitters, and polarizers. To obtain efficient device performance in such systems, the etched mirror facets must be extremely close to vertical (preferably, to less than 1 degree off-vertical) and extremely smooth (preferably, to better than a tenth of the wavelength) and without etching damage.

Waveguide devices utilizing multiple input/output ports require bends to separate interacting waveguides; total chip size is generally determined by bend requirements. Low-loss and compact bend structures are a key issue, and dry etching is a powerful technique for making compact waveguides. For GaAs and related alloys, reactive ion etching (RIE) in chloride is popular and produced good results. RIE of InP and related alloys, on the other hand, is often performed in a mixture of hydrocarbon gases. Since the first report of RIE of InP in CH_4/H_2 [28], many groups have reported similar results in methane/hydrogen [47,48], ethane/ hydrogen [49, 50], and methane/hydrogen/noble gas mixes [51]. One of the main problems with CH_4/H_2 RIE is the unavoidable polymer deposition on masked surfaces, and O_2 has therefore been used alternately with CH_4/H_2 to etch back the polymer before it builds up to an unacceptable level [52]. The etching of InGaAs/InAlAs MQW structures is similar to that of GaAs/AlGaAs systems and Cl_2 is used for MQW modulators. Electron-cyclotron-resonance (ECR) reactive ion etching is an especially powerful tool for etching off these materials because the semiconductor surface is irradiated with highly efficient ions [53]. A resonance in the microwave absorption of gas in the presence of a magnetic field occurs when the electron cyclotron frequency equals the microwave frequency (the ECR condition), and the efficient absorption allows dense plasmas to be generated at low gas pressures. In ECR etching apparutus the plasma is formed remotely from the etching chamber, and energy of incident ions to be changed. Under certain conditions ECR plasma ions reduce damage to below that produced by RIE.

It is often desirable to have an etching process with a high selectivity between certain layers, resulting in stopping vertical etch stop. Aluminium oxides are notoriously difficult to etch by RIE because of the involatility of AlO_x compounds. InP and InAlAs layers are both etched in CH_4/H_2 at similar rates. It is reported that the selectivity (defined here as the InP etch rate divided by the InAlAs etch rate) is only 1.4, whereas slight addition of O_2 increases the selectivity to 35 [54].

Another widely used technique for etch stopping on Al containing layers is the inclusion of a fluorine-containing feed gas during etching, since aluminium fluorine is very involatile. High selectivity has been achieved for GaAs/AlGaAs by using this techniques [54] .

11.3.7 Packaging and Modules

The light beam in usual semiconductor modulators as well as in other optical semiconductor devices is confined within an extreamly thin waveguide, so the sectional profile of the beam propagating through the waveguide is greatly distorted. Consequently, a large coupling loss occurs with single-mode optical fibers whose sectional beam profile is round and this large loss hinders the use of semiconductor modulators in practical optical transmission systems.

Furthermore, the space around the edge of the waveguide is usually occupied by the modulation circuit, which is set as close to the modulator chip as possible in order to reduce the time delay. Flexible optical circuit designs are therefore difficult to implement when fabricating semiconductor modulator modules.

A quasi-confocal lens combination configuration with aspherical lenses for optically coupling the waveguide to commercial fibers has been proposed and demonstrated [55,56]. Figure 11.14 shows the packaging design of the MQW-EA (electroaborption) modulator module. The first lens, placed near the waveguide, is an aspherical lens whose NA and focal length are 0.5 and 0.5 mm. The second lens, which is jointed to a polarization-maintaining pig-tail, has an NA of 0.2 and a focal length of 2.5 mm.

The data transmission line is an alumina substrate microstrip line whose surface is highly polished; its characteristic impedance is 50 Ω. The stripline is tapered so that it does not interrupt the optical beam between the waveguide and the aspherical lens. A 50 Ω thin-film resistor was evaporated onto the top edge of the tapered strip line to serve as the termination resistor. Because the heat genarated by the termination resistor severely affects the modulation characteristics of the EA modulator, a 100-pF thin-film integrated capacitor is connected in series to the termination resistor to block dc components. As a result, the mean insertion loss between fiber-to-fiber is 8 dB.

Another approach to packaging modulator modules uses specially designed micro-optics made at the cleaved endface of single-mode fiber pigtails [57]. These micro-optics consist of an apheric microlens fabricated on GRIN multimode fiber spliced to the single-mode fiber [58]. The focusing micro-optics exhibit a coupling loss ranging from 3 to 3.5 dB while the working distance is about 30 μm. To attach the fiber

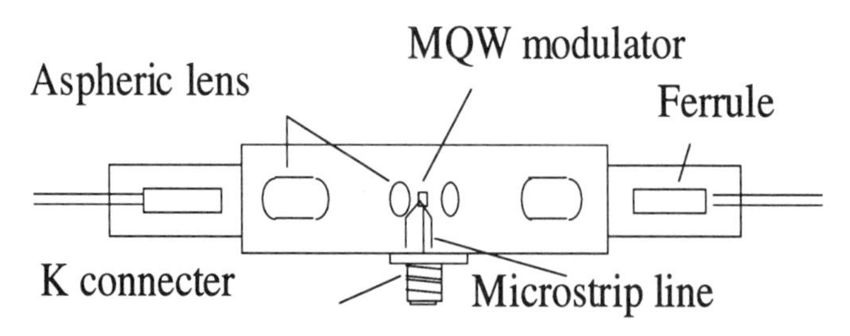

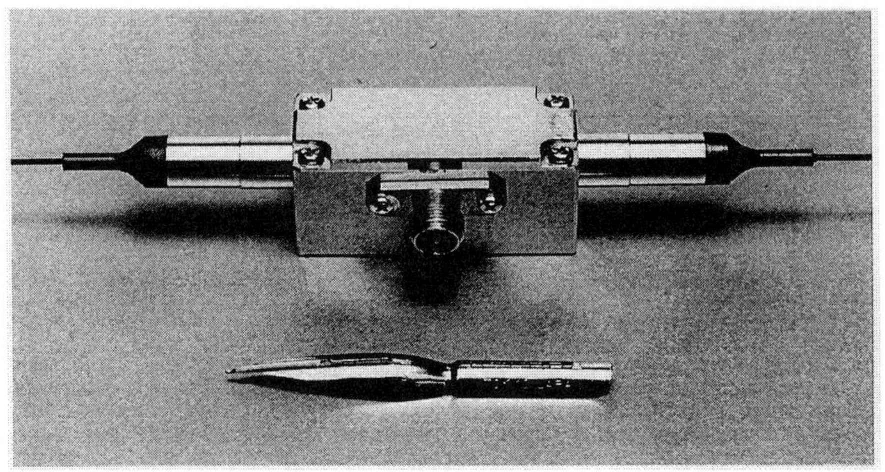

Fig. 11.14. Schematic diagram of modulator module and its photograph.

to the device and ensure long-term stability, a YAG laser welding technique with submicrometer accuracy is used. This assembling process yields typical packaging excess losses of 0.5 dB per facet. This approach has been used to make a polarization-independent MQW-EA module with fiber-to-fiber insertion loss of only 10 dB. This module assembling process is simpler than the one using the quasi-confocal lens configuration because each element need not be aligned and fixed separately.

11.4 References

[1] H. Reiss and J. O. McCladin (eds.), " *Progress in Solid State Chemistry*", Vol. 7, Pergamon Press, Oxford (1972).
[2] H. C. Casey, Jr., and M. B. Panish, "*Heterostructure Lasers, Part B; Materials and Operating Characteristics*", pp. 109-132, Academic

Press, New York (1978).
[3] W. T. Tsang (ed.),"*Semiconductors and Semimetals, Lightwave Communications Technology, Part A, Material Growth Technologies*", Vol. 22, pp. 1-94, Academic Press, Orlando (1985).
[4] G. B. Stringfellow, "*Organometallic Vapor Phase Epitaxy, Theory and Practice*", Academic Press (1989).
[5] M. Razeghi, "*The MOVPE Challenge*", Volume 1, A survey of GaInAsP-InP for photonic and electric applications, Adam Hilger (1989).
[6] K. Wakita, I. Kotaka, O. Mitomi, H. Asai, Y. Kawamura, and M. Naganuma, J. Lightwave Technol., vol. 8, no. 7, pp. 1027-1032, 1990.
[7] S. Kondo, L. Wakita, Y. Noguchi, N. Yoshimoto, M. Nakao, and K. Nakashima, J. Electronic Materials, vol. 25, no. 3, 1996.
[8] N. Yoshimoto, K. Kawano, Y. Hasumi, H. Takeuchi, S. Kondo, and Y. Noguchi, IEEE Photon. Technol. Lett., vol. 6, 208, 1994.
[9] T. Ido, H. Sano, D. A. Moss, S. Tanaka, and A. Takai, IEEE Photon. Technol. Lett., vol. 6, 127, 1994.
[10] J. E. Zucker, K. L. Jones, T. H. Chiu, B. Tell, and K. Broun-Goebeler, Integrated Photon. Res., paper Pd7, 1992.
[11] T. Kato, T. Sasaki, N. Kida, K. Komatsu, and I. Mito, Tech. Dig. ECOC '91, Paris, France, 1991, paper WeB7-1, 1991.
[12] M. Aoki, H. Sano, M. Suzuki, M. Takahashi, K. Uomi, and A. Takai, Electron. Lett.., vol. 27, no. 1, pp. 2138-2140, 1991.
[13] E. Colas, C. Caneat, M. Frei, E. M. Clausen, W. E. Quinn, Jun., and M. S. Kim, Appl. Phys. Lett.., vol. 59, pp. 2019-2021, 1991.
[14] M. Aoki, M. Suzuki, M. Takahashi, H. Sano, T. Ido, K. Kawano and A. Takai: Electon. Lett.., vol. 28, no.12, pp. 1157-1158, 1992.
[15] M. Nakao, K. Sato, M. Oishi, Y. Itaya, and Y. Imamura, J. Appl. Phys., vol. 63, no. 5, pp. 1722-1728, 1988.
[16] R. G. Walker, Electron. Lett.., vol. 21, 581, 1985.
[17] T. H. Wood, Appl. Phys. Lett., vol. 48, p. 1413-1415, 1986.
[18] M. K. Chin, Photon. Technol. Lett., vol.4, no. 8, pp. 866-869, 1992.
[19] K. Wakita and S. Matsuo, Jpn. J. Appl. Phys., 23, L556, 1984.
[20] S. Matsuo and M. Kiuchi, Jpn. J. Appl. Phys., 22, L210, 1983.
[21] A. T. Murrel and R. C. Grimwood, IEEE GaAs IC Symp. Proc., 173, 1992.
[22] W. D. Laidig, Appl. Phys. Lett., 38, 776, 1981.
[23] J. J. Coleman, P. D. Dapkus, C. G. Kirkpatrick, M. D. Camras, and N. Holonyak, Appl. Phys. Lett., 40, 904, 1981.

[24] D. G. Deppe and N. Holonyak, J. Appl. Phys., 64, R93, 1988. Detailed references are therein.
[25] R. L. Thornton, R. D. Burnham, T. L. Paoli, N. Holonyak, and D. G. Deppe, Appl. Phys. Lett., 48, 7, 1986.
[26] K. Ishida, K. Matsui, T. Fukunaga, T. Takamori, J. Kobayashi, and H. Nakashima, 13th Int. Symp. GaAs & Related Compounds, 1986 .(Institute of Physcs, 1987), p. 361.
[27] W. Xia, S. C. Lin, S. A. Pappert, C. A. Hewett, M. Fernandes, T. T. Yu, P. K. L. Yu, and S. S. Lau, Appl. Phys. Lett., 55, 2020, 1989.
[28] T. Miyazawa, H. Iwamura, O. Mikami, and M. Naganuma, Jpn. J. Appl. Phys., 28, L1039, 1989.
[29] D. G. Deppe, L. G. Guido, N. Holonyak, Jr., K. C. Hsieh, R. D. Burnham, R. L. Thornton, and T. L. Paoli, Appl. Phys. Lett., 49, 510, 1986.
[30] J. Y. Chin, E. S. Koteles, and R. P. Holnstrom, Appl. Phys. Lett., 53, 2185, 1988.
[31] Y. Suzuki, H. Iwamura, T. Miyazawa, and O. Mikami, Appl. Phys. Lett., 56, 19, 1990.
[32] Y. Suzuki, H. Iwamura, T. Miyazawa, and O. Mikami, Appl. Phys. Lett., 57, 2745, 1990.
[33] T. Miyazawa, H. Iwamura, and O. Mikami, IEEE Photon. Technol. Lett., 3, 421, 1991.
[34] A. Ramdane, P. Krauz, E. V. K. Rao, A. Hamoudi, A. Ougazzaden, O. Robein, A. Gloukhian, J. Landreau, M. Carre, and A. Mircea, Techn. Dig. CLEO '94, p. 239, 1994.
[35] J. E. Zucker, K. L. Jones, B. Tell, K. Brown-Goebeler, C H. Joyner, B. I. Miller, and M. G. Young, Electron. Lett., 28, 853, 1992.
[36] J E. Zucker, K. L. Jones, B. I. Miller, M. G. Young, U. Koren, B. Tell, and K. Brown-Goebeler, J. Lightwave Technol., l0, 924, 1992.
[37] J. E. Zucker, K. L. Jones, M. A. Newkirk, R. P. Gnall, B. I. Miller,M. G. Young, U. Koren, C. A. Burrus, and B. Tell, Electron. Lett., 28, 1888, 1992.
[38] J. E. Zucker, M. D. Divino, T. Y. Chang, and N. J. Sauer, IEEE Photon. Technol. Lett., 6, 1105, 1994.
[39] J. C. Cambell, F. A. Blum, D. W. Shaw, and K. L. Lawkey, Appl. Phys. Lett., 27, 202, 1975.

[40] P. A. Kirkby, P. R. Selway, and L. D. Westbrook, J. Appl. Phys., 50, 4567, 1979.
[41] L. D. Westbrook, P. N. Robson, and A. Majerfeld, Electron. Lett., 15, 99, 1979.
[42] L. D. Westbrook, and P. J. Fiddyment, Electron. Lett., 16, 170, 1980.
[43] T. M. Benson, T. Murotani, P. A. Houston, and P. N. Robson, Electron. Lett., 17, 237, 1981.
[44] L. S. Yu, Z. F. Guan, W. Xia, Q. Z. Liu, F. Deng, S. A. Pappert, P. K. L. Yu, S. S. Lau, L. T. Florez, and J. P. Harbison, Appl. Phys. Lett., 62, 2944, 1993.
[45] S. A. Pappert, W. Xia, X. S. Jiang, Z. F. Guan, B. Zhu, Q. Z. Liu, L. S. Yu, A. R. Clawson, P. K. L. Yu, and S. S. Lau, J. Appl. Phys., 75, 4352, 1994.
[46] U. Niggerbrugge, M. Klug, and G. Garus, Inst. Phys. Conf. Ser. No. 79, 367, 1985.
[47] T. R. Hayes, M. A. Driesbach, P. M. Thormas, W. C. Dautremont-Smith, and A. L. Heimbrook, J. Vac. Sci. Technol., B7,1130, 1989.
[48] J. Werking, J. Schramm, C. Nguyen, E. L. Hu, and H. Kroemer, Appl. Phys. Lett., 59, 2003, 1991.
[49] T. Matsui, H. Sugimoto, T. Ohishi, Y. Abe, K. Ohtsuka, and H. Ogata, Appl. Phys. Lett., 54, 1193, 1989.
[50] S. J. Pearton, W. S. Hobson, F. A. Baiocchi, A. B. Emerson, and K. S. Jones, J. Vac. Sci. Technol., B8, 57, 1990.
[51] I. Adesida, E. Andideh, A. Ketterson, T. Brock, and O. Aina, Inst. Phys. Conf. Ser. No. 96, 425, 1989.
[52] I. Adesida, K. Nummila, E. Andideh, and J. Hughes, J. Vac. Sci. Technol., B8, 1357, 1990.
[53] C. Constantine, and D. Johnson, J. Vac. Sci. Technol., B8, 596, 1990.
[54] S. E. Hicks, C. D. W. Wilkinson, G. F. Doughty, A. L. Bur-ness, I. Henning, M. Asghan, and I. White, Proc. ECI0'93, 2-36, 1993.
[55] S. Yoshida, Y. Tada, I. Kotaka, and K. Wakita, Electron. Lett.,.30, 21, pp. 1795-1796, 1994.
[56] K. Yamada, H. Murai, K. Nakamura, H. Satoh, Y. Ozeki, and Y. Ogawa, OFC'95, San Diego, Technical Digest, paper TuF4, pp. 24-25, 1995.
[57] N. Kalonji, J. Semo, F. Devaux, J. Tanniou, F. Foucher, and J. Saulnier, Proc. 21st Eur. Conf. on Opt. Comm. (ECOC'95- Brussels),

Th. B. 2.4, pp. 901-904, 1995.
[58] N. Kalonji, J. Semo, J. Tanniou, and F. Foucher, Electron. Lett., vol.30, no.11, pp. 892-893, 1994.

Chapter12

New Applications

12.1 Introduction

In the last chapter, we discuss the new applications and future trends using semiconductor modulators and their monolithic integrated light sources. In the first two sections, we present new applications of electroabsorption modulators for optical soliton light sources. In section 3 we describe ultra high speed modulation with a modulation bandwidth of over 40 GHz that operates at 1 V peak-to-peak large signal modulation.

12.2 Short Optical Pulse Generation and Modulation

Ultrashort optical pulses synchronized to an electrical clock are required in ultrahigh bit rate soliton transmission and optical signal processing systems. The word "soliton" is combined by "solitary" with the suffix "on" which is commonly used for describing particles such as electrons or photons. A soliton can be generated by in a nonlinear dispersive medium where the nonlinearity is the self phase modulation and the dispersion is group velocity dispersion. Optical solitons exist in a single-mode fiber and can propagate without distortion over long distances.

The generation of transform-limited optical pulses at a 1.55-μm wavelength is necessary for next generation time-division multiplexed communication systems and optical soliton transmission systems [1] that operate at a high bit rate and long haul distance. Methods such as active mode-locking [2,3] and gain-switching[4] have been used to produce optical pulses for laboratory experiments. Although these methods use chips without cavities, they are not suitable for commercial use because their repetition frequency cannot be altered. This is because fixed-length resonators are used in the mode-locking method. The large frequency chirping provided by the gain-switching method requires an optical filter, which results in a lack of flexibility and stability.

Use of a sinusoidally-driven InGaAsP bulk electroabsorption (Franz-Keldysh) modulator with an RF signal of 6.5-V peak-to-peak [5] has been reported to generate a transform-limited optical soliton pulse with a variable repetition rate. This is a simple and highly stable way of generating pulses. The response nonlinearity in time between transmitted light intensity and driving voltage was first reported in MQW modulators [6] but it was used only for a vertical configuration, where the optical interaction length was too short to obtain a sufficient on/off ratio (at most 4 dB) and nonlinearity. Response characteristics in the time and frequency domain of optical output from MQW electroabsorption modulators have been analyzed for waveguided MQW structures with a large on/off ratio and nonlinearity [7]. It is noted through calculations that the frequency response of electroabsorption modulators can be improved under large signal modulation, because of their nonlinear attenuation response to the applied voltage, and can thus shorten the light transmitted in the time domain [7]. It is reported that the modulated light spectra are broadened (full width at half maximum of ~50-GHz, corresponding to an optical pulse width of 7-ps) by a 30-GHz and 15-dBm signal even though chirp remains small [8]. It is also noted that the extinction ratio nonlinearity for the applied voltage is larger than that of a Franz Keldysh modulator [9], which results in a lower driving voltage (~1/2) and a shorter optical pulse (1/2~1/4) [10].

Additionally, an MQW DFB laser/modulator monolithically integrated light source [11], followed on a bulk DFB laser/modulator [12], has been reported to function as a single-chip pulse generator showing stable operation with low insertion loss and a low driving voltage, where a 20-GHz pulse train with a width of 7-ps is obtained with a 14-dBm (3.2-V peak-to-peak) RF signal [11]. A 10 Gbit/s soliton generated by the above integrated light source has been achieved that recirculates transmissions over 7200-km at BER=10^{-9} [13].

On the other hand, previous soliton light sources are only optical light generators and have no function of modulation. A tandem-integrated InGaAsP bulk modulator with two discrete modulators has been fabricated that produces 39-ps width pulses and modulates them at 5 Gbit/s by a 10-V peak-to-peak driving voltage [14]. However, this device has not been integrated with a light source and has the large insertion loss inherent in semiconductor modulators. Monolithic integration of a multi-section electroabsorption modulator and a DFB laser using strained-InGaAsP MQWs that choose, as a goal, the solution of above problems has been demonstrated. This single-chip device has a simple structure and provides a pulse train with a width of less than 17-ps and a repetition frequency of 10- GHz. It can encode a pseudo-random signal at a bit rate of 10 Gbit/s with a 2-V peak-to-peak swing.

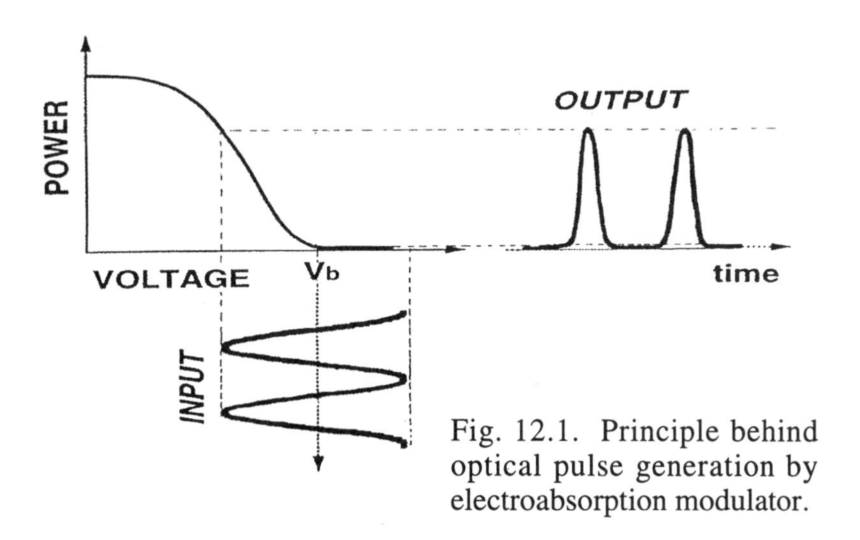

Fig. 12.1. Principle behind optical pulse generation by electroabsorption modulator.

12.2.1 Operation Principle

Figure 12.1 shows the operation principle of optical pulse generation. The intensity of light transmitted through the electroabsorption modulator can be expressed in approximate terms using voltage dependence factor **a** [9]: [transmitted light intensity] $\propto \exp\{(V/V_0)^{\mathbf{a}}\}$, where V is the applied voltage, and V_0 is the voltage where the on/off ratio is $\mathbf{e}^{-1}$. According to the on/off ratio against applied voltage, factor **a** is 2~6 for MQW modulators, while it is around 1 for bulk modulators. This significant voltage dependence in MQW structures is caused by a highly efficient electroabsorption effect (quantum-confined-Stark-effect: QCSE). When we apply a large signal to the modulator, the transmitted light is narrower than half the width of the driving sinusoidal wave, and short optical pulses can thus be obtained under a high-frequency operation. The width depends on the repetition frequency and driving voltage to the extent that the modulator can respond to the driving frequency.

When we integrate another electroabsorption modulator to drive tandem independently, we can easily control the optical pulse shape and encode as shown in Fig. 12.2. Generally, semiconductor modulators have large insertion loss, which is mainly due to their small spot size, resulting in mode mismatch between optical fibers and them. This has been accomplished by integrating a DFB laser and modulators monolithically [11,12].

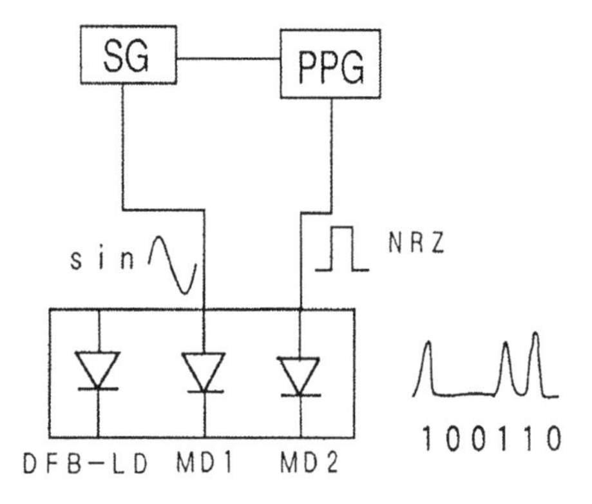

Fig. 12. 2. Principle behind optical pulse generation and encoding by a multi-section electroabsorption modulator/ DFB laser monolithic device. SG and PPG indicate a signal generator and a pulse pattern generator, respectively.

12.2.2 Sample Structure

Figure 12.3 shows a schematic diagram of the device. The device we used was grown by low-pressure MOVPE and is similar to that reported in Reference [15] except for the addition of another electroabsorption modulator. The modulator section consists of a p-i-n structure with an i-region consisting of eight periods of an 8.9-nm-thick InGaAsP well (bulk λ_{PL} =1.75μm) and a 5.0-nm-thick InGaAsP barrier (λ_{PL} =1.15μm), p-doped and n-doped InP cladding layers, and a p-doped InGaAsP capping layer. The laser section consists of two MQW guide layers: a laser active layer and a modulator core layer. The upper active layer consists of four periods of a 6.7-nm-thick InGaAsP well (λ_{PL} =1.75μm) and a 5.0-nm-thick InGaAsP barrier (λ_{PL} =1.15μm). The well was compressively

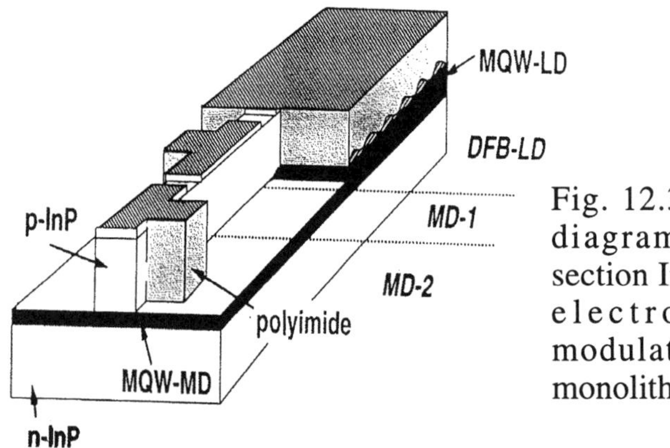

Fig. 12.3. Schematic diagram of a multi-section InGaAsP MQW electroabsorption modulator/DFB laser monolithic device.

strained with a magnitude of strain of $\varepsilon = 0.8\%$. Introducing a compressive strain to the wells increases the conduction-band offset (ΔE_C), which increases the exciton oscillator strength. The barriers are composed of InGaAsP with a 1.1 μm photoluminescence wavelengths.

First-order gratings were formed on the top of the guide layer at the laser section. The MQW-LD was selectively etched with a chemical etchant and a modulator section was formed. Next, a p-type InP cladding layer and an InGaAsP cap layer were grown. This configuration makes it possible to produce an integrated light source using simple crystal growth, compared with other monolithically integrated light sources. Light is confined laterally by using a conventional rib waveguide structure 2-μm wide and 450-μm long for the laser section and 220-μm long for each modulator section. Both separation regions between the laser and modulator and between the two modulators are 50-μm long with a resistance of over 30 kΩ. The total sample length is around 1-mm. To minimize stray capacitance, polyimide was spin-coated under the bonding pad over 50 μm-square area. The exciton resonance wavelengths for the modulator and the laser are 1.48~1.50 and 1.56 μm. The facet of the modulator was coated with an antireflection coating of less than 0.2% and that of the DFB laser was coated with high-reflectivity of around 90%.

12.2.3 Device Characteristics

The current vs. output power from the facet of the modulator is shown in Fig. 12.4. The CW threshold current is 18-mA, and single longitudinal operation is observed with a side-mode suppression ratio of more than 40-dB. The output from the modulator at zero bias voltage is 4-mW at

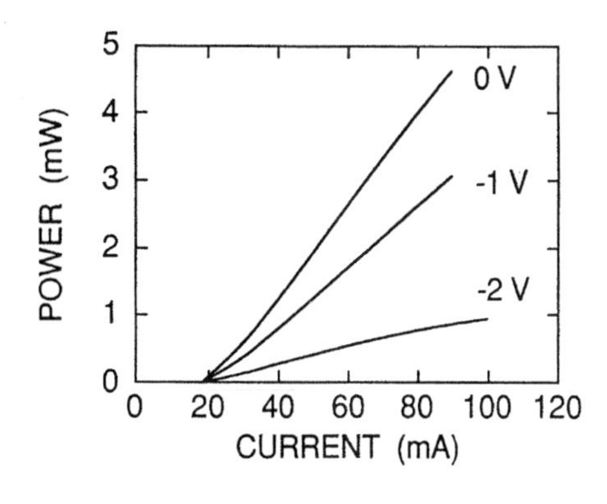

Fig. 12.4. Current vs. output power from the facet of the modulator.

an injection current of 80-mA. This is reduced to 0.75-mW as the reverse bias increases.

The on/off ratio of each modulator, monitored with a single-mode fiber, was over 30-dB at -3-V applied bias when reverse bias voltages were applied to one of the modulators and the other was not biased. The injection current to the laser was 80-mA. The optical coupling efficiency between the laser and the modulator was estimated at over 90%. The estimation was done by comparing the light output from the coupled single-mode optical fiber to the direct output from the modulator. A large-diameter Ge photodetector was used for the monitoring.

The small signal response in each modulator was measured and 3-dB bandwidth over 12-GHz was obtained for both sections at the injection current of 80-mA and the applied bias of -1-V.

One modulator (MD2) was used for pulse generation and driven by a 10-GHz and 16-dBm RF signal with a DC bias of -3-V using a synthesizer, whereas the other modulator was zero biased. The DFB laser was CW operated with an 80-mA current. An Er-doped fiber amplifier was used to measure the pulse. The optical pulse was detected by a high-speed streak scope (Hamamatsu Photonics C4334). Figure 12.5 shows the optical pulses generated from this integrated light source. The full width at half maximum (FWHM) is around 20-ps. The time resolution of the streak camera was 10-ps, which yields a 17-ps pulse

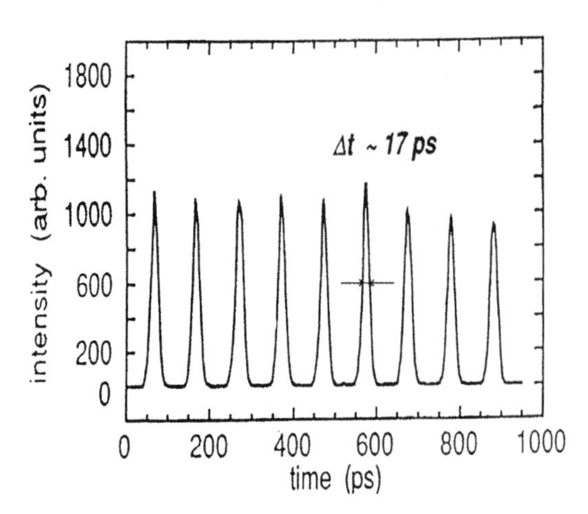

Fig. 12.5. Generated optical pulse train observed with a streak camera. The repetition frequency is 10-GHz.

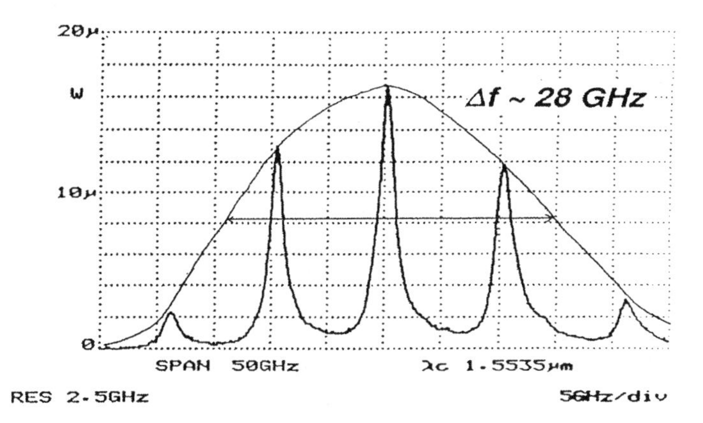

Fig. 12.6. Optical spectrum of the pulses driven by a 16-dBm RF signal and -3 V DC bias. This corresponds to that of Fig. 12. 5

width by assuming the Gaussian response. The waveform calculated based on the extinction characteristics was close to the Gaussian response. The optical spectrum for a generated optical pulse train is shown in Fig. 12.6. First, second, and third harmonics were observed for the optical pulse spectrum measured with an optical spectrum analyser. The FWHM is about 28-GHz, yielding a time-bandwidth product of 0.48. This value is similar to the transform-limited value of 0.44 for the Gaussian shape. This approximate transform-limited optical pulse indicates small

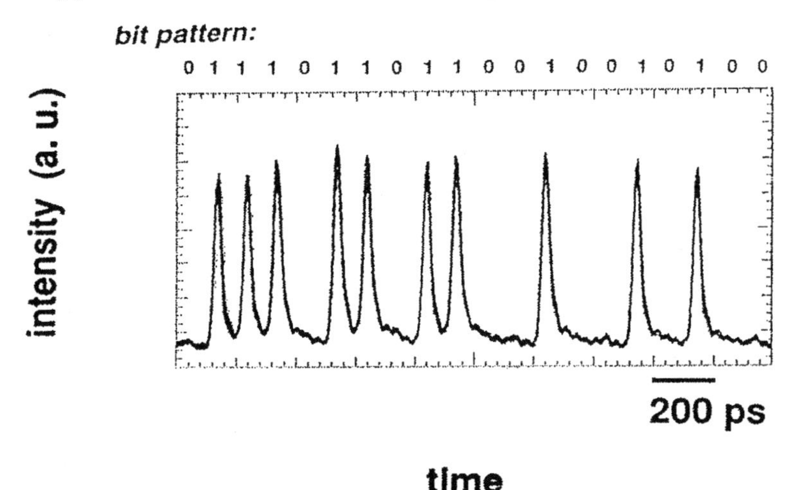

Fig. 12.7. Encoded optical pulses with a 10 Gbit/s pseudo-random bit signal.

linewidth increases and a small chirping factor in this MQW modulator.

Figure 12.7 shows an optical bit pattern when modulator MD1 is modulated with a 10 Gbit/s pseudo-random 2^{11}-1 NRZ signal synchronized with the RF signal driving modulator MD2. The MD1 is driven with a 10 Gbit/s signal with a 2-V peak-to-peak voltage at a -1.5-V bias voltage, while the MD2 is driven for optical pulse generation. A high-speed p-i-n photodetector with 3-dB bandwidth of 40-GHz (NEW FOCUS 1011) and a digitizing oscilloscope with 3-dB bandwidth of 50-GHz (HP 54120B) are used. The pulse train can be observed to be encoded at 10 Gbit/s. The fall time of the waveforms for random pattern shown in Fig. 12.7 is larger than that for (1,1) fixed pattern shown in Fig. 12.5. This is due to the poor response of the detecting system used in Fig. 12.7.

In summary, short optical pulses by using a strained MQW DFB laser/optical modulator monolithically integrated light source and applying low-voltage RF signals are generated and simultaneous data encoding is demonstrated using a multi-section electroabsorption modulator. The generated pulses show an approximate transform-limited value whose time-bandwidth product is 0.48. Even shorter pulses can be obtained by operating at higher frequencies and larger biases. This low-voltage, high-frequency operation with simultaneous pulse genaration and pulse modulation will give us a powerful pulse light source for optical solitons.

12.3 Active Mode Locking

12.3.1 Introduction

The previous section showed that electroabsorption modulators have various advantages such as compactness, stability and variability of repetition rate, but the emitted optical power is small due to the large insertion loss, and pulse width is not short relatively. To reduce optical pulse width, the repetition rate has to be increased. Proposed methods for reducing pulse width include combining the low-frequency gain switching of a DFB laser with synchronous high-frequency EA modulation [16], improving the electrical driver [17], and using tunable-dispersion chirped-fiber Bragg grating [18]. However, the emitted optical power is still small yet. This section presents another method for generating short optical pulses: active mode locking using high-speed MQW modulator.

12.3.2 Active Mode Locking Principle

The conventional approach to active mode locking is to use a semiconductor laser with a saturable absorber that acts as a pulse-shortening gate. Unfortunately, this results in poor reproducibility and reliability. Optical frequency chirping is also a problem in direct modulation of injection current to the gain sections. Pulse width tunability is essential for ultra-high bit rate soliton transmission and optical signal processing. Electroabsorption modulators have a fast nonlinear response to applied voltage and can gate lightwaves with a variable duty factor. This section briefly reviews active mode locking using intensity modulators.

Mode-locked semiconductor lasers provide optical pulses with high output. The key to obtaining stable and reliable mode locking in such lasers is to monolithically integrate the gain section, passive waveguides, and modulator section [19-23]. Active mode locking is the most promising way to generate short electrical-clock-driven optical pulses because it offers high emitting power and good pulse width tunability.

Mode locking was initially performed in a gas laser in order to generate ultrashort and repetitive optical pulses for, for instance, chromatic dispersion measurements. Figure 12.8 shows mode locking in a gas laser with an external cavity schematically to aid in understanding the mode-locking mechanism. If we remove the modulator, then the figure shows a typical gas laser.

The range of possible wavelengths is determined from the gain curve of the active material. The resonant condition inside the cavity defines the fine structure of the spectrum: the cavity is nothing more than a Fabry-Perot resonator. A resonant condition exists only for integer numbers of wave cycles in the cavity as shown in Fig. 12. 9. This condition leads to the mode spacing: $\Delta\lambda = \lambda^2 / 2nL$ (for wavelength) and $\Delta f = c / 2nL$ (for frequency) with $\Delta f = c\Delta\lambda / \lambda^2$, where c is the speed of light (3×10^{10} cm/s), n is the refractive index of the cavity medium, L is length of the cavity, and λ is wavelength in air.

The combination of cavity modes and gain curve provides the emitted spectrum. The total pulse width is approximately two times the inverse of the total width of the gain curve such that [25]

$$\Delta\tau = 2 / B_f = 2\lambda_0^2 / B_\lambda c \quad \text{with } B_f = cB_\lambda / \lambda^2,$$

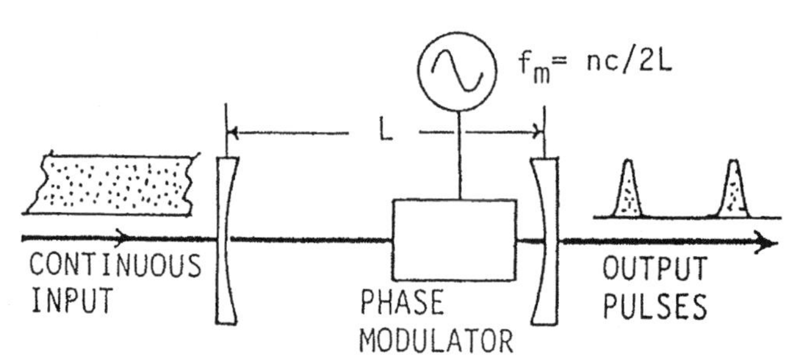

Fig. 12.8. Schematic diagram of mode-locking using gas laser.

where $\Delta\tau$ is the total pulse width, i. e., the time between the first zeros, B_f is the total width of the gain curve in units of frequency, and B_λ is the total width of the gain curve in units of wavelength.

In the configuration of Fig. 12.8, we have to manipulate the optical axis of the mirrors and the modulators, gain medium, and the stability is not good. Semiconductor lasers are compact, reliable and easy to use and have been used as sources of short optical pulses. The monolithic integration of a laser, a modulator and external mirrors is a powerful method for overcoming the above problems. There are two ways in

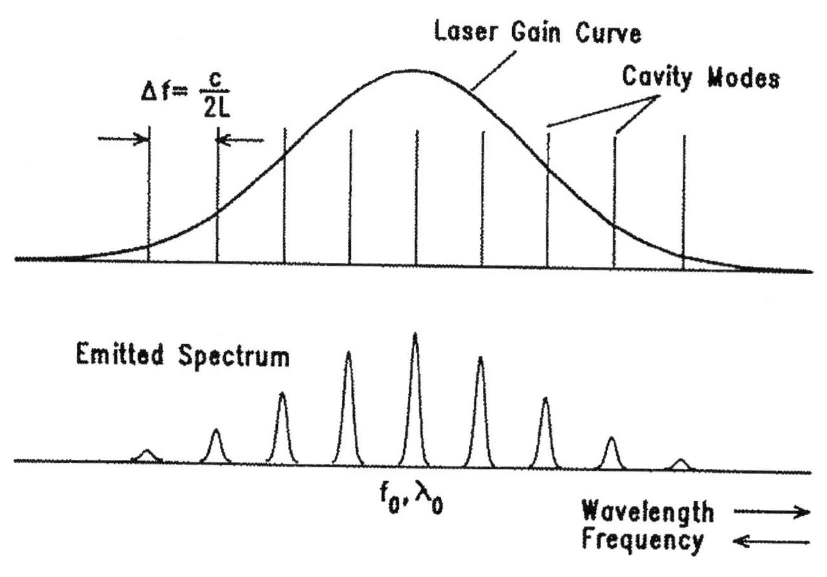

Fig. 12.9. Working principle of a Fabry-Perot interferometer and mode spacing.

which such a structure can be operated as an active mode-locked laser. One is to modulate the laser diode directly and to have the modulator operate as a saturable absorber as shown in Fig. 12. 10 [40]. The other is to use an external modulator. The former is simple and is commly used for active mode-locking. However, the emitted light broadens due to the chirping associated with direct modulation. The latter, a more recent development, enables us to obtain small chirping and jitter. This structure is shown in Fig. 12. 11 [25]. Using an optical pulse a few picoseconds in width has been obtained with a repetition rate of 20 GHz. Figure 12.12 shows the emitted optical pulse trains and their

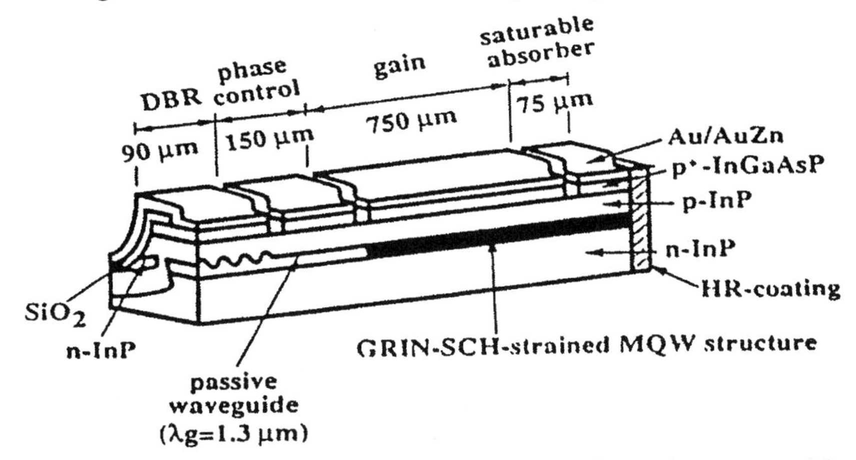

Fig. 12.10. Structure for an active mode locking using a saturable absorber [40].

spectra. The pulse width product is 0.6, which corresponds to the transform-limited theoretical value of 0.44. When a grating in the active medium that operates as a DFB laser or a DBR laser is used, the emitted light wavelength can be controlled.

These devices can be used for optical time-division multiplexing (OTDM) when an optical multiplexer is used. A shematic diagram of a 100-GHz transmitter using the above mode-locker is shown in Fig. 12.13. In this transmitter, a monolithic mode-locked laser with a repetition frequency of 20 GHz is used. Pulse trains of 20 GHz are multiplexed five times and modulated by the optical multiplexer [24]. The device consists of a multimode interference (MMI) splitter, five waveguides of different lengths, five electroabsorption (EA) modulators, and a taper-type combiner. Each path length difference corresponds to 50 ps delay

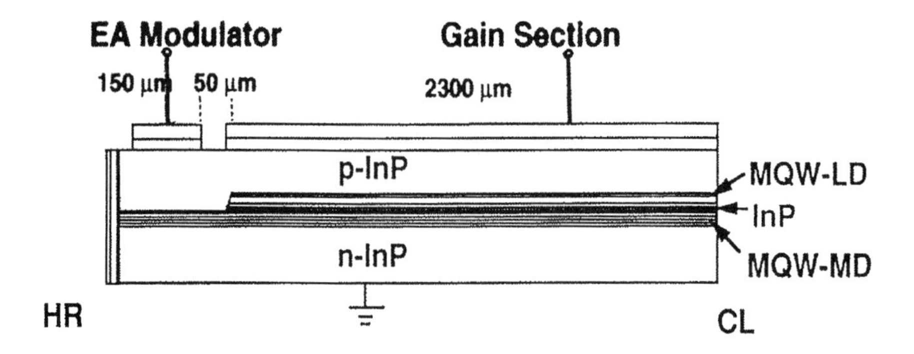

Fig. 12.11. Structure for an active mode locking using a semiconductor modulator.

so that 20 Gbit/s pulse train results in 100 Gbit/s pulses. By operating each modulator simultaneously, we can get a fully coded 100 Gbit/s optical pulse pattern.

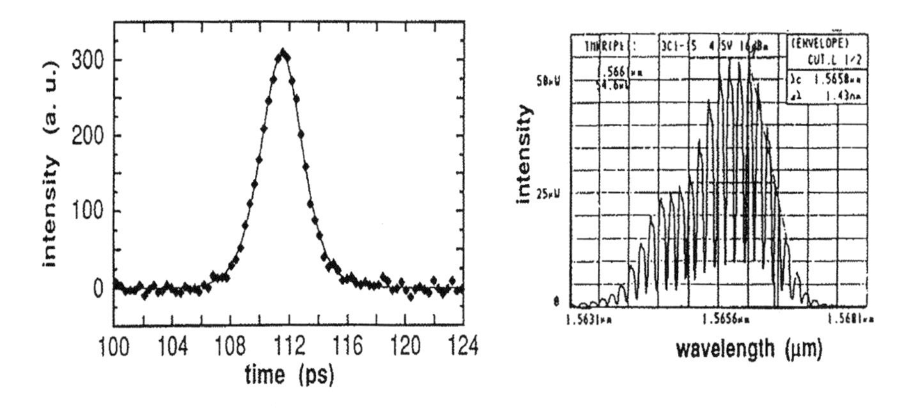

Fig. 12.12. 20-GHz pulse profile and spectrum from an actively mode-locked laser using MQW modulator.

Figure 12.14 shows a streak camera trace of 100 Gbit/s pulses with modulation on path 2. The pulse train propagating through path 2 clearly shows the pattern "010" by sinusoidal modulation. This InP-based optical multiplexer enables the modulation of 5 channels independently, leading a full 100-GHz modulated signal, and is well suited for optical time-division multiplexing.

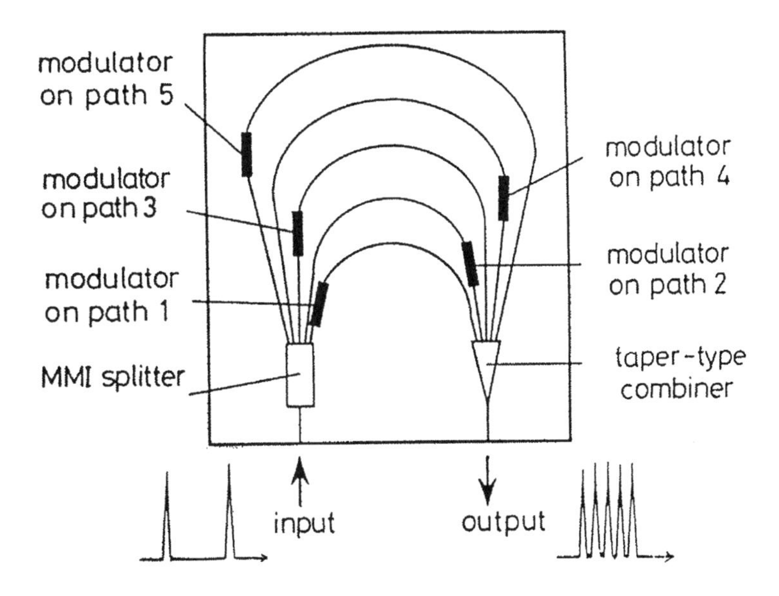

Fig. 12.13. Schematic diagram of the 100 Gbit/s transmitter.

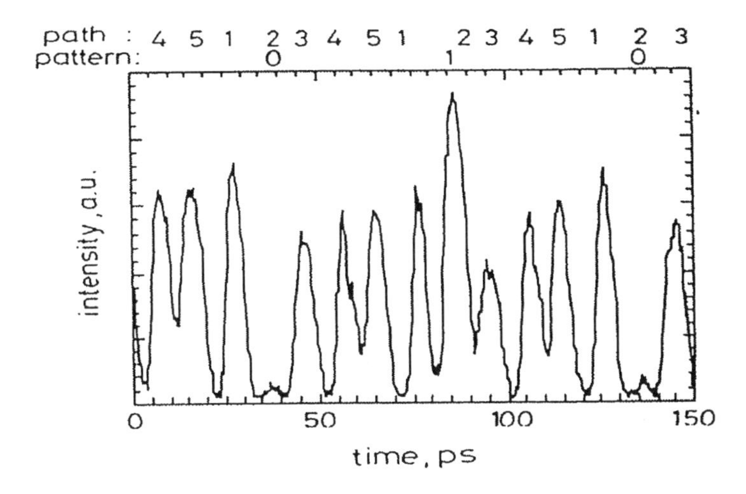

Fig. 12.14. Streak camera trace of 100 GHz pulses with modulation on path 2

12.4 Ultra-High-Speed Modulation

12.4.1 Introduction

High-speed, low-chirp modulators are needed to further take advantage of the capacity of optical fiber transmission systems. At present, high speed operation of optical modulators over 35 GHz has been achieved only under small signal modulation because of its large driving voltage [28-31]. Low-driving-voltage operation is the key point bringing such modulators into practical use because this eliminates the need for high-speed electrical amplifiers. There is generally a trade-off between modulator speed and driving voltage. Moreover, the reduction of the chirp parameter α for electroabsorption modulators is a hot topic in research on ultra high bit rate and long haul optical fiber transmission systems. It has been reduced, but the price has been increased insertion loss and reduced extinction ratio [32]. Consequently, it is necessary to optimize device characteristics for overall practical use. This section describes ultra-high-speed modulation over 40 GHz using semiconductor modulators and shows how the driving voltage can be reduced by using strain-compensated multiple quantum wells (MQWs) without the above trade-off. It also explains how the use of these MQWs provides polarization-insensitivity and low chirp.

12.4.2 Device Structure

In lumped modulators, device capacitance limits the bandwidth. Reducing the capacitance increases the bandwidth, but also increases the required operating voltage. One way to achieve high-speed without increasing voltage is to use short samples with small capacitance as shown in Fig. 12.15 [28]. This structure enables us to fabricate samples less than 100 μm long, but the extinction ratio decreases (see Fig. 12.16). The output voltage of electrical drivers usually decreases as speed increases. Table 12.1 shows the high-speed modulators with modulation bandwidth of over 40 GHz reported so far. Many modulators have been reported, but the modulation has been achieved only with a small-signal. This is because the driving voltages are relatively large and no drivers with relatively high-output have been built. Such a driver is required for modulators to achieve both high-speed modulation and low driving operation. As mentioned in Chapter 7, there is an optimum condition at which

Table 12. 1. Reported high-speed semiconductor modulators

year	3dB (GHz)		L (µm)	material	affiliat.	ref.
1990.5	40	20 dB,7V	100	InGaAs/InAlAs MQW	NTT	CLEO
1991.3	37	Vπ =7V 9V	1cm	GaAs/AlGaAs bulk	GEC-Marco.	IOOC/ECOC
1993.10	38	10dB,8V	1	GaAs/AlGaAs MQW	UCSB	DRC
1994.3	35	13dB,6V	3mm	InGaAsP/InP bulk	HHI	OFC
1994.8	35	19dB,4.2V	120	InGaAsP/InGaAsP MQW	CNET	EL,30, 1347
	28	10dB,2.5V	115			
1995.2	40	10dB,2.6V	50	InGaAs/InAlAs MQW	Hitachi	PTL,7,170
	30	20dB,2.6V	100			
1995.3	40 (70)	Vπ =25V 10V	2cm	GaAs/AlGaAs bulk	UCSB	OFC
1995.7	50	20dB,3.4V	63	InGaAs/InAlAs MQW	Hitachi	IOOC
	42	19dB,4.4V	75	(tensile)	CNET	IOOC
1995.9	42	20dB,2V	107		NTT	present

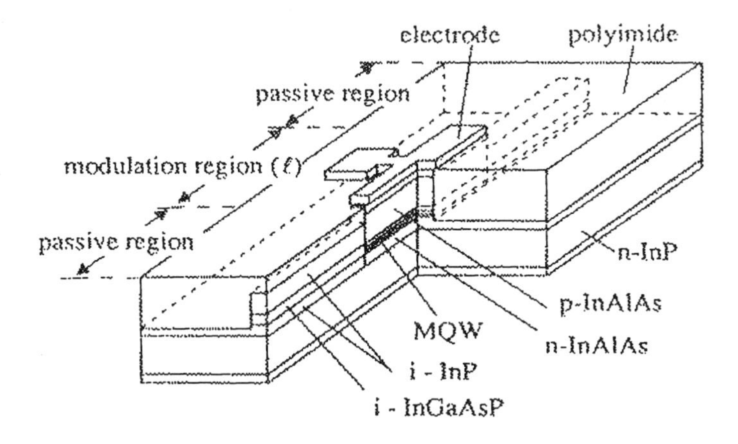

Fig. 12.15. Structure of an MQW modulator with integrated waveguides [28].

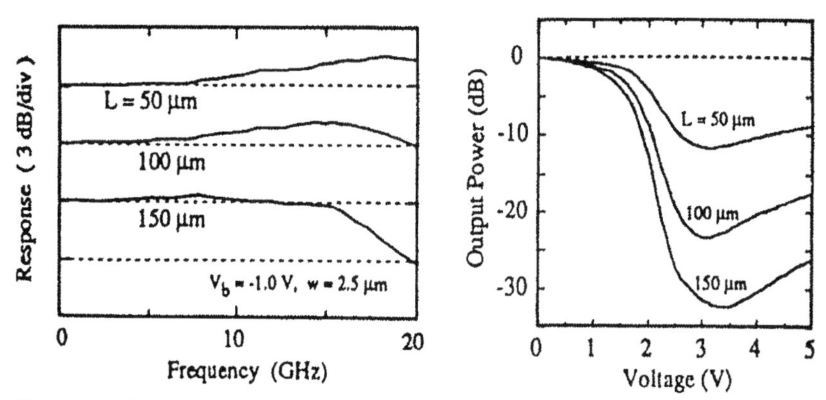

Fig. 12.16. Frequency characteristics for various sample lengths [28].

semiconductor modulators operate both at low driving and high speed, and the conclusion drawn was that well thickness of 12 nm is best. In order to improve the figure of merit, some stress is introduced in the wells and barriers so that the condition will be changed.

In what follows, the use of an MQW structure with quaternary InGaAlAs wells 11-15 nm thick and InAlAs barriers 5 nm thick with 8-10 periods is examined. Tensile strain of 0.05, 0.13, and 0.45% is introduced into the quantum wells, changing the well content/thickness so as to keep the absorption edge wavelength at 1.49 μm, whereas 0.5% compressive strain is introduced into the barriers to compensate the strain. The MQW layers are surrounded by 50- and 10-nm thick undoped InGaAsP layers. It is topped by 1.5-μm thick p-InP and 0.1-μm thick p-InGaAs contact layers. The high-mesa structure waveguides were etched down to the InP substrate by reactive ion etching, and low-capacitance contact pads were formed from polyimide dielectrics. The samples were 100 and 200-μm long and 2-μm wide. Some samples were assembled into modules, which were coupled with aspheric lenses in the confocal condition to single mode fibers. The mean insertion loss at 1.55 μm was 8 dB from fiber-to-fiber.

Figure 12.17 shows the fabricated modulators' figures of merit as a function of the detuning energy with tensile-strain magnitude as a parameter. The introduction of tensile strain enhances the electroabsorption resulting in reduced drive voltage. This is due to the heavy- and light-hole degeneracy, where the absorption of heavy-hole excitons transition coincides with that of the light-hole, and to the electric field effect difference in the absorption edge shift and oscillator strength

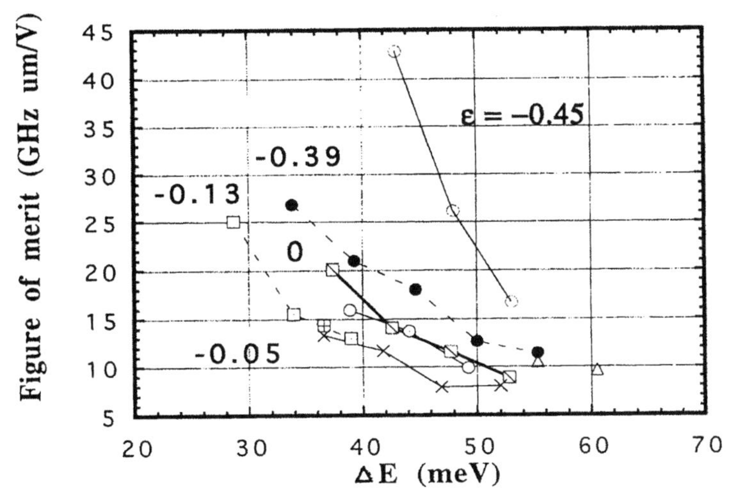

Fig. 12.17. Figure of merit as a function of detuning energy. The figure of merit is defined in the text. The number indicates tensile-strain magnitude.

decrease between them [34]. This enhances the oscillator strength at the zero bias condition and absorption coefficient change. Based on the obtained data, strain of -0.45% gives the best figure of merit.

12.4.3 Modulator Characteristics

A. On/off ratio. The transmitted optical power from the modulator with -0.45% tensile-strain is shown in Fig. 12.18 for incident light wavelengths of 1.54, 1.55 and 1.56 μm for both TE and TM polarization. Polarization insensitivity has been achieved in a 100-μm-long sample. The estimated extinction ratio is about 20 dB at the swing voltage of 1.0 V. The frequency response of this modulator was measured in small signal modulation. The modulated signal was detected with a 50 GHz photodiode and a network analyzer. The 3-dB bandwidth is around 40 GHz [31], as shown in Fig. 12.19. This value was fitted with a RLC model. The ratio of 3-dB bandwidth to drive voltage required for 20-dB on/off ratio is over 40 GHz/V. This value is, to our knowledge, the largest ever reported.

Figure 12.20 shows 40 Gbit/s eye diagrams obtained with a 100-μm-long modulator for a 0.9 V_{pp} pseudo-random modulation signal and 1.5 V DC bias. Clear eye opening is observed. This modulator was used in a 320-km long dispersion-shifted fiber transmission experiment with four channel wavelength division multiplexing, which was the first experiment of its kind [36]. No power penalty has been obtained. This indicates that this strained-compensated modulator can be used for ultra-high-speed modulation over 40 Gbit/s without a high-speed electrical amplifier.

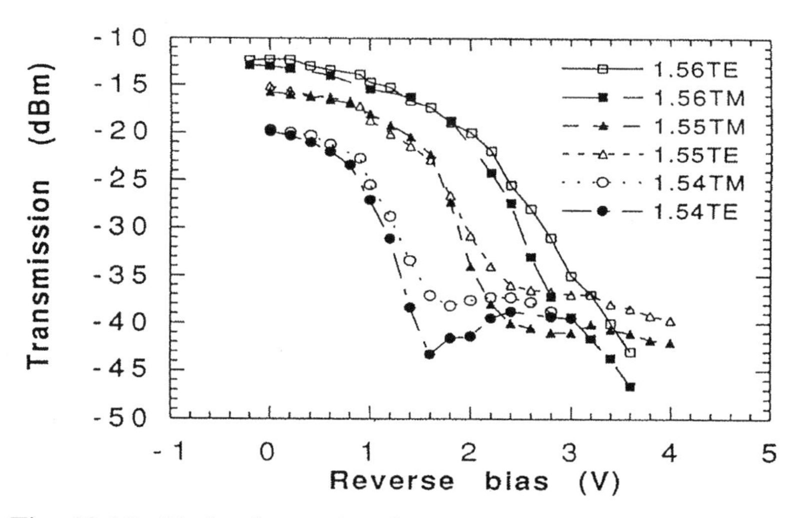

Fig. 12.18. Extinction ratio of a modulator comprising thick tensile strained quantum wells as a function of applied bias. Measured light is TE- and TM-polarized with wavelengths of 1.54, 1.55 and 1.56 μm.

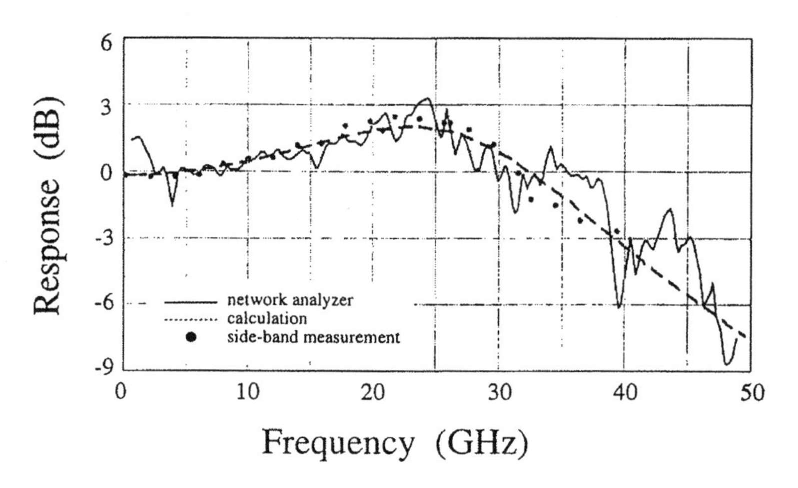

Fig. 12.19. Frequency response of the high-speed modulator.

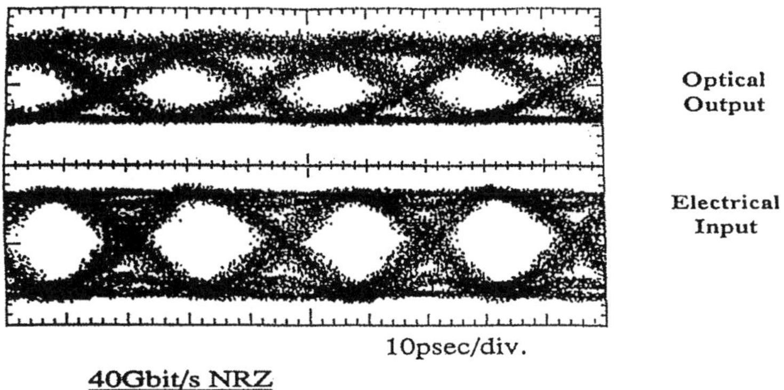

Fig. 12.20. 40 Gbit/s eye diagram recorded with a strain-compensated InGaAs/InAlAs modulator driven by a 1.6 Vpp pseudo-random modulation signal.

B. Chirp measurement. The chirp parameter α was measured by the frequency response peak method from the study of signal distortion after propagation in a standard fiber [37]. Applied bias dependence of α parameters for lattice-matched wells with 20-nm thick are shown in Fig. 12.21 with TE-polarized incident light wavelength as a parameter. The α parameters for 1.54-μm wavelength (the detuning energy of 33 meV) are almost zero even at zero bias. This is unlike those of modulators comprising lattice-matched [37] or compressively strained [38] InGaAsP thin quantum wells, in which abrupt increase in α parameters at small bias were observed. The magnitude of α parameters decreases as the well thickness increases. This indicates bulk-type modulators based on Franz-Keldysh effect will give negative α parameters. In fact, negative α parameters have been reported even with zero bias [39]; however, the detuning energy was small (about 28 meV) and insertion loss was large (about 15 dB). The fact that negative a parameters have generally been accompanied by increased insertion loss allowed us to investigate the relationship between α parameters and insertion loss with incident light wavelength as a parameter.

Figure 12.22 shows the dependence of chirp parameter α on the insertion loss with prebias. The parentheses indicates the magnitude of strain and the well thickness, respectively. The absolute values of α decrease as the insertion loss, which is associated with detuning energy and applied bias, increases and as the tensile strain increases. Note that short samples show a better relationship between α and the insertion

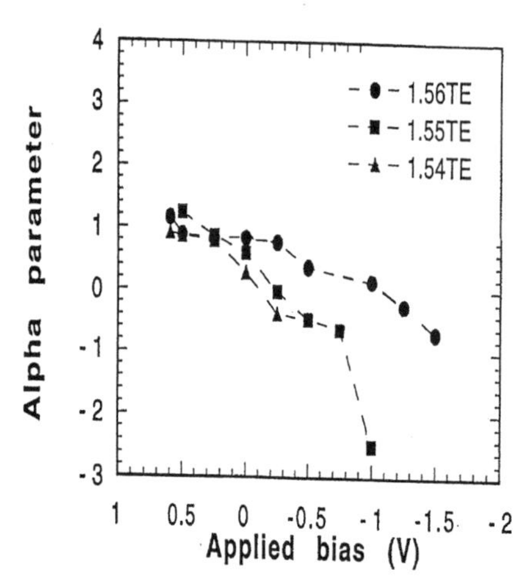

Fig. 12.21. Chirp parameter α as a function of applied bias. Measured light is TE-polarized with wavelengths of 1.54, 1.55, and 1.56 μm.

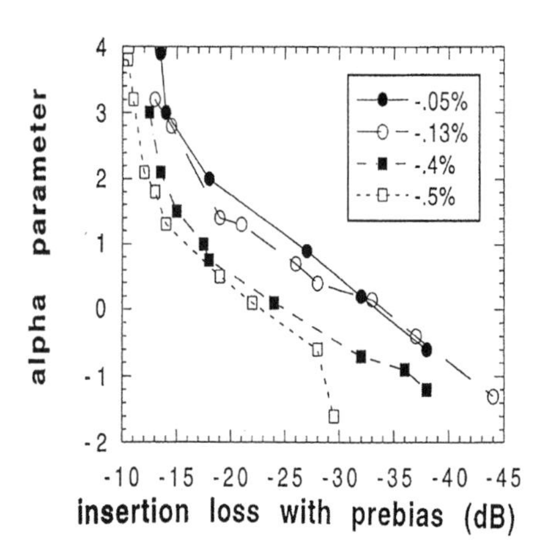

Fig. 12.22. Chirp parameter α as a function of insertion loss with prebias for incident wavelength of 1.55 μm. The numbers indicate the amount of tensile strain.

loss. This indicates that samples should be shortened as much as possible and still keep the required on/off ratio obtained. The short sample results in reducing device capacitance and high speed operation. Moreover, the insertion loss that gives zero α parameter is relatively small for strained modulators. This relatively small α parameter is due to the optimized structure, in which at nearly zero electric field, a strong exciton resonance exists that sharply reduces absorption, resulting in small insertion loss. When reverse bias is applied, the exciton resonance decreases at a relatively weak electric-field because there is no degeneracy of heavy- and light-hole excitons and the absorption-edge-energy shifts considerably due to the thick quantum wells.

The small magnitude of a at small bias (corresponding to the initial 3 dB of transmission) is expected to greatly improve the results of transmission experiments [37].

In summary, characteristics of strain-compensated InGaAlAs/InAlAs multiple quantum well (MQW) electroabsorption modulators have been investigated by changing the thickness and strain in the quantum wells over a wide range. Tensile strain of -0.45% and thick quantum wells 12-nm thick were found to give the best results: efficiency of 40 GHz/V and a 3 dB bandwidth of 40 GHz. We also investigated polarization insensitivity and low chirp operation using strain-compensation. Very low driving-voltage (<1 Vpp) modulators capable of 40-Gbit/s NRZ modulation have been fabricated using strained multiple quantum wells. The low drive operation eliminates the need for a high-speed electrical amplifier. Small chirp operation for low insertion loss with prebias was also demonstrated. The small magnitude of driving voltage and chirp parameter α at small bias (corresponding to the initial 3 dB of transmission) will greatly improve the results of future ultra high-speed and long haul optical fiber transmission experiments.

12.5 References

[1] M. Nakazawa, E. Yamada,, H. Kubota, and K. Suzuki, Electron. Lett., vol. 27, no. 14, pp. 1270-1272, 1991.

[2] P. B. Hansen,, G. Raybon,, M. -D. Chien,, U. Koren,, B. I. Miller, M. G. Young, J. -M. Verdiell, and C. A. Burrus, Photon. Technol. Lett., vol. 4, no. 5, pp. 411-413, 1992.

[3] M. C. Wu, Y. K. Chen, T. Tanbun-Ek, R. A. Logan, M. A. Chin, and G. Raybon, Appl. Phys. Lett., vol. 57, no. 8, pp. 759-761, Aug. 1990.

[4] M. Nakazawa, K. Suzuki, and Y. Kimura, Opt. Lett., vol. 15, no.12, pp. 715-717, June 1990.

[5] M. Suzuki, H. Tanaka, K. Utaka, N. Edagawa, and Y. Matsushima, Electron. Lett., vol. 28, no. 11, pp. 1007-1008, May 1992.

[6] T. H. Wood, C. A. Burrus, D. A. B. Miller, D. S. Chemla, T. C. Damen, A. C. Gossard, W. Wiegmann, IEEE J. Quantum Electron., vol. QE-21, pp.117-118, 1985.

[7] K. Kawano, O. Mitomi, K. Wakita, I. Kotaka, and M. Naganuma, IEEE J. Quantum Electron., vol. QE-28, no. 1, pp. 224-226, Jan. 1992.

[8] K. Wakita, I. Kotaka, O. Mitomi, H. Asai, and Y. Kawamura, Photon. Technol. Lett., vol. 3, no. 2, pp. 138-140, Feb. 1991.

[9] O. Mitomi, K. Kawano, K. Wakita, I. Kotaka, S. Nojima, and M. Naganuma, J. Lightwave Technol., vol. LT-10, no. 1, pp. 71-75, Jan. 1992.

[10] K. Wakita, I. Kotaka, O. Mitomi, H. Asai, and M. Asobe, Electron. Lett., vol. 29, no. 8, pp. 718-719, Apl. 1993.

[11] K. Wakita, K. Sato, I. Kotaka, M. Yamamoto, and M. Asobe, Photon. Technol. Lett., vol. 5, no. 8, pp. 899-901, Aug. 1993.

[12] M. Suzuki, H. Tanaka, and Y. Matsushima, Photon. Technol. Lett., vol. 4, no. 10, Oct. 1992.

[13] S, Kawai, K. Iwatsuki, K. Suzuki, S. Nishi, M. Saruwatari, K. Sato, and K. Wakita, Electron. Lett., vol. 30, no. 3, pp. 251-252, Feb.1994.

[14] H. Tanaka, S. Takagi,M. Suzuki, and Y. Matsushima, Electron. Lett., vol. 29, no. 11, pp. 1002-1003, May 1993.

[15] K. Sato, I. Kotaka, K. Wakita, Y. Kondo, and M. Yamamoto, Electron. Lett., vol. 29, no. 12, pp. 1087-1088, June 1993.

[16] M. J. Guy, S. V. Shernikov, and J. R. Taylor, Electron. Lett., 31, 1767,1995.

[17] N. M. Froberg, G. Raybon, A. Gnauck, Y. K. Chen, T. Tanbun-Ek, R. A. Logan, A. Tate, A. M. Sergent, K. Wecht, P. F. Sciortino, Jr., and A. M. Johnson, OFC'95 Technical Digest, TuI5, pp. 40-41, 1995.

[18] S. V. Shernikov, M. J. Guy, and J. R. Taylor, Opt. Lett., 20, 2399, 1995.

[19] R. S. Tucker, U. Koren, G. Raybon, C. A. Burrus, B. I. Miller, T. L. Koch, G. Eisenstein, and A. Shahar, Electron. Lett., 25, 622, 1989.

[20] P. A. Morton, J. E. Bowers, L. A. Koszi, M. Soler, J. Lopata, and D. P. Wilt, Appl. Phys. Lett., 56, 111, 1990.

[21] Y. K. Chen, M. C. Wu, T. Tanbun-Ek, R. A. Logan, and M. A. Chin, Appl. Phys. Lett., 58, 1253, 1991.

[22] P. B. Hansen, G. Raybon, U. Koren, P. P. Iannone, B. I. Miller, G. M. Young, M. A. Newkrik, and C. A. Burrus, Appl. Phys. Lett., 62, 1445, 1993.

[23] S. Arahira, Y. Matsui, T. Kunii, S. Oshiba, and Y. Ogawa, Electron. Lett., 29, 1013, 1993.

[24] A. Takada, K. Sato, M. Saruwatari, M. Yamamoto, Electron. Lett.,30, 899, 1994.

[25] K. Sato, K. Wakita, I. Kotaka, Y. Kondo, and M. Yamamoto, Appl. Phts. Lett., Appl. Phys. Lett., 65, 1, 1994.

[26] A. Yariv, "*Quantum Electronics*", Second edition, John Wiley & Sons, 1975.

[27] F. Zamkotsian, K. Sato, H. Okamoto, K. Kishi, I. Kotaka, M. Yamamoto, Y. Kondo, H. Yasaka,, Y. Yoshikuni, and K. Oe, Electron., Lett., 31, 578, 1995.

[28] T. Ido, S. Tanaka, M. Suzuki, and H. Inoue, in Proc. IOOC'95, 1995, pp. 1-2, paper PD1-1.

[29] F. Devaux, S. Chelles, J. C. Harmand, N. Bouadma, F. Huet, M. Carre, and M. Foucher, in Proc. IOOC'95, 1995, pp. 56-57, paper FB3-2.
[30] R. Weinmann, D. Baums, U. Cebulla, H. Haisch, D. Kaiser, E. Kuhn, E. Lack, K. Satzke, J. Weber, P. Wiedemann, and E. Zielinski, Photon. Technol. Lett., 8, pp. 891-893, 1996.
[31] K. Wakita, K. Yoshino, I. Kotaka, S. Kondo, and Y. Noguchi, in Proc. ECOC'96, Th. B.3.2, pp. 1011-1014, Brussels, Belgium.
[32] J. F. Fells, M. A. Gibbon, I. H. White, G. H. B. Thompson, R. V. Peny, C. J. Armistead, E. M. Kimber, D. J. Moule, and E. J. Thrush, Electron. Lett., vol. 30, no. 14, pp. 1168-1169, 1994.
[33] K. Wakita and I. Kotaka, Microwave and Optical Technol. Lett., vol. 7, pp. 120-128, 1994.
[34] B. N. Gomatam and N. G. Anderson, IEEE J. Quantum Electron. Lett., vol. 28, no. 6, pp. 1496-1507, 1992.
[35] M. Erman, "*GaAs and Related Compounds 1987*", *Inst. Phys. Conf. Ser. no. 91*, p. 33, 1988.
[36] S. Kuwano, N. Takachio, K. Iwashita, T. Otsuji, Y. Imai, and T. Enoki, K. Yoshino, and K. Wakita, in Proc. OFC'96, 1996, paper PD25.
[37] F. Dorgeuille and F. Devaux, IEEE J. Quantum Electron., 30, 2565, 1994.
[38] K. Morito, R. Sahara, K. Sato, Y. Kotaki and H. Soda, Electron. Lett., vol. 31, pp. 975-956, 1995.
[39] K. Yamada, K. Nakamura, Y. Matsui, T. Kunii, and Y. Ogawa, Photon. Technol. Lett., vol. 7, no. 10, pp. 1157-1158, 1995.
[40] S. Arahira, S. Oshiba, Y. Matsui, T. Kunii, and Y. Ogawa, Appl. Phys. Lett.,vol. 64, no. 15, pp. 1917-1919, 1994.

index

A

B

C

D

E

F

G

I

O

P

Q

R

S

T

V

W

X

Y

Z